Fluid Flow in Sedimentary Basins and Aquifers

Geological Society Special Publications

Series Editor K. COE

GEOLOGICAL SOCIETY SPECIAL PUBLICATION NO 34

Fluid Flow in Sedimentary Basins and Aquifers

EDITED BY

J. C. GOFF
British Petroleum
London

B. P. J. WILLIAMS
Department of Geology
Bristol University

1987

Published for
The Geological Society by
Blackwell Scientific Publications

OXFORD LONDON EDINBURGH
BOSTON PALO ALTO MELBOURNE

Published for
The Geological Society by
Blackwell Scientific Publications
Osney Mead, Oxford OX2 0EL
8 John Street, London WC1N 2ES
23 Ainslie Place, Edinburgh EH3 6AJ
52 Beacon Street, Boston, Massachusetts 02108, USA
667 Lytton Avenue, Palo Alto, California 94301, USA
107 Barry Street, Carlton, Victoria 3053, Australia

First published 1987

DISTRIBUTORS

USA and Canada
Blackwell Scientific Publications Inc.
PO Box 50009, Palo Alto
California 94303

Australia
Blackwell Scientific Publications
(Australia) Pty Ltd,
107 Barry Street, Carlton,
Victoria 3053

Typeset, printed and bound in Great Britain by
William Clowes Limited, Beccles and London

British Library Cataloguing in Publication Data

Fluid flow in sedimentary basins and aquifers.
—(Geological Society special publication,
ISSN 0305–8719).
1. Sediment transport 2. Setting basins
I. Goff, J. C. II. Williams, B. P. J.
III. Geological Society of London
IV. Series
552'.5 QE571

ISBN 0–632–01806–2 (46)

Library of Congress Cataloging-in-Publication Data

Fluid flow in sedimentary basins and aquifers.
(Geological Society special publication; no. 34)
Proceedings of a meeting held at the Geological Society in London in June 1985.
Includes index.
1. Sedimentation and deposition—Congresses.
2. Fluid dynamics—Congresses. 3. Groundwater flow—Congresses. 4. Aquifers—Congresses. 5. Petroleum—Geology—Congresses. I. Goff, J. C. (Jeremy C.)
II. Williams, B. P. J. III. Geological Society of London. IV. Series.
QE571.F62 1987 551.3'53 87–10261

ISBN 0–632–01806–2 (46)

Contents

Preface

This Special Publication of the Geological Society is the proceedings of a meeting on 'Fluid Flow in Sedimentary Basins and Aquifers' held at the Geological Society in London in June 1985. The meeting was organized by the Petroleum and Hydrogeological Specialist Groups of the Society and attended by 200 earth scientists. The aim of the meeting was to bring together geologists concerned with different aspects of fluid flow in sedimentary basins, to present case histories of flow in different types of basin, and to emphasize economically important aspects of fluid flow. It is hoped that this volume which includes thirteen of the papers presented at the meeting will be a useful reference on this complex and evolving subject in geology.

Further advances in our understanding of fluid flow will require accurate description and prediction of rock types, aquifer and seal geometry, and permeability in basins, and advances in modelling flow—particularly over long periods of geological time. The convenors, Jeremy Goff of the Petroleum Group and Brian Williams of the Hydrogeological Group, hope that this volume will stimulate further interest in this topic and encourage further research in this economically important aspect of geology. The convenors wish to express their sincere thanks to the Geological Society for their help and support in the organization of this meeting.

Introduction

Knowledge of fluid flow in sedimentary basins is of vital economic importance for understanding water resources, aquifer behaviour, petroleum migration, accumulation and production, geothermal resources and toxic waste disposal. The meeting on 'Fluid Flow in Sedimentary Basins and Aquifers' allowed workers in different fields of geology to compare and contrast their approaches to the study of fluid flow. Advances in the study of fluid flow by hydrogeologists, for example, may be significant for flow studies by petroleum geologists and engineers, and vice versa.

The papers in this volume discuss principles of fluid flow, and geologically important aspects of flow at all scales ranging from large scale basinal flow systems to flow in small aquifers and fracture systems. The volume is divided into four sections covering the following topics: fluid flow in compacting basins, fluid flow in the Western Canadian Sedimentary Basin, fluid flow in the United Kingdom Sedimentary Basin, and fluid flow in low permeability and fractured media.

In the first section, Chapman discusses the large-scale geological controls on fluid flow in compacting basins, the physical principles controlling fluid flow, and the rock properties which affect flow. He emphasizes how the stratigraphy (distribution of lithologies) in a basin controls the directions and pattern of fluid flow. He also comments on the development of abnormal pressures, and on the role of faults in vertical flow in basins. Magara attempts to quantify horizontal, compaction driven, fluid flow caused by sediment loading. He applies his theory to the sourcing of the oil fields in the Arabian Gulf region. He combines his fluid flow study with work on the maturation of the source rock and concludes that major horizontal oil migration has occurred.

The Western Canadian Basin has become an important area for studies of fluid flow and the second section of the volume includes four papers which cover this topic. Hitchon describes the recent research work being carried out at the Basin Analysis Group of the Alberta Geological Survey. Using hydrogeological studies of oil sand and of a deep waste disposal site as examples, he describes how fluid flow can be modelled in areas with extensive geological and hydrogeological data bases. He emphasizes the importance of identifying individual 'hydrostratigraphic units' (aquifers, aquitards and aquicludes), and of modelling the fluid flow within and between the identified aquifers. Corbett and Tóth discuss the evolution of fluid flow systems through geological time in part of the Western Canadian Basin. They recognize flow systems controlled by the present land surface, erosional rebound effects, and an ancient cross formational flow system which is not adjusted to the present land relief.

Jones and Majorowicz, using a large bottom hole temperature data base, show that redistribution of heat due to groundwater motion occurs in the Western Canadian Basin. Bradbury and Woodwell describe ancient fluid flow in the thrust belt flanking the Western Canadian Basin. They show how isotopic data can be used to help identify ancient flow systems. In Western Canada they identify ancient flow systems resulting from dewatering of thrust sheets with major flow in basal aquifer systems. In another example from the Southern Pyrenees they show that aquifer flow was largely confined to the crystalline basement of the foreland, with shear zones acting as important conduits for fluid flow.

The third part of the volume covers fluid flow in United Kingdom groundwater basins. Downing, Edmunds and Gale define seven groundwater provinces in the United Kingdom. They use geological, hydrochemical and heat flow data to interpret the patterns of regional flow. They demonstrate that the Permo–Triassic sandstones are major drains for subsurface flow from pre-Jurassic rocks. They show that other flow systems are controlled by the outcrop pattern, the small size of the drainage basins, and by the development of high permeability zones.

Bath, Milodowsky and Strong discuss fluid flow in the East Midlands Triassic sandstone aquifer. They show that the freshwater flow regime in this aquifer has existed for several tens of millions of years. They discuss diagenetic changes occurring in the aquifer due to the long period of freshwater influx. They suggest that diagenetic mineralogies in ancient aquifers can be used to determine palaeoflow regimes in clastic aquifers. Price describes the physical properties of the UK Chalk aquifer: an aquifer with high secondary permeability due to fractures and a low matrix permeability. He shows that enlargement of fractures by dissolution to form major key permeable zones is necessary for the Chalk to form a major aquifer, and also discusses the importance of outcrop weathering in forming shallow permeable zones. Wilson and Luheshi model the convective transport of heat by moving groundwater in the Carboniferous rocks of the East Midlands Basin. They show that the thermal anomaly over the East Midlands Eakring oil field could be due to water movement.

In the final section of the volume, Alexander, Black and Brightman discuss the role of low permeability rocks in controlling regional flow in basins. They show that although cross formational flow rates in low permeability rocks are low, major quantities of fluid are involved because of the large surface areas through which flow occurs. They use chemical evidence to suggest that pore waters recovered from thick mudstone sequences are not original formation water or compaction water but may represent groundwater moving in cross formational flow systems. They show that the ratio of the hydraulic conductivity of beds of high permeability to beds of low permeability is an important factor controlling the extent of cross formational flow systems.

The last two papers in the volume are concerned with flow through fracture systems. Black discusses an end member of flow in sedimentary basins: flow in crystalline rocks which bound sedimentary basins. He concludes that groundwater flow in crystalline rocks is probably concentrated in specific systems of interconnected channels with hydraulic conductivities five orders of magnitude greater than the bulk rock. Continuing the theme of flow in fracture systems, Brown describes the role of fracture systems in fluid flow during production from the fractured chalk reservoir of the Ekofisk Field. He argues from laboratory experiments that imbibition of water from the fracture network into the rock matrix (where most of the oil is stored) will control the displacement of oil and thus influence the amount of oil recovered from the field. A field water injection project has been set up to test this conclusion.

The papers in this volume thus illustrate the wide ranging applications of studies of fluid flow in sedimentary basins. It is hoped that this book will be a useful reference for geologists interested in the principles and geological controls of fluid flow in basins, as well as to practising hydrogeologists, petroleum geologists and researchers.

J. C. Goff
B. P. J. Williams

Section 1
Fluid Flow in Compacting Basins

Fluid flow in sedimentary basins: a geologist's perspective

R. E. Chapman

SUMMARY: Water begins to flow from higher-energy to lower-energy positions as soon as sediment begins to accumulate in a new sedimentary basin, and flow continues during the sedimentary development of the basin. Water expelled by compaction from the compactible less-permeable beds displaces water in the less-compactible more-permeable beds. Compaction flow also causes subsidence that contributes to the sedimentary development of a basin.

The pattern of flow is determined by the stratigraphy. In transgressive sequences, it is downwards in mudstone to the basal permeable unit and then lateral towards the land of the time. In regressive sequences, it is upwards and downwards in mudstones towards the intercalated sandstones and then lateral in these towards the land. In compacting sequences of alternating sandstones and mudstones, the sandstones drain distinct isolated hydraulic units that are related to the stratigraphy.

When a sedimentary basin ceases to accumulate sediment and compaction has reached equilibrium, the flow patterns change and the main energy comes from elevation of the intake areas.

In rift basins the main source of compaction water is the thick mudstone in the post-unconformity–disconformity sequence. Abnormal pressures in the pre-unconformity–disconformity sequence indicate lack of flow in this part of the basin.

There are many reasons, both scientific and practical, for interest in the flow of fluids in sedimentary basins. Petroleum, a material of geological as well as commercial interest, flowed through water in the pores of sedimentary rocks before it arrived in the position in which we find it. Some base metals reached their site of concentration as a result of transport in flowing pore water. Large areas of the N American, African and Australian continents are made habitable by fresh-water flow in artesian basins. Ground water flow is essential for the flow of rivers, and its smoothing effect on the irregularities of rainfall, particularly in times of drought, makes life more comfortable for land animals. Most of our sedimentary rocks have suffered diagenesis to some extent as a result of water flow through the pores, and the processes of metamorphism must include fluid flow.

Compaction of water-saturated sediments involves water flow, and compaction of thick mudstones eventually expels huge quantities of water. For example, 1 km^3 of mudstone compacting from 50% to 15% porosity expels 0.4 km^3 of water, and its thickness is reduced to 60% of its original thickness. This compaction allows the accumulation of a commensurate thickness of new sediment and so contributes to the development of a sedimentary basin.

The geological consequences of failure of water to flow and failure of sediment to compact are also important to geology because they include early deformation of sedimentary basins by virtue of abnormally high pore pressures generated in thick mudstones.

Methods of study of water flow in sedimentary basins are extraordinarily difficult and diverse. The rates of water movement in the sub-surface are normally so slow (of the order of metres per year) that direct studies using chemical or physical tracers cannot be made, and theoretical extrapolations from small-scale work to sedimentary basins must be used. There is a beautiful but little-known example of sub-surface water flow that has been thoroughly studied for more than a century using chemical and physical tracers: the flow of water from the River Danube on the Malm outcrop to the great spring at Aach (mainly), and from there into the Bodensee (Lake Constance) and so to the Rhine (Fig. 1). In 1877, in what was perhaps the first use of tracers to study groundwater flow, oil was introduced into a sink in the right bank of the Danube in the afternoon of Saturday, 22 September, and detected in the spring at Aach from 6 a.m. on the morning of Monday, 24 September (see Käss 1969, 1972; Käss & Hötzl 1973; Chapman 1981, pp. 98–101). On Tuesday, 9 October 1877, an organic dye, sodium fluorescein (uranin), was put into the water at the sink and was detected at the spring at Aach at dawn on 12 October. The sub-surface velocities over the straight-line distance of 13.3 km were 0.09 m s^{-1} in September and 0.06 m s^{-1} in October. This flow through fissures in the limestone is, from time to time, greater than the flow of the Danube onto the outcrop, so that the Danube runs dry over a short length. In contrast, the rates of interstitial flow in the Great Artesian Basin of Australia are less than 5 m a^{-1} (less than 1.6×10^{-7} m s^{-1}) (Habermehl 1980).

From GOFF, J. C. & WILLIAMS, B. P. J. (eds), 1987, *Fluid Flow in Sedimentary Basins and Aquifers,* Geological Society Special Publication No. 34, pp. 3–18.

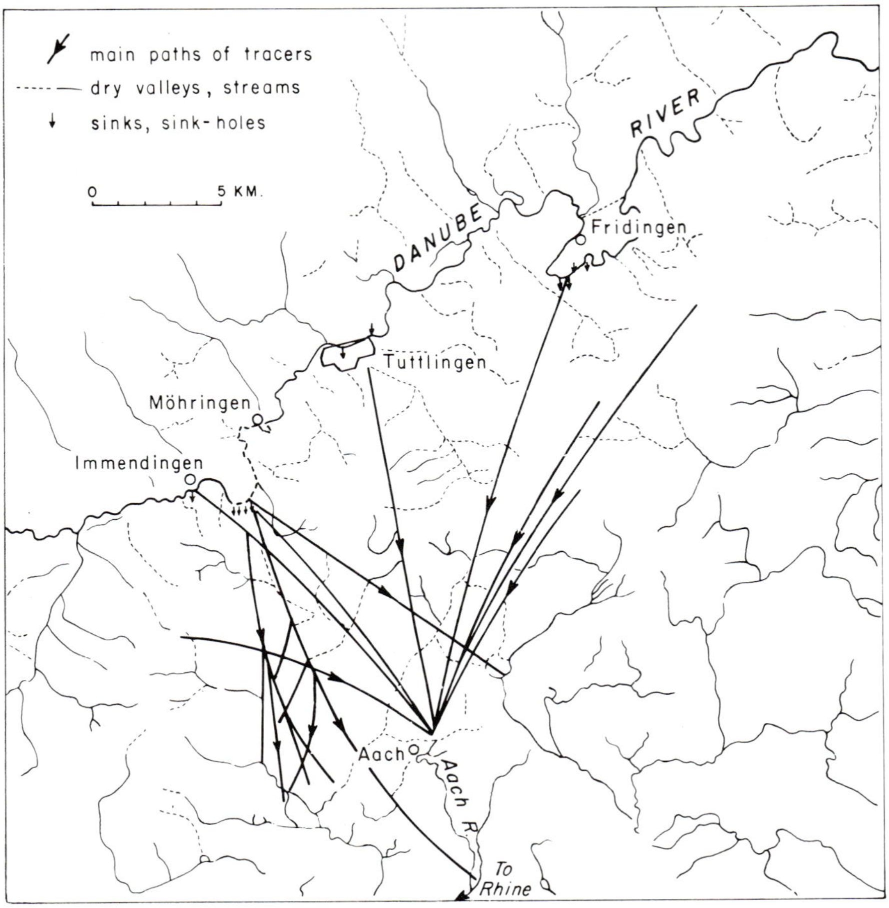

FIG. 1. Sub-surface flow directions in the Malm in the area between the Danube and the springs near Aach, Baden–Württemberg (after Käss & Hötzl 1973).

There is, as yet, no reliable method of age-dating water beyond the ^{14}C limit of about 30 000 years, but there is hope that ^{36}Cl, with a half-life of 308 000 years, will become a reliable tool for dating old water.

One great difficulty that has hampered study of fluid flow has been widespread misunderstanding and confusion over the physical principles of fluid flow and its terminology. The basic physics of fluid flow in porous media does not require abstruse mathematical treatment for its proper understanding. In the *Annual Report of the U.S. Geological Survey* for 1897–98, there are two papers on groundwater flow. King (1899) reveals a deep understanding of the subject without using any mathematics, and his paper is well worth reading today. Slichter's (1899) paper is full of mathematics and is rightly ignored today. However, old work should not always be ignored. Versluys (1917) understood fluid flow and introduced the term 'potential'. Amongst other things, he clearly understood the nature of the Ghyben–Herzberg lenses under the coastal dunes of Holland (Versluys 1919). Hubbert (1940, 1969) developed a more complete theory of groundwater motion, but it remains a topic of difficulty that is widely misunderstood (even by some mathematicians and physicists). We must, of course, seek proper mathematical modelling of some processes, such as water-flooding of oil reservoirs, but it is not essential for geologists in their daily work.

Physical principles of liquid flow

The basic physical principles of liquid flow on a macroscopic scale are relatively simple and can be expressed as an axiom with two corollaries.

Axiom. Liquid that is flowing loses energy.

Corollary. The free upper surface of a body of liquid at rest is horizontal.

Corollary. If the free upper surface of a body of liquid is not horizontal, the liquid is in motion. All the liquid is moving: liquid under the elevated free surface is moving towards liquid under the depressed free surface, and this movement will continue until equilibrium is achieved and the free surface is horizontal.

The axiom and its corollaries apply to water in porous and permeable rocks, but there is a scale effect. On a scale comparable with the grains of a sedimentary rock, the air–water interface is deformed near the water–solid interfaces by surface tension, and the air–water interface can be raised some metres in fine-grained rocks (i.e. rocks with small pores) by capillary pressure. This capillary rise is not free water or it would not be elevated. It has been known for several centuries that if a length of pipe is filled with fine sand and one end is placed in a pan of water, the water will rise in the sand. If it were free water, a hole could be drilled through the wall of the pipe and water allowed to flow out and so to do work while returning to the pan. This, as Perrault pointed out more than 300 years ago, would be perpetual motion (Perrault 1674 (English translation 1967)).

The axiom and corollaries apply to the water beneath a crude oil accumulation, and they can be extended to notional surfaces above confined aquifers, i.e. the *potentiometric surface* (less desirably, the *piezometric* surface). This is the surface passing through the points to which water would rise in a vertical manometer if one could be inserted into the aquifer. It is a notional surface that is related directly through its elevation to the *energy* of the water in that position in the aquifer. The higher the surface, the greater the energy of the water in that position. Water flows from positions of higher energy to positions of lower energy, and movement involves loss of energy.

The basic principles can be illustrated by an hypothetical cross-section through an artesian basin, and this will also introduce the proper terms and their meanings. Figure 2 is a schematic cross-section through an elevated intake area and a confined aquifer. Two boreholes provide data from which the potentiometric surface can be sketched. From these data we infer the following.

(a) The water is in motion, so the artesian aquifer is leaking.

(b) The water is flowing in the direction down the slope of the potentiometric surface.

(c) The potentiometric surface can be contoured, the contours being also lines of equal potential or energy, and these lines are normal to the flow lines.

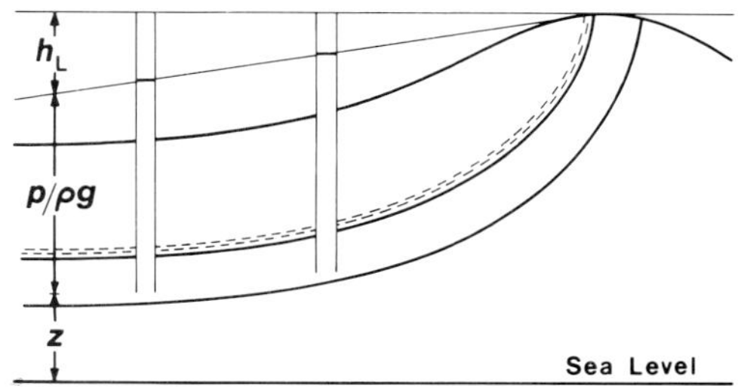

FIG. 2. Schematic cross-section through a leaking artesian basin showing the elevation head z, the pressure head $p/\rho g$ and the head loss.

If (a) and (b) were not true, the water would be at rest and the potentiometric surface would be horizontal and level with the intake area. Inference (c) is logically satisfying and can be proved analytically (but it is not without difficulty because tracers from a point source diverge downstream from that source).

Terminology

The terminology of groundwater is very confused (indeed, it is unsafe to use the second edition of the AGI *Glossary of Geology* (1972) for any hydrogeological term). The energy of water in an aquifer is proportional to its *total head* (Fig. 3), which is the algebraic sum of its *pressure head* $p/\rho g$ and its *elevation* above (+) or below (−) an arbitrary horizontal datum such as sea-level (all with the dimension of length).

We frequently use *gradients* of quantities. The *gradient of total head*, or *hydraulic gradient*, is the loss of total head per unit macroscopic length along the aquifer (not the horizontal distance between the two points of measurement). *Pressure gradient* has two meanings, pressure divided by depth (p/z) and the rate of change of pressure with depth ($\Delta p/\Delta z$), so we must specify the one meant.

Normal hydrostatic pressures are the pressures that would be measured in an unconfined body of water at rest and, by extension, pressures that are consistent with a free-water table near the land surface onshore or near sea-level offshore. A *normal hydrostatic pressure gradient* is the weight density ρg of the water, which can be expressed in units of pressure per unit of vertical depth.

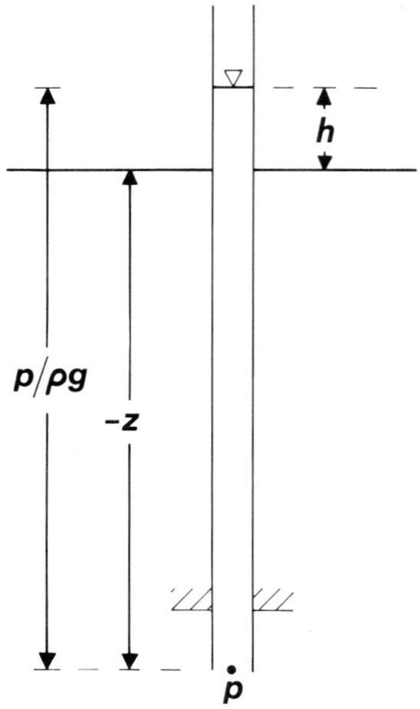

FIG. 3. The total head h is the algebraic sum of the elevation head z and the pressure head $p/\rho g$. The elevation head and the total head are measured from an arbitrary horizontal datum.

Abnormal pressures are pressures that are not normal hydrostatic pressures, but this term is usually used for pressures above hydrostatic, with pressures below hydrostatic being called *subnormal*.

The term *connate water* is commonly used for pore water that is supposed to be the water that was originally trapped in the sediment when it accumulated. It is not at all clear that this is a useful term, and if it applies strictly it applies only to abnormally-pressured under-compacted mudstones. Even then, diagenesis has probably altered the chemistry of the water (but there are great difficulties in obtaining true water samples from mudstones).

The flow of pore water is the medium for much of the diagenesis of sedimentary rocks, particularly the growth of authigenic clay minerals. The main evidence for this lies in petroleum reservoirs where diagenesis is either retarded relative to the water-bearing part of the same rock unit or is absent (e.g. Philipp *et al.* 1963a, b; Campbell & Smith 1982; Lambert-Aikhionbare 1982). This is also one of the lines of evidence that petroleum accumulations are totally impermeable to water flow. An extension of this process is that water flow affects the geothermal gradient, and so does the presence of petroleum fields. Heigold *et al.* (1971) found that the geothermal gradients in the Fairfield basin, Illinois, U.S.A., were greater in the areas of crude oil accumulation than in the areas barren of oil. They attributed the relatively higher temperatures around oil fields to the deflection of ground water around them on account of their 'low water conductivity'. (This is a nice example of ambiguity in natural science because it could be argued that the petroleum was generated because of the higher geothermal gradients.)

Factors that affect fluid flow

Properties of sedimentary rocks

Mean grain size is a property that is regarded as affecting fluid flow, but it is obviously an indirect measure of pore size, which is what is wanted. It can be shown (see Chapman 1981, pp. 52*ff*.) that the harmonic mean grain diameter is a better measure of pore size than the geometric mean. Better still would be a characteristic dimension that embraced both size and shape, which we shall consider after the hydraulic radius.

Hydraulic radius is the total volume of effective pore space divided by the wetted area of solids (the area of the liquid–solid interface). It is a very good measure of both size and shape for flow in pipes and channels, but it is a difficult quantity to measure in sedimentary rocks. It is a very small number (with the dimension of length) in sedimentary rocks, independent of the volume of rock through which the fluid flows (except very small volumes), so it can be regarded as the characteristic dimension of the pores, taking both size and shape into account.

Reverting to the problem of *mean grain size*, we can define a characteristic dimension for the solids that is similar to the hydraulic radius of the pore space of a sedimentary rock. This characteristic dimension is the total volume of solids divided by their bounding surface area (and I agree with Winkelmolen (1982) that *shape* has the dimension of length). We shall call this the *characteristic grain dimension* (or simply characteristic dimension when there is no ambiguity) and denote it by D; it combines the size and shape of the grains into one number with the dimension of length. Such a characteristic grain dimension may have value beyond this hydraulic use.

When this characteristic dimension is used, the expression for the hydraulic radius of the pore space of sedimentary rocks becomes

$$R = \frac{fD}{1-f}$$

where f is the fractional effective porosity and $f/(1-f)$ is the void ratio.*

Tortuosity is the ratio of the true mean flow-path length through the rock to the macroscopic length of rock traversed (i.e. it is greater than unity and would be equal to unity if the pores were straight capillaries). It is dimensionless, but it is a vector quantity (with magnitude and direction) and accounts for the anisotropy of permeability in many sedimentary rocks. Tortuosity can be estimated from the formation resistivity factor, which is the ratio of the resistivity of a rock saturated with an electrolyte to the resistivity of the electrolyte:

$$F = \frac{R_o}{R_w}$$

It can be shown (see Chapman 1983, pp. 113–114) that the tortuosity T is given by

$$T = (Ff)^{1/2}$$

where f is, as before, the fractional porosity. The *pendular-ring* part of the pore space around point contacts probably has virtually no effect on permeability and fluid flow. Füchtbauer (1967, p. 359) found that ordinary cementation had very little effect on permeability.

Permeability

There are two measures of permeability, the coefficient of permeability (also called the hydraulic conductivity), and intrinsic permeability. The first is a measure of the ease with which water flows through the material, and the second is a property of the rock that is independent of the nature of the fluid in it.

The coefficient of permeability K has the dimensions of a velocity, but it is a notional velocity in sedimentary rocks because only part of the cross-sectional area is available for fluid flow and tortuosity increases the flow-path length. Included in the coefficient of permeability is a component that takes the fluid properties into account, and one that takes the pore size and shape and their hydraulic properties into account. It is not, therefore, a constant—even for the same materials.

The intrinsic permeability k has the dimensions of an area, and the units are usually square centimetres. In SI Units, the unit is square micrometres. The *darcy* is also a measure of intrinsic permeability† and, for practical purposes, it is equal to 10^{-8} cm^2 and 1 μm^2. Intrinsic permeability, being a property of the rock that is independent of the fluid in it, is the coefficient of permeability without those quantities ascribable to the fluid flowing. Hence $K = k\rho g/\eta$, where ρ and η are the mass density and coefficient of viscosity of the fluid and g is the acceleration due to gravity.

The intrinsic permeability k is a function of measurable parameters of sedimentary rocks. Using the characteristic grain dimension, as defined above, we obtain

$$k = cfT^{-3}R^2$$
$$= \frac{cT^{-3}f^3D^2}{(1-f)^2}$$
$$= \frac{cF^{-1.5}f^{1.5}D^2}{(1-f)^2}$$

where c is a true constant with a value probably equal to or close to 0.5, F is the formation factor, f is the fractional porosity, T is the tortuosity and R is the hydraulic radius with D as the characteristic grain dimension (cf. Chapman 1981, pp. 55–57, 65).

Viscosity

Viscosity is the property of a fluid's internal resistance to flow (the greater the resistance, the larger the viscosity). There are two measures of viscosity: the coefficient of viscosity (also known as absolute viscosity or dynamic viscosity) and kinematic viscosity.

The coefficient of viscosity η has the dimensions $ML^{-1}T^{-1}$ and is defined as the ratio of shear stress to shear strain. The unit is newton second per square metre (N s m^{-2}) or the *poise* (dyne second per square centimetre or grams per centimetre second (g cm^{-1} s^{-1})).

The kinematic viscosity ν is the ratio η/ρ of dynamic viscosity to mass density of the fluid

* The practical problem of measuring the bounding areas needs re-examination. Gas adsorption techniques are well established for large specific surfaces (surface area of solids divided by bulk volume (dimension L^{-1})) and could be used routinely and cheaply (e.g. Barrer & Barrie 1952; see also the *Specialist Periodical Reports in Colloid Chemistry*, Chemical Society, London). Statistical techniques on photomicrographs could also be used (Corrsin, 1955).

† It is one of the tragedies of science, as Hubbert has pointed out on several occasions, that Henry Darcy's name should be given to a unit that was improperly defined (Wyckoff *et al.* 1933, 1934). Their definition, which is inconsistent with their description of the apparatus for measuring the darcy, is physically erroneous in that it ascribes the flow to a pressure differential, and improper in that it involves pressure in *atmospheres*. In the first paper, Darcy's name was incorrectly spelt (an error that was fortunately not carried into the name of the unit).

which appears in many mathematical expressions in fluid mechanics. It has the dimensions of a length multiplied by a velocity [L^2T^{-1}], and it gets its name because it concerns motion without reference to force. The unit of kinematic viscosity is square metres per second in SI Units, or the *stokes* (which is square centimetres per second).

The viscosity of water is largely a function of temperature, and so it is a variable in most geological contexts (Fig. 4).

Darcy's law

Darcy's law, which is derived from the experiments of Henry Darcy (1856), can now be written

$$q=\frac{K\Delta h}{l}$$

$$=\frac{k(\rho/\eta)g\Delta h}{l}$$

$$=\frac{cF^{-1.5}f^{1.5}D^2}{(1-f)^2}\frac{(\rho/\eta)g\Delta h}{l}$$

where q is the *specific discharge* [LT^{-1}], K is the coefficient of permeability [LT^{-1}], k is the intrinsic permeability [L^2], ρ is the mass density [ML^{-3}] of the liquid flowing and η is its absolute viscosity [$ML^{-1}T^{-1}$]. The term $g\Delta h$ is an energy per unit of mass [L^2T^{-2}] where Δh is the difference of total head between two points in the aquifer separated by distance l, which is the macroscopic length of flow along the aquifer. The other quantities are given on pp. 6–7.

Darcy's law applies to liquids. No lower limit has yet been demonstrated. There is an upper limit, but natural ground water flow is probably always well within the realm of Darcy's law. Darcy's law must be modified for gases because mass density is not constant but a function of pressure (see Hubbert 1940, pp. 823–826; Klinkenberg 1942).

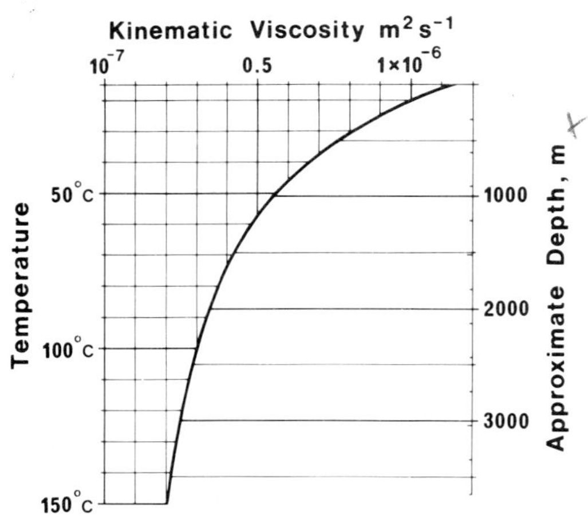

FIG. 4. Approximate kinematic viscosity of water under sub-surface conditions with a geothermal gradient of 36°C km^{-1} and a normal hydrostatic pressure gradient (interpolated from data of Gray, 1957, p. 2–138, Table 2n–1, and p. 2–209, Table 2v–6).

Sedimentary basins

A sedimentary basin is an area in which sediments have accumulated during a particular time span at a significantly greater rate, and so to a significantly greater thickness, than in surrounding areas. The essence of a sedimentary basin is the *accumulation* of sediment relative to the surrounding areas, and its relative rather than absolute thickness. The concept of a sedimentary basin is distinct from that of a physiographic basin. A physiographic basin is a depression in the surface of the land or sea-floor that may or may not fill with sediment. Part of the area of a physiographic basin is an area of erosion, providing materials that will form sediment in other areas of the physiographic basin. In those areas in which sediment accumulates, the upper surface of the sediment does not necessarily form a depression: it may be physiographically indistinguishable from neighbouring areas that are not accumulating sediment. Sedimentation is one thing: sediment accumulation is another. The primary control on sediment accumulation is subsidence (assuming sediment supply) because without subsidence the sediment is prone to dispersion.

The lateral continuity of sediments and sedimentary rocks in the stratigraphic record, which is important for fluid flow in sedimentary basins, is a matter of three dimensions of space and one of time. A rock unit may be discontinuous over an area for a number of reasons: (a) the sediment was not distributed over the whole area; (b) the sediment was distributed but did not accumulate over the whole area; (c) the sediment was distributed and accumulated over the whole area, but accumulation was only temporary over parts of the area owing to changing energy patterns.

Lateral continuity of rock units is not just a matter of environments, and the margin of a discontinuous rock unit in a sedimentary basin is not necessarily the margin of that environment in the physiographic basin. The sequence and areal distribution of rock types in a sedimentary basin constitute an incomplete record in space of the variations of the environments in the physiographic basin over that area in time. (For a fuller discussion of sedimentary basins see Chapman 1983, pp. 1–40.)

Transgressions and regressions

When subsidence relative to base level in a physiographic basin exceeds the supply of sediment and the area has a capacity to accumulate a larger volume of sediment than is supplied, the sea tends to deepen over the depositional surface and the facies tend to migrate towards the land (Fig. 5). This is a *transgression*. Sedimentary basins tend to be enlarged during transgressions, but the enlargement is only permanent if subsidence relative to base level is permanent. We are concerned here with the dominant first-order effects, and not with the ephemeral effects that are important for rock-unit distribution. If the sea over part of a physiographic basin becomes shallower, the facies migrate seawards: this is a *regression*. By extension of these concepts, sedimentary sequences in which deeper-water sediments accumulated on top of shallower-water sediments are called *transgressive sequences*, and those in which shallower-water sediments accumulated on top of deeper-water sediments are called *regressive sequences*. In general, it is wrong to infer that regressive sequences indicate uplift and the beginning of orogeny in the sedimentary basin because there is much evidence from petroleum geology that subsidence continues during the accumulation of regressive sequences (Chapman 1983). Regression differs from transgression in the volumes of sediment supplied to the sedimentary basin, and this is more important than changes of sea level. The significance of orogeny is that it creates mountains that generate increasingly large volumes of sediment outside the sedimentary basin, and much of this sediment accumulates in the subsiding neighbouring sedimentary basin. Figure 6 illustrates this schematically.

There are also lithological associations in sedimentary sequences. There is an association between carbonates/evaporites and transgressive sequences, and between regressive sequences and sand/sandstones. Transgressive sequences may contain carbonates or sandstones, but regressive sequences do not normally contain carbonates. The explanation for these associations appears to be that carbonates accumulate where sands are absent during transgressions over subdued topography, but that the other associations result from an adequate supply of sand from a more pronounced topography. In other words, it is the rate of sediment supply that determines whether a

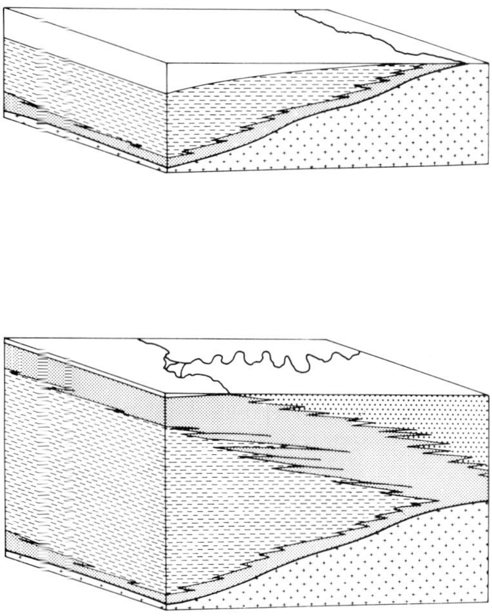

FIG. 5. Schematic block diagram showing a transgressive sequence (above) and a transgressive sequence followed by a regressive sequence (below): neritic facies, dashes; neritic–paralic, fine dots; paralic–terrestrial, coarse dots.

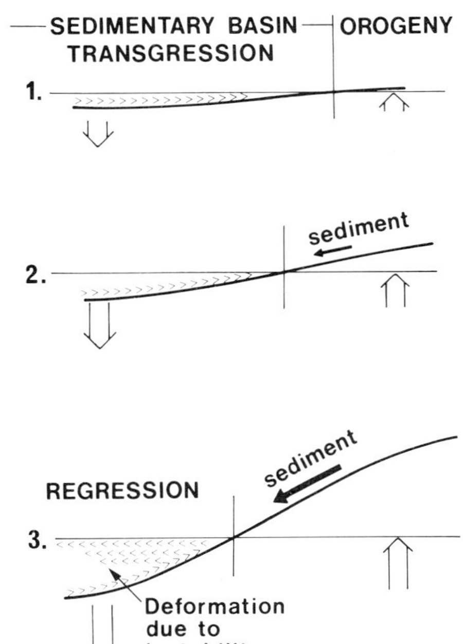

FIG. 6. Schematic representation of the relationship between sedimentary basin development and orogeny.

sedimentary basin preserves the record of a transgression or a regression in the development of the physiographic basin, and it also determines whether carbonates or sands shall accumulate in the transgressive sequence.

There are also structural associations with sedimentary sequences. Transgressive sequences tend to be undeformed, with few small faults and small regional dips. Regressive sequences tend to be folded and faulted. We shall return to these associations in the context of petroleum accumulations.

Stratigraphy

Stratigraphy controls water movement on account of the variable permeability of various rock types. When water flows, it flows mainly through the permeable beds (sandstones and some carbonates). However, water also flows on account of compaction, so a distinction must be made between the pre-orogenic development of sedimentary basins and those that have suffered orogeny and reached some measure of maturity and mechanical equilibrium. This must not be interpreted as a pre-deformation stage and a post-deformation stage because there is a great deal of evidence from petroleum geology that sedimentary basins suffer deformation during the accumulation of the sediment in them (Chapman 1983, pp. 23–38).

As a broad but useful generalization, permeable beds are less compactible than the relatively impermeable beds, so water flow is generated during the pre-orogenic stage by compaction of compactible beds and takes place largely in the less compactible beds. Sandstones and mudstones are clearly in this category; limestones and dolomites tend to be permeable, and marls tend to be relatively impermeable. Once compaction is complete, or near completion, the hydraulic properties of the rock units may change because of the development of fractures that create permeability where little existed before.

Fluid flow in pre-orogenic sedimentary basins

Sedimentary basins fall into two main classes. Basins that typically begin with a transgression and end with a regression are the common sort, to be found on all continents and in some continental shelves. In contrast, rift basins appear to begin with a dominantly regressive phase followed by a dominantly transgressive phase with fine-grained sediments and sedimentary rocks (mudstones, marls and carbonates). These are perhaps characteristic of aseismic continental margins, but they are not exclusive to them. The ordinary sedimentary basin develops stratigraphic patterns that have a profound influence on the flow of fluids in them, but the patterns of rift basins are not so clear.

We shall first consider ordinary sedimentary basins with a transgressive–regressive cycle.

Transgressive phase

Water begins to flow as soon as sediment begins to accumulate and compact in a new sedimentary basin. The pattern of this flow is governed by the stratigraphy, which is essentially a pair of diachronous rock units, either mudstone on carbonate or mudstone on sandstone. The deeper-water facies lies on the shallower-water facies, the more compactible on the less compactible, and the less permeable on the more permeable. The basal permeable unit is the conduit for water of compaction, and the migration direction is towards the land of the time.

Mudstone on carbonate

In a new sedimentary basin both these rock units accumulate at the same time but in different environments, and both are continuously in 'outcrop' in the areas of their environments. As the transgression proceeds, so the environments (and their rock units) migrate towards the land and the mudstone comes to overlie the carbonate. Compaction of the mudstone expels pore water both upwards and downwards, with the downward flow inducing lateral flow in the carbonate towards the land of the time. This is the natural direction of flow in a permeable rock unit that terminates near sea level.

If organic reefs form on the sea-floor, they may act as vents for pore-water flow until they are sealed by mudstone (Fig. 7). This happens if the rate of subsidence exceeds the capacity of the reef-building organisms to maintain their position near sea-level, or when the organic reef is smothered by mud and exterminated. Of the two, the latter is probably the common cause because rates of subsidence are normally so very slow. Once the reef is covered with mud, it ceases to be a vent and becomes part of the larger hydraulic system by receiving pore water that is expelled downwards from the overlying compacting mudstone. This is the beginning of a development that may lead to its becoming an attractive petroleum prospect.

Within the compacting mudstone there exists a surface that separates downward from upward

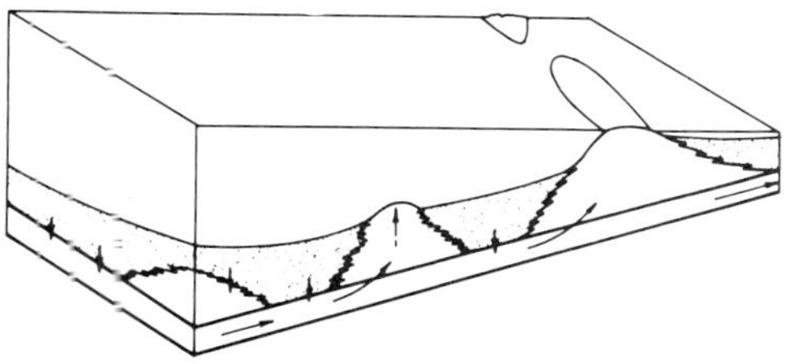

FIG. 7. Organic reefs may act as vents for pore water until they are sealed by mudstone.

flow, and this surface is probably a little below the level of half-thickness in the unit. As the mudstone thickens, this surface tends to rise stratigraphically (there are many possible combinations of compaction rate, subsidence rate and accumulation rate that give variations on this theme).

The downward flow in the mudstone is a consequence of the energy imparted to the pore water by compaction and the relatively small permeability of the unit. These influences tend to raise pore pressures above the hydrostatic within the mudstone, while those in the more permeable member of the diachronous pair remain hydrostatic because the pores are connected to the sea. There is therefore an energy gradient downwards that induces flow within the mudstone.

It is a characteristic of transgressive sequences in ordinary sedimentary basins (and, indeed, in rift basins) that there is very little deformation. Typically we find transgressive carbonates with regional dips of less than 5° (e.g. middle and upper Devonian of Western Canada, Cretaceous of NE Mexico and Palaeocene of Libya). Faults are usually rare, so this fluid flow is essentially in a single rock-stratigraphic unit that is porous and permeable (perhaps with fracture porosity in joints).

Mudstone on sandstone

The same principles apply to transgressive sequences with sandstone as to those with carbonates—downward flow of pore water in the lower part of the mudstone to the more permeable member of the diachronous pair followed by lateral flow towards the land of the time.

An important variation on this theme is the inundation of a pre-existing topography from which sand was being generated. The hill becomes an island, and soon after the island has been submerged sand generation ceases and the whole may be covered with mud. Once so sealed, pore water will continue to flow through the porous and permeable sands (and also the basement, if it is permeable), ultimately reaching the sea near sea level. This is an important class of petroleum trap: the palaeogeomorphic trap. Water flows round any accumulation of petroleum.

Petroleum in transgressive sequences

As a generalization, primary migration of petroleum in transgressive sequences (both arenaceous and calcareous) is downwards in the mudstone of the diachronous pair, and then secondary migration is lateral in the more permeable unit, towards the land of the time. These directions are also those of water migration, but the secondary migration of petroleum is assisted by buoyancy. The petroleum source rock accumulates in a deeper-water environment that is almost certainly not contiguous with that in which the permeable facies accumulates. Because transgressive sequences are not commonly deformed, there are usually no structural traps for the petroleum, so that the habitat of petroleum in transgressive sequences is in stratigraphic traps (reefs and palaeogeomorphic traps) rather than structural traps. Because the relief on the permeable unit of a carbonate transgression is usually small, with regional dips of 5° or less, the flow of pore water may affect crude-oil accumulations significantly, giving rise to hydrodynamic traps (Hubbert 1953; Dahlberg 1982).

Gussow's differential entrapment theory

Silurian reefs in the Michigan basin, U.S.A. (Gill 1979, p. 614, Fig. 4), have a petroleum pattern that eloquently confirms Gussow's (1954) princple and also supports the theory that petroleum accumulations affect water flow. Prior to the accumulation of petroleum, water flow in a reef system on a carbonate platform (which seems to be a necessary feature of the stratigraphy if reef provinces are to become important petroleum provinces) involves movement of all the free water in the system. However, as soon as petroleum begins to accumulate in a trap, pore water is displaced downwards and the accumulation becomes totally impermeable to the flow of water as soon as the irreducible water saturation has been reached (Chapman 1982). In those areas where Gussow's principle is evident, secondary migration paths may be quite long—from below the deepest gas accumulation to the highest oil accumulation—and removal of water-soluble components may be a cause of the observed increase in the specific gravity of crude oils in successive accumulations up-dip.

Regressive phase

The stratigraphy of a sedimentary basin is largely determined by the progress of orogeny just outside it. During the transgressive phase, nearby orogeny is not apparent and subsidence creates a volume that is greater than the volume of sediments supplied, and the water tends to deepen with time. As orogeny begins to create mountains, so the volume of sediment generated and supplied increases until it becomes greater than the volume created by subsidence. Sediments then prograde, environments tend to migrate seawards and the corresponding rock units form a three-component diachronous unit of outer neritic mudstones, inner neritic mudstones and sandstones, and paralic sandstones and mudstones.

The stratigraphy of regressive sequences comprises alternating sandstones and mudstones, with the sand/shale ratio increasing upwards. Nearly all these sandstones are connected to the surface and have pore water at normal hydrostatic pressures. Of course, the neritic sediments accumulated in water.

The neritic mudstones tend to compact, expelling a commensurate part of their pore water. This water flows upwards and downwards to sandstones of considerable lateral extent, and further flow in these is lateral towards the land of the time (Fig. 8). This direction is imposed by the geometry of the sandstones, which tend to wedge out away from the land and to be in contact with the sea near the land.

In detail, flow paths are inferred to be close to vertical in mudstones because compaction is induced by gravity, and in each of those volumes of mudstone intercalated between two sandstones there is a surface that separates upward from downward flow. The downward potential (or energy) gradient above each sandstone creates a perfect barrier to upward migration of petroleum (Chapman 1972, p. 791). Each sandstone (indeed, each permeable layer with sufficient lateral continuity) acts as a drain, and the surfaces within the mudstones act as total physical and chemical barriers to the migration of fluids. In this manner, isolated hydraulic units exist within the stratigraphy of regressive sequences, each drained by a sandstone (Fig. 9).

These alternations of mudstone and sandstone can be regarded as second-order transgressions and regressions superimposed on the dominant first-order regression, possibly from sea-level

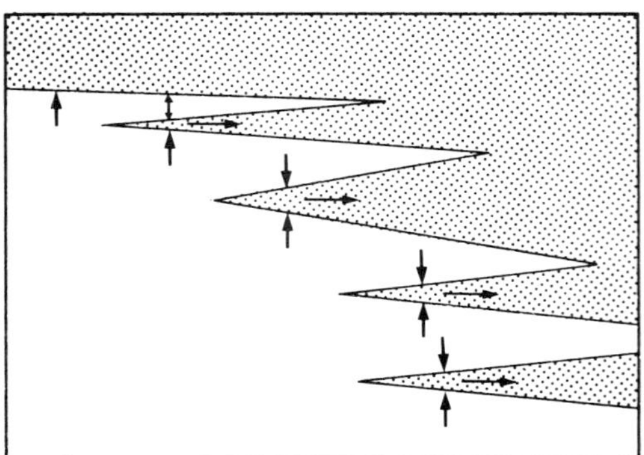

FIG. 8. Schematic cross-section through a regressive sequence showing pore-water flow directions imposed by the stratigraphy.

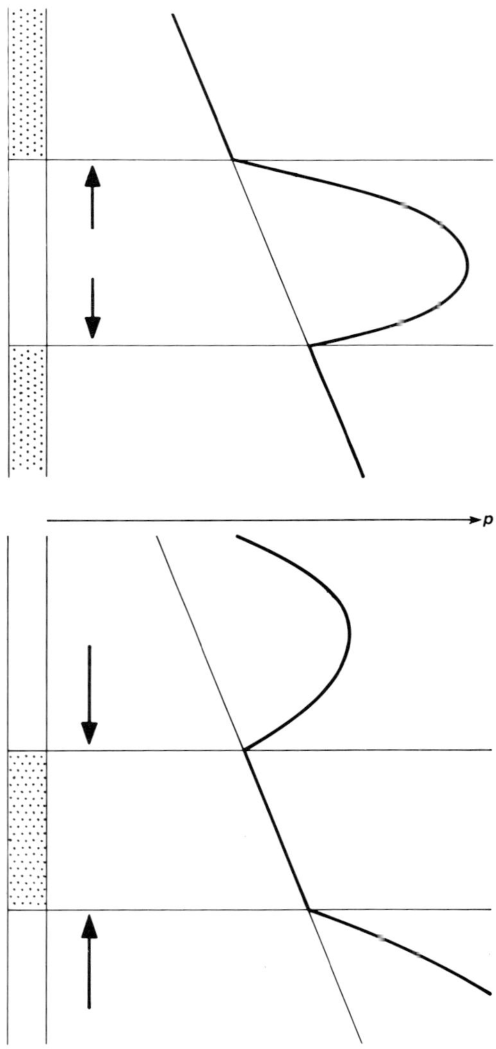

FIG. 9. Pressure–depth plots illustrating the directions of water flow in a compacting mudstone intercalated between two permeable sandstones. A perfect barrier is created in the mudstone (upper diagram), and an isolated stratigraphic unit is drained by the intercalated permeable sandstones (lower diagram).

fluctuations or possibly from redistribution of sand on the sea-floor. Ephemeral transgressions may be very important for petroleum geology because they are more favourable for the development of petroleum source rocks than ephemeral regressions on account of their more rapid accumulation.

Abnormal pressures

Many Tertiary sedimentary basins around the world have, in the massive mudstone unit that intervenes between the permeable transgressive and the permeable regressive units, pore pressures that greatly exceed the normal hydrostatic value. In some areas they are so large that they approach the pressure exerted by the total overburden. (For fuller discussion of abnormal pressures see Fertl 1976, Chapman 1983 and Gretener & Feng 1985.)

There is little doubt that most of such abnormality is caused by gravitational loading of a thick, relatively impermeable, compactible rock unit. Many over-pressured mudstones are undercompacted and have physical properties that would be expected in a mudstone at a much shallower depth. Sedimentary rocks can only compact if the pore fluids can be compressed or a commensurate part expelled, so an important cause of abnormal pressures in mudstone is insufficient permeability of the mudstone so that pore water does not flow sufficiently fast for normal compaction to take place.

Abnormal pressures are stratigraphically controlled. When they occur below the base of regressive sandstones, the top of abnormal pressures is a diachronous surface (as Dickinson 1951, 1953, so clearly saw in Louisiana, U.S.A.). There is a structural style (onshore, at least) associated with such abnormal pressures: long narrow sinuous anticlines separated by broad gentle synclines. There is also a marked association between regressive sequences, abnormal pressures, growth faults, growth anticlines and petroleum fields in structural traps with multiple reservoir sandstones (commonly more than 50) with the crude oil tending to be progressively lighter in stratigraphically deeper, or older, reservoirs. (Abnormal pressures in rift basins will be discussed later.)

Although abnormally-pressured mudstone is a zone in which fluid flow is retarded, slow compaction of such a large volume leads to important and protracted water flow, in permeable beds above and below it, towards the land from which the regression came. The growth faults overlying abnormally-pressured mudstone are also stratigraphically controlled. They are

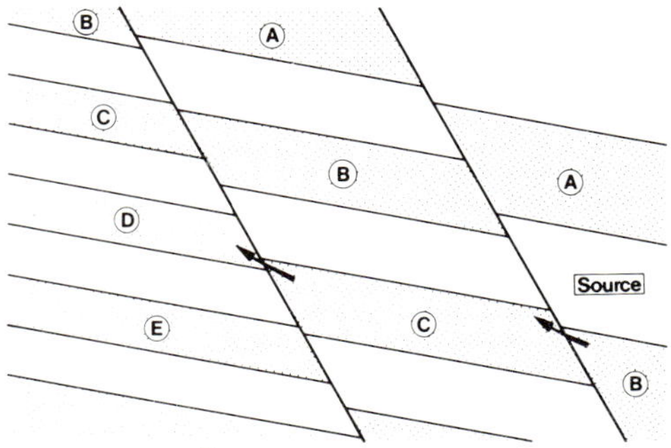

FIG. 10. Migration of fluid *through* a growth fault tends to transfer the fluid to a stratigraphically deeper sandstone.

normal faults with the down-thrown side away from the land from which the regression came, and younger faults are on the down-thrown side of the older faults. Therefore fluid flow in the sandstones is interrupted and transferred to other sandstones as they become juxtaposed (Fig. 10). When a sandstone finally becomes juxtaposed to the massive mudstone, it too may become overpressured.

Role of faults in fluid flow

The role of faults as conduits for petroleum migration has been debated for at least 50 years, and Weeks's account is still well worth reading (Weeks 1958, p. 32). It is my view that the current popularity of faults as conduits follows, as a consequential hypothesis, from geochemical postulates of source rocks much deeper than their petroleum reservoirs.

There is no true general statement of the form: 'Faults do/do not act as conduits for fluid flow'. Such statements must be qualified to take the nature and strength of the faulted rocks into account. There is no doubt that some faults at shallow depth in young sedimentary rocks, and faults in well-consolidated rocks, can act as conduits for fluids. It is most unlikely that open fractures can exist in Tertiary mudstones below depths of a kilometre or so because the cohesive strength of the mudstone is too small. There are several petroleum fields in which the data strongly suggest that the faults not only are sealing (otherwise there would be no petroleum field) but that they always have been (see, e.g., Chapman 1983, p. 245).

Growth faults not only cut unconsolidated sediments (as indeed they must) but also move very slowly. For example, assume that rapid 'accumulation' of 1 km of growth fault with a throw of 100 m took 500 000 years. This is a movement of 0.2 mm a^{-1} on average, and it is hard to conceive of open fractures resulting in such material under these conditions.

The observation that faults in consolidated rocks can form conduits suggests that the role of fractures in fluid flow in sedimentary basins changes with time, and that by the time the sedimentary history comes to an end and compaction to mechanical equilibrium has been achieved, fractures may form important paths for fluid flow.

As regards flow of fluids *through* faults from one aquifer to a juxtaposed aquifer, there is direct field evidence that crude oil has flowed through faults from one reservoir to another under oil-field conditions (Smith 1980), and it seems quite certain that faults are not normally barriers to water flow when there are juxtaposed permeable beds.

Petroleum in regressive sequences

Much of the world's known petroleum occurs in regressive sequences, and the habitat of this petroleum is structural traps (anticlinal and fault traps) which are typically growth structures. This habitat of petroleum is, of course, a direct consequence of the geology of regressive sequences, and it is in strong contrast with that of transgressive sequences. Stratigraphic traps form an insignificant part of the known petroleum reserves of regressive sequences, and they are confined to sand lenses and pinch-outs.

Petroleum-bearing anticlines are typically culminations of long sinuous trends. This is the structural style of young regressive sequences with long narrow sinuous anticlines separated by broad gentle synclines. There are two expressions of this style: the independent anticlines, and the roll-over anticlines in the down-thrown blocks of growth faults (as found, for example, in the U.S. Gulf Coast and the Niger delta). The independent anticlines appear to have a slightly different character offshore from that onshore; the latter are usually steeper and more deformed than the former. Part of the explanation for this difference is probably that both are due to the mechanical instability that is inherent in a young regressive sequence, and that the younger structures on the seaward side of the older structures have had less time to develop.

Petroleum fields in regressive sequences also tend to have many reservoirs. In the Handil field (Kalimantan, Indonesia), for example, there are 150 individual reservoirs (Verdier *et al.* 1980). In the Niger delta there are perhaps fewer reservoirs in fields, but there are many sandstones in most fields.

Faulting is important in many fields in regressive sequences, and some fields, such as Seria field in Brunei (Schaub & Jackson 1958; James 1984), are fault traps rather than anticlinal traps.

Petroleum migration is governed largely by stratigraphy, and stratigraphy has also caused the structures that began to develop while sediment was still accumulating. It is thus clear that secondary migration paths are relatively short, being limited to the dominant wavelength of the deformation (typically 15–25 km).

Geological arguments for the position of source rocks and the paths of primary migration differ significantly from the geochemical arguments. Evidence of interstratified source rocks is the variability of crude-oil and water compositions between different reservoirs of the same field. This argument was used by Fränkl & Cordry (1967) and Short & Stäuble (1967) for fields in the Niger delta. If the petroleum source rock is a stratigraphic facies of a regressive sequence, primary migration is vertical and relatively short, being limited to about half the mudstone thicknesses.

If the geochemical arguments are correct, and the sources of petroleum in the U.S. Gulf Coast, the Niger delta and many fields in Indonesia lie much deeper than their reservoirs, then the difficulties of accounting for dry sands between source and accumulation must be addressed, apart from the problems of flow up faults.

Rift basins on aseismic continental margins

It is not at all clear that rift basins and half-graben associated with aseismic continental margins should be treated as pre-orogenic sedimentary basins; nevertheless, fluid flow in them during development will be different from that which will obtain when stability and maturity have been reached.

Rift basins and half-graben on aseismic continental margins have, as Kent (1975a, b, 1977, 1978) has pointed out, a remarkably similar sequence of events in their development. There are two stages.

1 At first, normal faulting creates a graben that rapidly fills with sediment (largely non-marine in most basins) from the margins, and the graben tends to be enlarged by a series of sub-parallel normal faults. (These are growth faults: they moved while the sediment accumulated and this affected the thickness of rock units.)

2 The normal faulting ceases, the environment changes and fine-grained sediments accumulate that are virtually devoid of structure. One or two unconformities or disconformities occur between the two components of the stratigraphy, and faults are virtually confined to the pre-unconformity sequence.

The timing of these events is typically Permian or Triassic for the inception, late Jurassic or early Cretaceous for the unconformities or disconformities and the cessation of fault movement, followed by late Cretaceous and Tertiary accumulation of fine-grained sediments—a span of 250 Ma or so. The so-called Cretaceous Transgression is part of this global development.

For example, the stratigraphy and structure of the NW shelf of Australia is remarkably similar in style and timing to the North Sea rift basin. However, not all such basins share the timing. Along the western and southern margins of Australia the same sequence of events took place, but progressively later in an anticlockwise direction; in the Gippsland basin in the SE corner of the continent, the unconformity–disconformity is Eocene, separating a dominantly non-marine faulted sequence from the fine-grained unfaulted sequence of mudstones and marls (Threlfall *et al.* 1976).

The deeper parts of most rift basins are beyond our reach but, in general, the sandstone bodies probably tend to conduct water of compaction to the margins of the rift basin in the pre-unconformity–disconformity sequence. As with growth faults in regressive sequences, the contemporaneous movement of the faults may mean that such flow is temporarily interrupted and may resume through another sandstone.

The fine-grained post-unconformity–disconformity sequence is commonly abnormally pressured, and these zones have a great effect on the flow of fluids in the basin. The zone of abnormal pressures may extend well down into the pre-unconformity–disconformity sequence (as in the North Sea). When permeable rock units are abnormally pressured, it is evidence that they do not act as effective conduits for fluids to the surface.

The top of abnormal pressures in mudstones that do not have an overlying permeable sequence is not as definite as that commonly found in regressive sequences. In the latter, the overlying permeable sandstones act as a filter-press and so compact the top of the thick mudstone unit, thus tending to preserve the high pore pressures. In the absence of a porous and permeable overburden abnormal pressures are simply due to the thickness of the mudstone and its relatively rapid accumulation, and no sharp top is seen (e.g. Van den Bark & Thomas 1980, p. 220, Fig. 24).

Petroleum in rift basins

Structural style in rift basins is related to position. In the pre-unconformity–disconformity sequence the habitat of petroleum is unconformity and fault traps beneath closed highs on the unconformity–disconformity surface (see, for example, the many fields of the North Sea described by Woodland (1975) and Illing & Hobson (1981)). The mudstone on the unconformity–disconformity surface is commonly a prolific petroleum source rock, and primary migration takes place downwards from it into permeable subcrops.

In the post-unconformity–disconformity transgressive sequence, absence of structure is the rule, and the habitat of petroleum is stratigraphic traps, such as submarine fans near the graben margin (e.g. Frigg in the North Sea (Heritier *et al.* 1979)) and in permeable drains to the abnormally-pressured mudstones, or potential drains, such as the Ekofisk field in the North Sea (Van den Bark & Thomas 1980) with structure that is perhaps induced by salt diapirism.

There is too much room for doubt about the petroleum source rocks of the North Sea in general, and these fields in particular, to discuss the flow of petroleum fluids in these geological contexts in detail. Nevertheless, the extensive and varied abnormal pressures in the North Sea, both in the pre-unconformity–disconformity sequence and in the overlying mudstones, must impose barriers to and drains for petroleum migration, and these must also be largely stratigraphically controlled.

In the question of faults as conduits for petroleum migration, there is a matter of interest in rift basins. Secor's (1965) argument for deep open fractures appears to be valid for rocks of relatively small tensile strength only if there are abnormally high pore pressures in a stress field in which the deviatoric stress is very small. It is therefore possible that the abnormally-pressured mudstones overlying the unconformity–disconformity surface have small open fractures. If such fractures do exist, we must seek to understand how the abnormal pressures have been maintained.

Downward flow of fluids from mudstone into the subcrops of porous and permeable units presents no great difficulties. Prudhoe Bay field (N slope of Alaska) which, if not in a rift basin, has all the features of being in one, has several reservoirs of different ages, but the crude oil was sourced almost certainly partly from the Lower

Cretaceous marine mudstones on the unconformity (Jones & Speers 1976), but not necessarily within the area of the field.

It is worth making the general point that, when petroleum reservoirs are abnormally pressured, the source rock must be in a position where it has or had higher energy than the reservoir fluids and the migration path is or was one of continuously decreasing energy towards the reservoir.

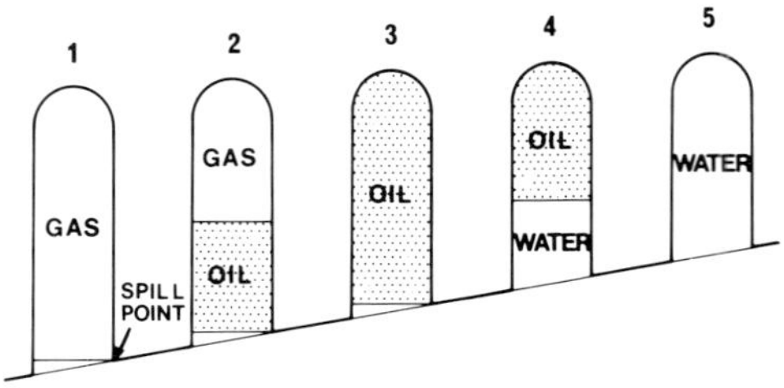

FIG. 11. Stylized distribution of gas and crude-oil accumulations in reefs in hydraulic continuity according to Gussow's (1954) differential entrapment theory.

Conclusion

During the development of sedimentary basins, a vast volume of water is eventually redistributed, and this flows mainly from the thick neritic mudstones downwards and upwards into the more permeable parts of the transgressive and regressive sequences and, in these, laterally towards the land of the time. Petroleum may also be generated in large volumes, and this too will tend to flow in porous and permeable carrier beds towards the land of the time. In both cases stratigraphy has a strong influence on the flow paths.

These statements are plausible and consistent with the evidence, but we do not have any real proof of water-flow paths in sedimentary basins in general. They are inferred from what we understand about the compaction of mudrocks and the flow of water through porous and permeable sedimentary rocks. The observed diagenesis of sandstones and carbonates is evidence of water flow, but not usually of the paths of flow.

The case for petroleum migration, however, rests on stronger evidence, the strongest point of which involves petroleum in fossil coral reefs. Two principal arguments indicate the direction of migration of petroleum to fossil coral reefs.

1 The reef environment itself is most unlikely to be favourable for the generation and preservation of petroleum because it is richly oxygenated and has relatively high energy, and because most of the organic matter is in the food chain.
2 Patterns of gas and crude oil accumulations that give credence to Gussow's (1954) differential entrapment theory are also evidence that the main gas and oil source rocks are not local, but down-dip from the deepest gas accumulation of the series (Fig. 11).

This evidence for petroleum migration, which is supported by the general evidence that reservoir rocks are usually too poor in organic matter and too inert chemically to be likely source rocks for petroleum, serves to strengthen our belief that the inferences we draw about water flow are generally valid. Buoyancy is not the only force acting on petroleum.

The other side of that coin is that, in the sedimentary basins in which water has evidently been unable to flow freely, we find mudstones that have pore pressures well above normal hydrostatic, and a good case can be made that the consequences of this lack of flow are important to geology as a cause of deformation in young sedimentary basins (mainly diapirism and sliding). Again, there is as yet no *proof* that this water has been retained, but the indirect evidence is strong in those parts where geophysical evidence is consistent with under-compaction of the mudstones.

The main geological problem requiring elucidation is the role of faults in fluid flow in sedimentary basins. In many parts of the world petroleum is supposed to have migrated *up* faults from a deep petroleum source rock to a shallower reservoir. It is easy to make such assertions; it is very difficult to substantiate them. It is equally easy to assert that faults are not conduits for petroleum, and it is difficult to substantiate such an assertion. Many fields around the world have dry sands interposed between the supposed source rock and the known reservoirs, and the evidence they afford must not be ignored. There may be good physical reasons why petroleum did not enter them—or there may be none. It is of immense importance for both 'pure' and applied geology that credible answers to the various questions be found.

References

BARRER, R. M. & BARRIE, J. A. 1952. Sorption and surface diffusion in porous glass. *Proc. R. Soc. Lond., Ser.* A, **213**, 250–65.

CAMPBELL, I. R. & SMITH, D. N. 1982. Gorgon 1—southernmost Rankin Platform gas discovery. *J. Aust petrol. explor. Assoc.* **22** (1), 102–11.

CHAPMAN, R. E. 1972. Clays with abnormal interstitial fluid pressures. *Bull. Am. Assoc. Petrol. Geol.* **56**, 790–5.

—— 1981. *Geology and Water: An Introduction to Fluid Mechanics for Geologists*, Martinus Nijhoff/Dr W. Junk, The Hague.

—— 1982. Effects of oil and gas accumulation on water movement. *Bull. Am. Assoc. Petrol. Geol.* **66**, 368–78.

—— 1983. *Petroleum Geology*, Elsevier, Amsterdam.

CORRSIN, S. 1955. A measure of the area of a homogeneous random surface in space. *Q. Appl. Math.* **12**, 404–8.

DAHLBERG, E. C. 1982. *Applied Hydrodynamics in Petroleum Exploration*, Springer-Verlag, Berlin.

DARCY, H. 1856. *Les Fontaines Publiques de la Ville de Dijon*, Victor Dalmont, Paris.

DICKINSON, G. 1951. Geological aspects of abnormal reservoir pressures in the Gulf Coast region of Louisiana, U.S.A. *Proc. 3rd World Petroleum Congr.*, Vol. 1, pp. 1–16.

—— 1953. Geological aspects of abnormal reservoir pressures in the Gulf Coast region of Louisiana. *Bull. Am. Assoc. petrol. Geol.* **37**, 410–32.

FERTL, W. H. 1976. *Abnormal Formation Pressures*, Elsevier, Amsterdam.

FRÄNKL, E. J. & CORDRY, E. A. 1967. The Niger delta oil province: recent developments onshore and offshore. *Proc. 7th World Petroleum Congr.*, Vol. 2, pp. 194–209.

FÜCHTBAUER, H. 1967. Influence of different types of diagenesis on sandstone porosity. *Proc. 7th World Petroleum Congr.*, Vol. 2, pp. 353–70.

GILL, D. 1979. Differential entrapment of oil and gas in Niagaran pinnacle-reef belt of northern Michigan. *Bull. Am. Assoc. petrol. Geol.* **63**, 608–20.

GRAY, D. E. (ed.) 1957. *American Institute of Physics Handbook*, McGraw-Hill, New York.

GRETENER, P. E. & ZENG-MO FENG 1985. Three decades of geopressure—insights and enigmas. *Bull. Verein. schweiz. Petrol. Geol. Ing.* **51** (120), pp. 1–34.

GUSSOW, W. C. 1954. Differential entrapment of oil and gas: a fundamental principle. *Bull. Am. Assoc. petrol Geol.* **38**, 816–53.

HABERMEHL, M. A. 1980. The Great Artesian Basin, Australia. *Bureau of Mineral Resources, Australia, J. Aust. Geol. Geophys.* **5**, 9–38.

HEIGOLD, P. C., MAST, R. F. & CARTWRIGHT, K. 1971. Temperature distributions and ground-water movement associated with oil fields in the Fairfield basin Illinois. *Ill. State Geol. Surv., Ill. Petrol.* **95**, 127–40.

HERITIER, F. E., LOSSEL, P. & WATHNE, E. 1979. Frigg field—large submarine-fan trap in Lower Eocene rocks of North Sea Viking graben. *Bull. Am. Assoc. petrol. Geol.* **63**, 1999–2020.

HUBBERT, M. K. 1940. The theory of ground-water motion. *J. Geol.* **48**, 785–944.

—— 1953. Entrapment of petroleum under hydrodynamic conditions. *Bull. Am. Assoc. petrol. Geol.* **37**, 1954–2026.

—— 1969. *The Theory of Ground-water Motion and Related Papers*, Hafner, New York.

ILLING, L. V. & HOBSON, G. D. (eds) 1981. *Petroleum Geology of the Continental Shelf of North-West Europe*, Heyden, London.

JAMES, D. M. G. (ed.) 1984. *The Geology and Hydrocarbon Resources of Negara Brunei Darussalam*, Muzium Brunei, Kota Batu & Brunei Shell Petroleum, Seria, Brunei.

JONES, H. P. & SPEERS, R. G. 1976. Permo–Triassic reservoirs of Prudhoe Bay field, North Slope, Alaska. *In:* BRAUNSTEIN, J. (ed.) *North American Oil and Gas Fields, Mem. Am. Assoc. petrol. Geol.* **24**, 23–50.

KÄSS, W. 1969. Schrifttum zur Versickerung der oberen Donau zwischen Immendingen und Fridingen (Südwestdeutschland). *Steir. Beitr. Hydrogeol.* **21**, 215–46.

—— 1972. Die Versickerung der oberen Donau, ihre Erforschung und die Versuche 1969. *Geol. Jahrb., Reihe C*, **2**, 13–18.

—— & HÖTZL, H. 1973. Weitere Untersuchungen im Raum Donauversickerung-Aachquelle (Baden-Württemberg). *Steir. Beitr. Hydrogeol.* **25**, 103–16.

KENT, P. E. 1975a. The tectonic development of Great Britain and the surrounding seas. *In:* WOODLAND, A. W. (ed.) *Petroleum and the Continental Shelf of North-west Europe*, Vol. 1, *Geology*, pp. 3–28, Applied Science, Barking, Essex.

—— 1975b. Review of North Sea basin development. *J. geol. Soc. Lond.* **131**, 435–68.

—— 1977. The Mesozoic development of aseismic continental margins. *J. geol. Soc. Lond.* **134**, 1–18.

—— 1978. Mesozoic vertical movements in Britain and the surrounding continental shelf. *In:* BOWES, D. R. & LEAKE, B. E. (eds) *Crustal Evolution in Northwestern Britain and Adjacent Regions, Geol. J. Spec. Issue*, **10**, 309–24.

KING, F. H. 1899. Principles and conditions of the movements of ground water. *Ann. Rep. U.S. geol. Surv. (1897–98)*, **19** (2), 59–294.

KLINKENBERG, L. J. 1942. The permeability of porous media to liquids and gases. *Am. petrol. Inst. Drill. Product. Practice 1941*, pp. 200–213.

LAMBERT-AIKHIONBARE, D. O. 1982. Relationship between diagenesis and pore fluid chemistry in Niger delta oil-bearing sands. *J. petrol. Geol.* **4** (3), 287–98.

PERRAULT, P. 1674. *De l'Origine des Fontaines*, pp. 154–56, Pierre le Petit, Paris.

—— 1967. *On the Origin of Springs*, p. 78, Hafner, New York (translated by A. LaRocque).

PHILIPP, W., DRONG, H. J., FÜCHTBAUER, H., HADDENHORST, H.-G. & JANKOVSKY, W. 1963a. Zur

Geschichte der Migration im Gifhorner Trog. *Erdöl Kohle Erdgas Petrochem.* **16**, 456–68.

——, ——, ——, —— & —— 1963b. The history of migration in the Gifhorn Trough (NW-Germany). *Proc. 6th World Petroleum Congr.*, Vol. 1, pp. 457–78.

SCHAUB, H. P. & JACKSON, A. 1958. The northwestern oil basin of Borneo. *In:* WEEKS, L. G. (ed.) *Habitat of Oil*, pp. 1330–1336, American Association of Petroleum Geologists, Tulsa, OK.

SECOR, D. T. 1965. Role of fluid pressure in jointing. *Am. J. Sci.* **263**, 633–46.

SHORT, K. C. & STÄUBLE, A. J. 1967. Outline of geology of Niger delta. *Bull. Am. Assoc. petrol. Geol.* **51**, 761–79.

SLICHTER, C. S. 1899. Theoretical investigation of the motion of ground waters. *Annu. Rep. U.S. geol. Surv. (1897–98)*, **19** (2), 295–384.

SMITH, D. A. 1980. Sealing and nonsealing faults in Louisiana Gulf Coast salt basin. *Bull. Am. Assoc. petrol. Geol.* **64**, 145–72.

THRELFALL, W. F., BROWN, B. R. & GRIFFITH, B. R. 1976. Gippsland basin, off-shore. *In:* LESLIE, R. B., EVANS, H. J. & KNIGHT, C. L. (eds) *Economic Geology of Australia and Papua New Guinea*, Vol. 3, *Petroleum, Aust. Inst. Min. Metall., Monogr. Ser.* **7**, 41–67.

VAN DEN BARK, E. & THOMAS, O. D. 1980. Ekofisk: first of the giant oil fields in western Europe. *In:* HALBOUTY, M. T. (ed.) *Giant Oil and Gas Fields of the Decade 1968–1978, Mem. Am. Assoc. petrol. Geol.* **30**, 195–224.

VERDIER, A. C., OKI, T. & SUARDY, A. 1980. Geology of the Handil field (East Kalimantan—Indonesia). *In:* HALBOUTY, M. T. (ed.) *Giant Oil and Gas Fields of the Decade 1968–1978, Mem. Am. Assoc. petrol. Geol* **30**, 399–422.

VERSLUYS, J. 1917. De beweging van het grondwater. *Water*, **1**, 23–25, 44–46, 74–76, 95.

—— 1919. De duinwater-theorie. *Water*, **3** (5), 47–51.

WEEKS, L. G. 1958. Habitat of oil and some factors that control it. *In:* WEEKS, L. G. (ed.) *Habitat of Oil*, pp. 1–61, American Association of Petroleum Geologists, Tulsa, OK.

WINKELMOLEN, A. M. 1982. Critical remarks on grain parameters, with special emphasis on shape. *Sedimentology*, **29**, 255–65.

WOODLAND, A. W. (ed.) 1975. *Petroleum and the Continental Shelf of North-west Europe*, Vol. 1, *Geology*, Applied Science, Barking, Essex.

WYCKOFF, R. D., BOTSET, H. G., MUSKAT, M. & REED, D. W. 1933. The measurement of the permeability of porous media for homogeneous fluids. *Rev. Sci. Instrum., New Ser.* **4**, 394–405.

——, ——, —— & —— 1934. Measurement of permeability of porous media. *Bull. Am. Assoc. petrol. Geol.* **18**, 161–90.

RICHARD E. CHAPMAN, University of Queensland, St Lucia, Queensland, Australia 4067.

Fluid flow due to sediment loading—an application to the Arabian Gulf region

Kinji Magara

SUMMARY: Fluid pressure in excess of hydrostatic pressure can be generated during sediment loading. The rate of increase in the excess fluid pressure (psi Ma^{-1}) can be calculated from the sedimentation rate (ft Ma^{-1}) of a formation and its average bulk density. Further, the inferred directions of horizontal migration of the compaction fluid can be shown by means of a contour map based on the rate of increase in the excess fluid pressure, the first derivative of which gives the intensity of the horizontal motion of the compaction fluid (psi Ma^{-1} $mile^{-1}$). The horizontal fluid movement may be essential for driving hydrocarbons towards a trapping position during the primary stages of migration.

Application of this technique using the regional isopach maps of the Jurassic, Cretaceous and Tertiary formations in the Arabian Gulf region suggests that the major oil fields are concentrated in areas of relatively strong horizontal fluid movement. The area N of the Qatar Peninsula, where horizontal migration of fluid was relatively weak during the Cretaceous period when both the Jurassic and Cretaceous source rocks reached their early stages of oil generation, seems to have relatively low concentrations of oil. At present there are relatively few oil fields in this part of the region.

There are basically two different kinds of water moving in a sedimentary basin. Magara (1978) defined their characteristics as follows.

Sediment-source (or compaction) water

1 The movement of sediment-source water takes place in any part of a sedimentary basin (deep or shallow).
2 The principal direction of small-scale movement is primarily from shale or clay to a sandstone or other permeable bed.
3 The direction of large-scale movement is from the centre to the edges of the basin or from the deeper to the shallower levels.
4 The amount of water is limited because the amount of sediment in a basin is usually limited.
5 Movement of this type of water is probably important in the primary migration of hydrocarbons.
6 Most movement of this type of water took place in the geological past.

Meteoric water

1 Movement of meteoric water is important in the relatively shallow intervals of a sedimentary basin.
2 The direction of small-scale movement can be either from sandstone to shale or from shale to sandstone. However, most movement of this type of water may take place in sandstone or other permeable rock only.
3 The direction of large-scale water movement is from the edge to the centre of the basin or from shallower to deeper levels.
4 The amount of water is unlimited.
5 Movement of this type of water is probably unimportant in primary migration, but it may affect the conditions for trapping hydrocarbons in a pool.
6 Movement of meteoric water is mainly taking place at present and may or may not have developed in the geological past.

Items 5 and 6 in the above lists are of particular interest to petroleum geologists. The movement of sediment-source water is mostly a past event and was probably a major influence in primary hydrocarbon migration, whereas that of meteoric water persisted for a relatively short recent geological period and may have affected late secondary migration and conditions for the entrapment of hydrocarbons.

Chiarelli (1973) reported the present potentiometric surface of the Arabian Gulf region and inferred directions of movement of meteoric water in Cretaceous formations (Fig. 1). No major work on sediment-source (or compaction) water has been reported so far.

The purpose of this paper is to present the results of an analysis of the movement of sediment-source water in an attempt to understand the primary hydrocarbon migration which took place in the region in the geological past.

Excess fluid pressure due to sediment loading

Let us consider a sediment sequence in which the sediment has reached a compaction equilibrium and the fluid pressure is hydrostatic (Fig. 2, stage A). Such a compaction equilibrium for shales has

From GOFF, J. C. & WILLIAMS, B. P. J. (eds), 1987, *Fluid Flow in Sedimentary Basins and Aquifers*, Geological Society Special Publication No. 34, pp. 19–28.

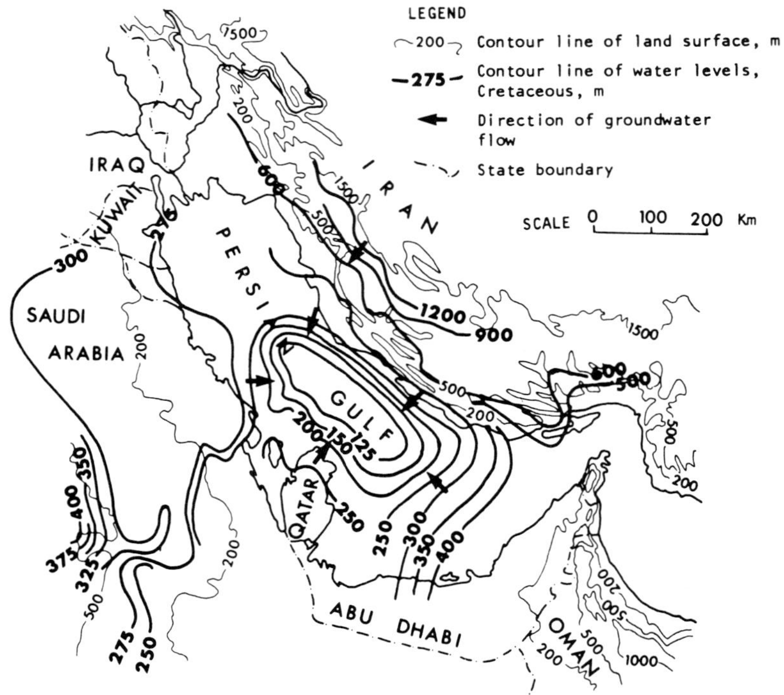

FIG. 1. Topography, potentiometric surface and lateral flow directions of groundwater in Cretaceous formations in the Arabian Gulf region. (From Chiarelli 1973.)

been established in many sedimentary basins. According to Hubbert & Rubey (1959), who studied the Palaeozoic shales of Oklahoma analysed by Athy (1930), the compaction equilibrium condition can be described by an exponential function which is represented by a straight line on semi-logarithmic paper (porosity on the logarithmic scale and depth on the arithmetic scale) (Fig. 2, stage A).

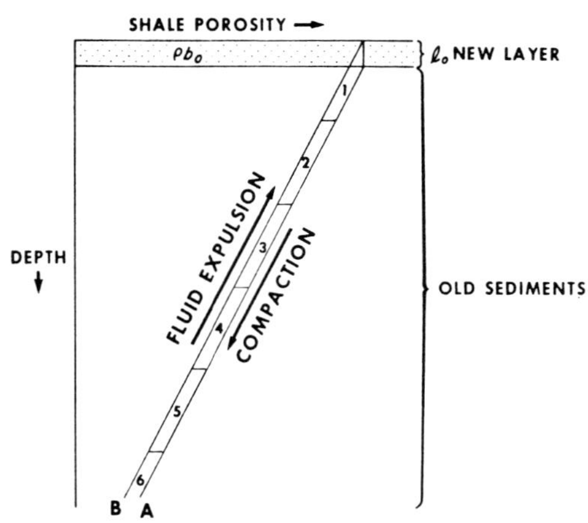

FIG. 2. Schematic diagram showing the stepwise compaction of shales and fluid expulsion due to the loading of a new layer. (From Magara 1978.)

Additional sediments of thickness l_0 are deposited above the sequence under water. If the entire sequence reaches a new equilibrium condition of compaction within a time interval t, a porosity distribution such as that shown in Fig. 2, stage B, is established. Suppose that the outlet for fluid expulsion exists only at the surface and that the fluid is expelled vertically upwards. In this case compaction of the sediment sequence from stage A to stage B occurs from the shallower to the deeper level in an approximately stepwise manner (Fig. 2).

The increase in fluid pressure p due to the instantaneous loading of a new sediment of thickness l_0 (original thickness before compaction) is given by

$$p = \rho_{b_0} g l_0 \qquad (1)$$

where ρ_{b_0} is the water-saturated original bulk density (before compaction) of the new sediment layer l_0 and g is the acceleration due to gravity. When the increase in fluid pressure is resolved into two components (Hubbert & Rubey 1959) we obtain

$$p = p_n + p_a \qquad (2)$$

where p_n is the increase in the normal or hydrostatic pressure and p_a is the increase in excess pressure. In this case p_n is given by

$$p_n = \rho_w g l_0 \quad (3)$$

Rearranging equation 2 and substituting equations and 3, we obtain

$$p_a = (\rho_{b_0} - \rho_w) g l_0 \quad (4)$$

If the thickness of the new sediment layer is not constant but changes horizontally at a rate of dl_0/dx (see Fig. 3), the rate of increase dp_a/dx of the horizontal excess pressure, which controls the horizontal fluid migration, is governed by the rate of change in the horizontal thickness of the new sediment layer:

$$\frac{dp_a}{dx} = (\rho_{b_0} - \rho_w) g \frac{dl_0}{dx} \quad (5)$$

However, most sediments in the sub-surface are not new but have been compacted, so that a correction for the compaction effect must be applied if the horizontal fluid migration which took place in the geological past is being considered.

The relationship between the thickness l_0 of a layer at the time of deposition (before compaction) and the thickness l at present (after compaction) is

$$l_0(1 - \phi_0) = l(1 - \phi) \quad (6)$$

where ϕ_0 is the porosity at the time of deposition (before compaction) and ϕ is the porosity at present (after compaction). The porosity can be expressed in terms of the density as follows:

$$\phi_0 = \frac{\rho_m - \rho_{b_0}}{\rho_m - \rho_w} \quad (7)$$

and

$$\phi = \frac{\rho_m - \rho_b}{\rho_m - \rho_w} \quad (8)$$

Where ρ_b is the bulk density at present (after compaction) and ρ_m is the matrix (or grain) density of the sediment. Substituting equations 7 and 8 into equation 6 gives

$$(\rho_{b_0} - \rho_w) l_0 = (\rho_b - \rho_w) l \quad (9)$$

or

$$(\rho_{b_0} - \rho_w) \frac{dl_0}{dx} = (\rho_b - \rho_w) \frac{dl}{dx} \quad (10)$$

From equations 5 and 10 we obtain

$$\frac{dp_a}{dx} = (\rho_b - \rho_w) g \frac{dl}{dx} \quad (11)$$

An advantage of using equation 11 rather than equation 5 is that the horizontal fluid migration which occurred in the geological past can be estimated directly from the change dl/dx in the thickness of the compacted sediment layer.

The above discussion is based on the increase in fluid pressure due to loading (increase in hydraulic pressure). If the geothermal gradient is constant in the study area, the increase in the aquathermal pressure (Barker 1972; Magara 1974) would be almost proportional to the increase in fluid pressure due to loading. In other words, aquathermal fluid migration can also be inferred from the study of sediment loading.

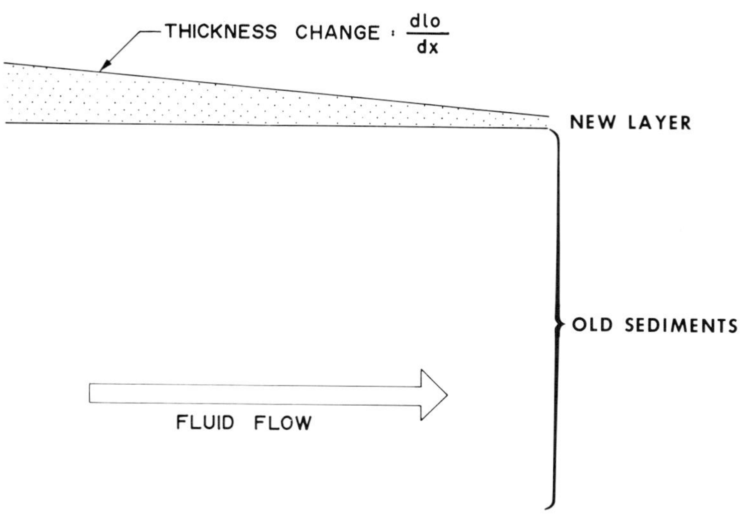

FIG. 3 Schematic diagram showing the direction of fluid flow due to wedge-shaped sedimentary loading.

Significance of horizontal fluid migration for hydrocarbon accumulation

During secondary hydrocarbon migration in a reservoir, globules of oil or gas are able to move from a structurally lower point to a higher point if the buoyancy is high enough to overcome the capillary restrictions of the reservoir rock. In other words, the accumulation of oil or gas is dependent on structural development or trap formation during secondary migration.

During the primary migration stages, however, the buoyancy is usually considered to be too low to overcome the relatively high capillary pressure of most fine-grained source rocks. The presence of structured water and both hydraulic and aquathermal forces may push oil or gas towards a trapping position (Dickey 1975; Magara 1978, 1981). The structured water, which is highly viscous or semi-solid, is particularly important in enabling hydrocarbons to overcome relative permeability by pore water and capillary pressure in the source rock. Under such conditions the concentration of oil or gas may be possible if there are strong horizontal hydraulic and aquathermal forces related to sediment loading.

Generation of excess fluid pressure in the Arabian Gulf region

Excess fluid pressures due to sediment loading were estimated using the isopach maps of the Arabian Gulf region constructed by Kamen-Kaye (1970). The graphs in Figs 4 and 5 show the change in the average rate of increase in excess fluid pressure (psi Ma^{-1}) for two selected locations (A and B in Fig. 6).

At location A in the southeastern part of the region the rate gradually increased throughout the geological period since the Permian. In contrast, at location B, which lies in Iran, the average rate fluctuated in the geological past and reached its highest value (about 60 psi Ma^{-1}) during the Tertiary period.

The broken curves in Figs 4 and 5 show the estimated vitrinite reflectance values R_0 at the bases of the Jurassic (J), Cretaceous (K) and Tertiary (T) formations (this method is described by Waples 1980). According to the estimated reflectance values, the basal Jurassic and Cretaceous formations at locations A and B reached their early stages of petroleum maturation ($R_0 >$ 0.5%) by the end of the Cretaceous period. Primary hydrocarbon migration would have taken place at that time if the conditions described above had existed in this region.

Figures 6, 7 and 8 show the calculated rates of increase in fluid pressure (psi Ma^{-1}) due to loading of the Jurassic, Cretaceous and Tertiary formations respectively. The arrows in these figures show the inferred directions of fluid movement. As stated previously, the horizontal change in the increase in excess fluid pressure (psi Ma^{-1} $mile^{-1}$), which is the first derivative of the values shown in Figs 6–8, may be one of the factors controlling the formation of an effective oil/gas accumulation.

Figures 9, 10 and 11 show maps of the horizontal changes in the increase in excess fluid

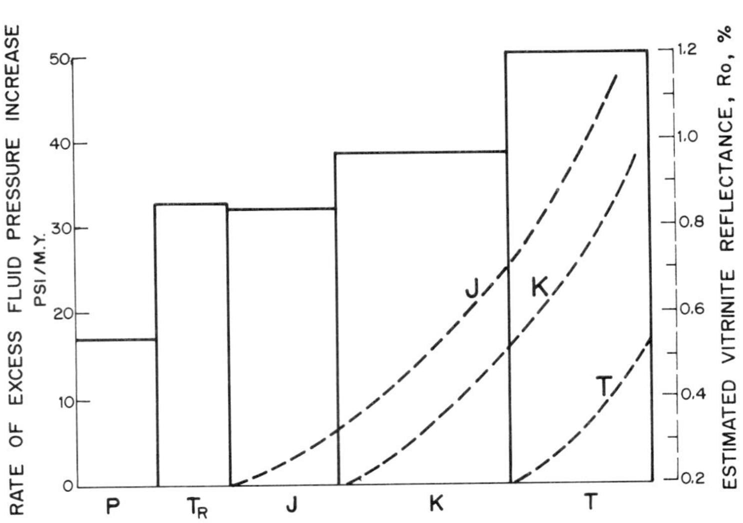

FIG. 4. Average rate of increase in excess fluid pressure (psi Ma^{-1}) for location A in Fig. 6: --- estimated vitrinite reflectance R_0 at the bases of Jurassic (J), Cretaceous (K) and Tertiary (T) formations.

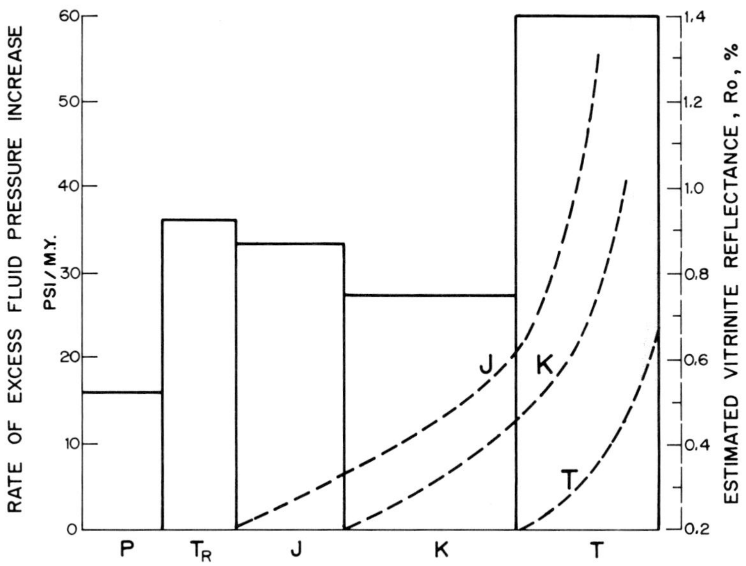

FIG. 5. Average rate of increase in excess fluid pressure (psi Ma^{-1}) for location B in Fig. 6: --- estimated vitrinite reflectance, R_0 at the bases of Jurassic (J), Cretaceous (K) and Tertiary (T) formations.

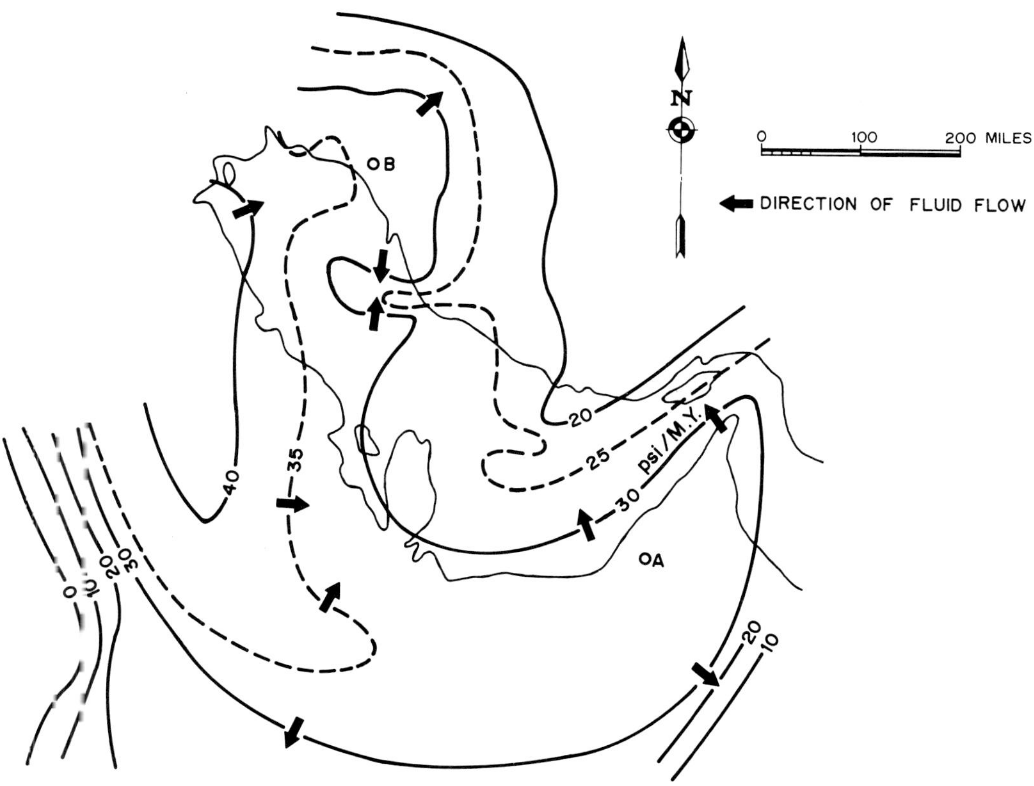

FIG. 6. Contour map showing the calculated rate of increase in excess fluid pressure (psi Ma^{-1}) due to the loading of Jurassic formations. The arrows show the inferred directions of fluid flow.

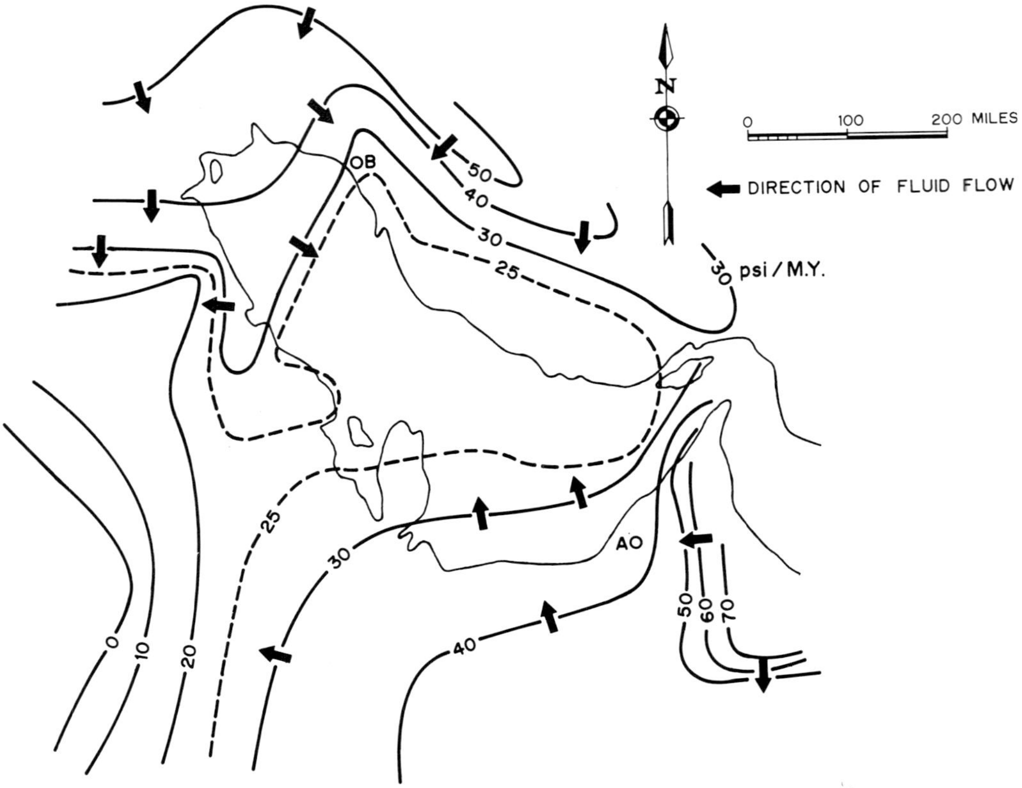

FIG. 7. Contour map showing the calculated rate of increase in excess fluid pressure (psi Ma^{-1}) due to the loading of Cretaceous formations. The arrows show the inferred directions of fluid flow.

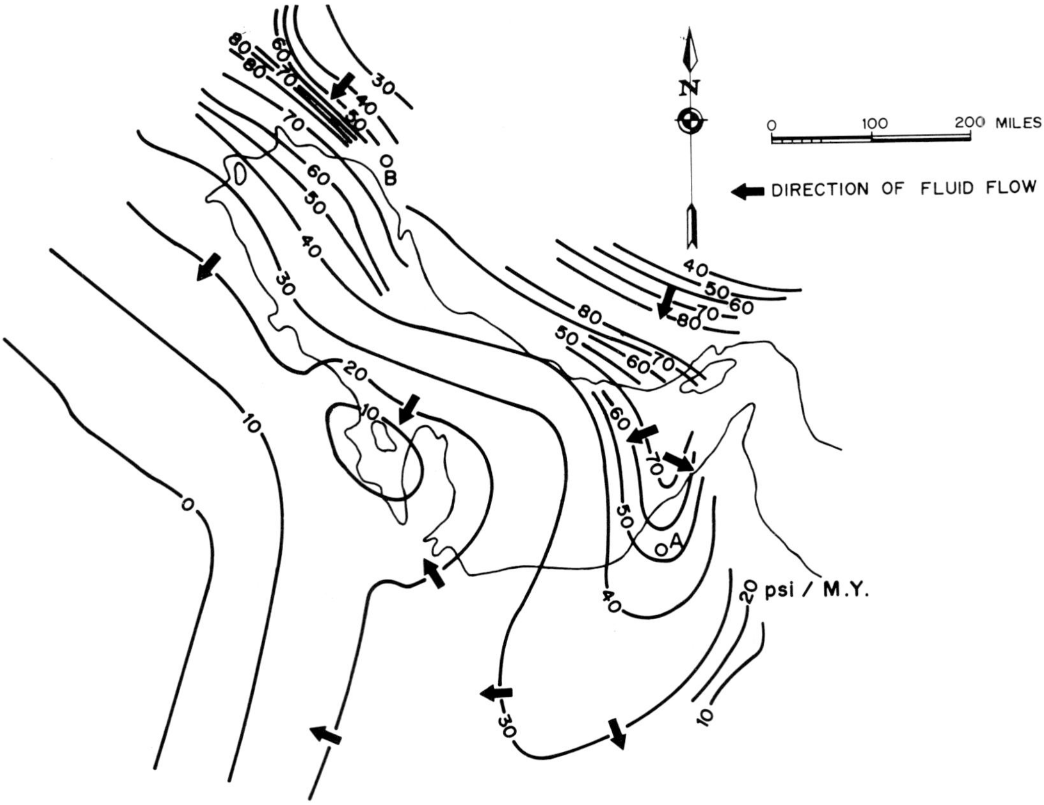

FIG. 8. Contour map showing the calculated rate of increase in excess fluid pressure (psi Ma^{-1}) due to the loading of Tertiary formations. The arrows show the inferred directions of fluid flow.

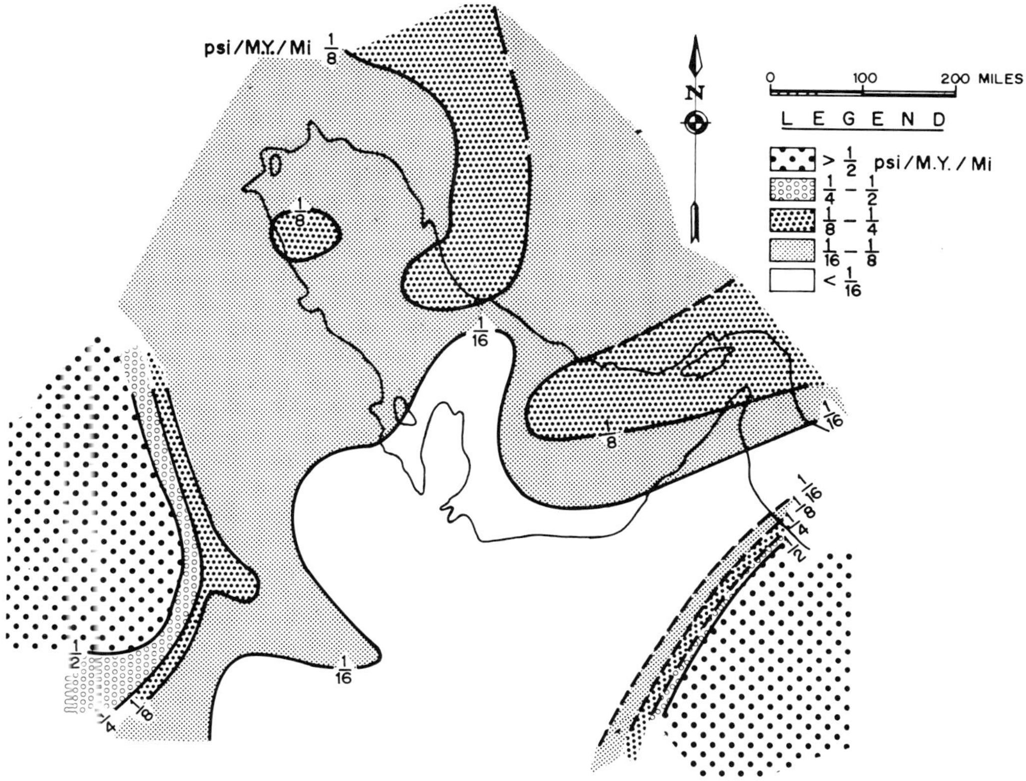

FIG. 9. Contour map showing the horizontal change in the increase in excess fluid pressure (psi Ma^{-1} $mile^{-1}$) by Jurassic formations.

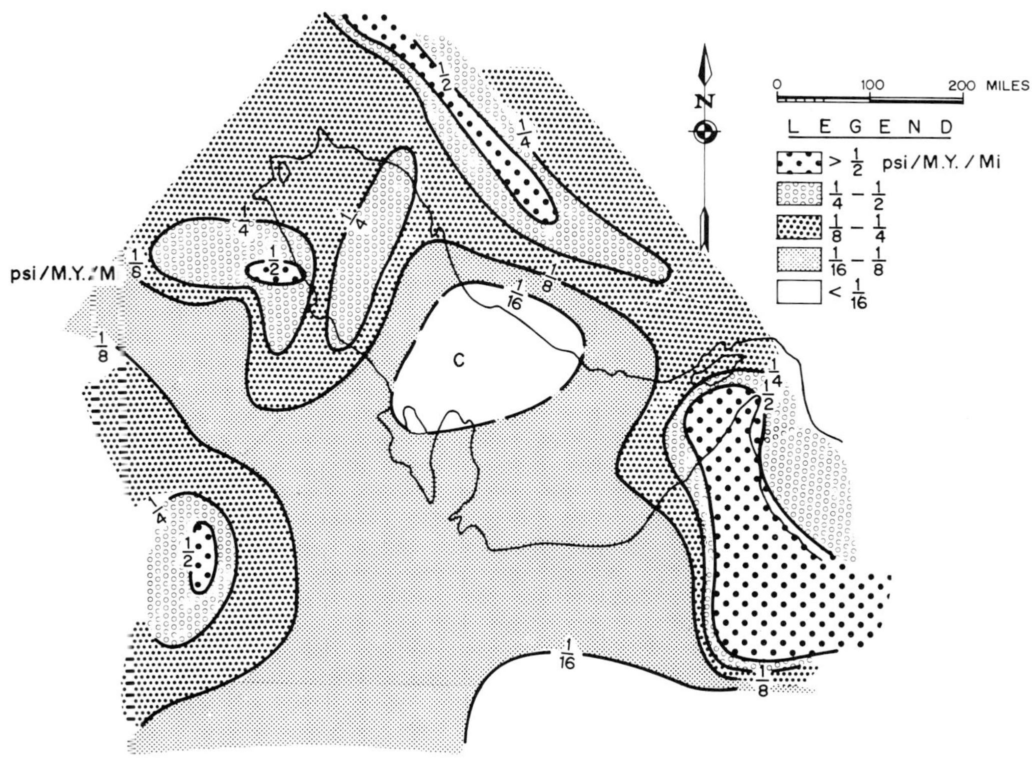

FIG. 10. Contour map showing the horizontal change in the increase in excess fluid pressure (psi Ma^{-1} $mile^{-1}$) by Cretaceous formations.

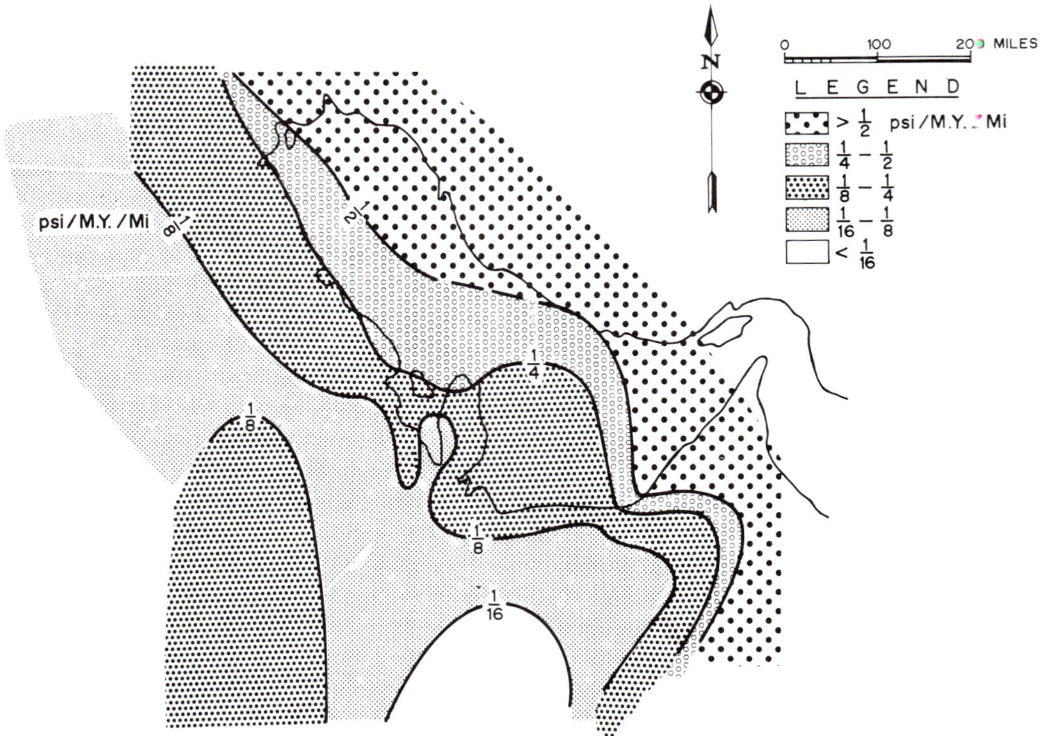

FIG. 11. Contour map showing the horizontal change in the increase in excess fluid pressure (psi Ma^{-1} $mile^{-1}$) by Tertiary formations.

pressure by the Jurassic, Cretaceous and Tertiary formations respectively. An interesting observation is that the values N of the Qatar Peninsula were relatively low during the Cretaceous period (Fig. 10, region C). The horizontal changes in the excess fluid pressure in the surrounding areas were relatively large, suggesting effective concentration of the generated oil or gas into traps during the primary stages of migration.

It should be noted that the early stages of petroleum maturation for both the basal Jurassic and Cretaceous formations were reached by the end of the Cretaceous period (Figs 4 and 5). This is why horizontal fluid migration during the Cretaceous period is important in the discussions in the remainder of this paper.

Petroleum generation, maturation and migration in the Arabian Gulf region

Sail & Magara (1985) constructed regional maps of hydrocarbon-generation zones based on Kamen-Kaye's (1970) regional isopach maps and Waples's (1980) method of estimating vitrinite reflectance. Figure 12 shows such a map for the basal Jurassic formations superimposed on the oil-field distribution map for the Jurassic reservoirs. Figure 13 is a similar composite map of the Cretaceous formations. These maps show excellent agreement between the estimated oil-generation zones and the oil-field distributions, except that relatively few oil fields have been discovered in the area N of the Qatar Peninsula (Figs 12 and 13, region C). This area is a northern extension of the Qatar Arch where the sedimentation rate was relatively slow. In fact some reservoir sections are missing owing to both erosion and non-deposition. This may partially explain the absence of oil fields in this area. However, there are also some porous sections of both Jurassic and Cretaceous ages which are water saturated (Schlumberger 1981). Relatively weak horizontal fluid movement as shown in Fig. 10 may be a reason for the relative scarcity of oil fields in this area.

Figure 14 shows a similar composite map of the hydrocarbon-generation and oil-field distribution for the Tertiary formations. In this case there is a significant mismatch because most oils in the Tertiary reservoirs were derived from the older source rocks (Young *et al.* 1977).

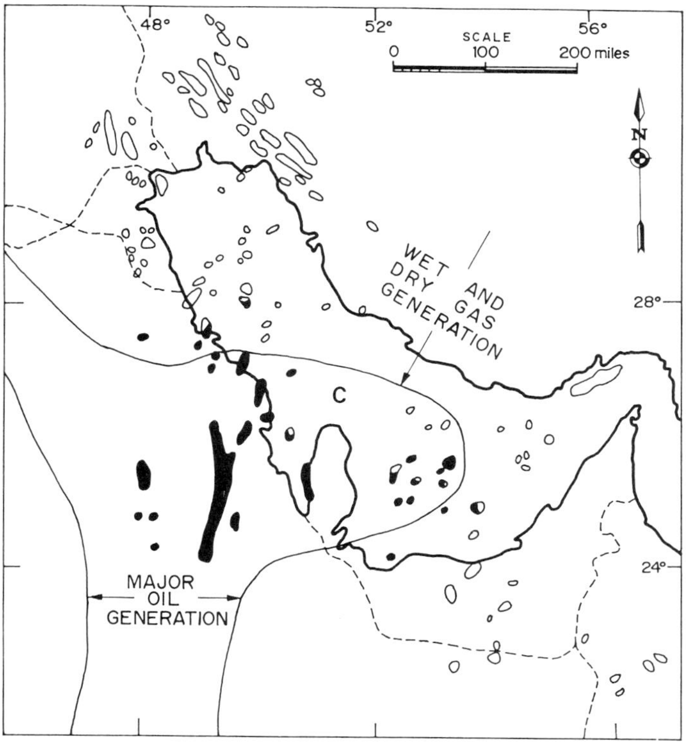

FIG. 1[illegible]. Map showing Jurassic oil-generation zones and oil fields: ●, Jurassic oil fields; ○, other oil fields. (From Sail & Magara 1985.)

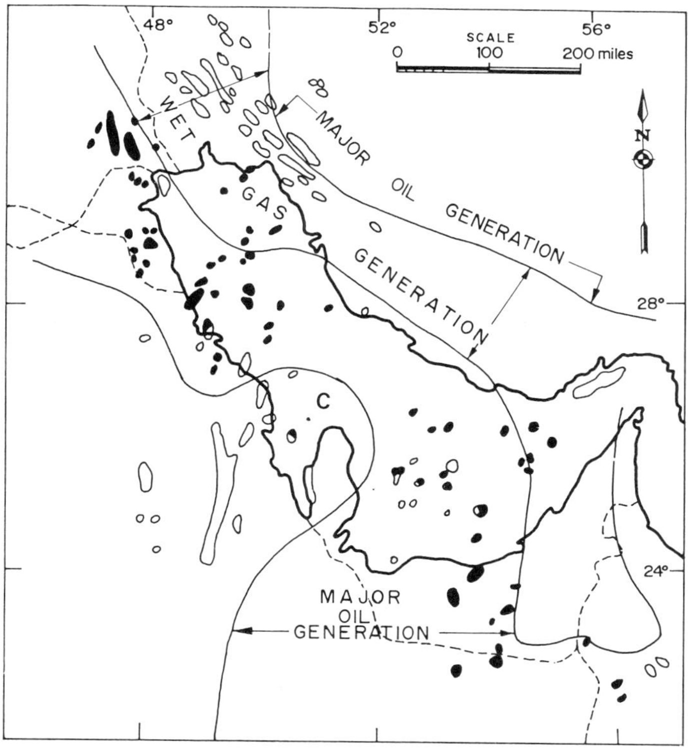

FIG. 13. Map showing Cretaceous oil-generation zones and oil fields: ●, Cretaceous oil fields; ○, other oil fields. (From Sail & Magara 1985.)

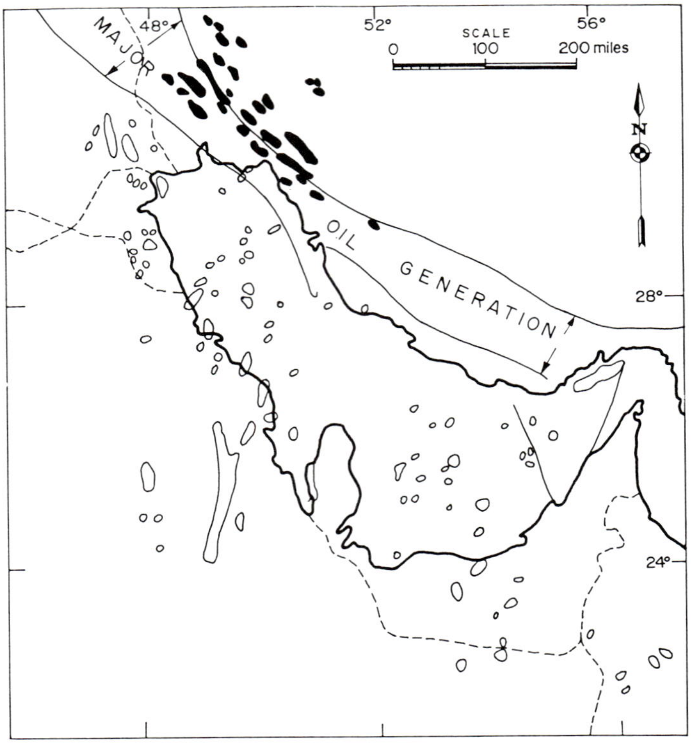

FIG. 14. Map showing Tertiary oil-generation zones and oil fields: ●, Tertiary oil fields; ○, other oil fields. (From Sail & Magara 1985.)

References

ATHY, L. F. 1930. Density, porosity and compaction of sedimentary rocks. *Bull. am. Assoc. petrol. Geol.* **14**, 1–24.

BARKER, C. 1972. Aquathermal pressuring—role of temperature in development of abnormal-pressure zone. *Bull. am. Assoc. petrol. Geol.* **56**, 2068–71.

CHIARELLI, A. 1973. Étude des nappes aquifères profondes. Contribution de l'hydro-géologie à la connaissance d'un bassin sédimentaire et à l'exploration pètrolière. *D.Sc. Thesis No. 401*, Université de Bordeaux I, Bordeaux, 187 pp.

DICKEY, P. A. 1975. Possible primary migration of oil from source rock in oil phase. *Bull. am. Assoc. petrol. Geol.* **59**, 337–45.

HUBBERT, M. K. & RUBEY, W. W. 1959. Role of fluid pressure in mechanics of overthrust faulting. *Geol. Soc. Am. Bull.* **70**, 115–206.

KAMEN-KAYE, M. 1970. Geology and productivity of Persian Gulf synclinorium. *Bull. am. Assoc. petrol. Geol.* **54**, 2371–94.

MAGARA, K. 1974. Aquathermal fluid migration. *Bull. am. Assoc. petrol. Geol.* **58**, 2513–6.

—— 1975. Importance of aquathermal pressuring effect in Gulf Coast. *Bull. am. Assoc. petrol. Geol.* **59**, 2037–45.

—— 1978. *Compaction and Fluid Migration—Practical Petroleum Geology*, Elsevier, Amsterdam, 319 pp.

—— 1978. The significance of expulsion of water in oil-phase primary migration. *Can. petrol. Geol. Bull.* **26**, 123–31.

SAIL, I. A. N. & MAGARA, K. 1985. Petroleum generation and maturation of the Arabian Gulf region. *Faculty of Earth Sciences Bulletin*, King Abdulaziz University, Jeddah, Saudi Arabia, in press.

SCHLUMBERGER 1981. *Well Evaluation Conf., United Arab Emirates/Qatar*, Vol. 1, *Geology*, Schlumberger Middle East S.A., 271 pp.

WAPLES, D. W. 1980. Time and temperature in petroleum formation: application of Lopatin's method to petroleum exploration. *Bull. am. Assoc. petrol. Geol.* **64**, 916–26.

YOUNG, A., MONAGHAN, P. H. & SCHWEISBERGER, R. T. 1977. Calculation of ages of hydrocarbons in oils. *Bull. am. Assoc. petrol. Geol.* **61**, 573–600.

K. MAGARA, Faculty of Earth Sciences, King Abdulaziz University, Jeddah, Saudi Arabia.

Section 2
Fluid Flow in the Western Canadian Sedimentary Basin

Dynamic basin analysis: an integrated approach with large data bases

Brian Hitchon, S. Bachu, C. M. Sauveplane & A. T. Lytviak

SUMMARY: Basin analysis must include an appreciation of both the rocks and the processes affecting their contained fluids. Therefore the objective of dynamic basin analysis is to synthesize and evaluate the present geological, hydrogeological, thermal and geochemical situation of sedimentary basins as well as their past history. For sedimentary basins such as that in western Canada, where petroleum exploration has resulted in a vast amount of information, synthesis and evaluation can only be achieved through an integrated multidisciplinary computer-oriented approach. This requires the development of a data-base management system and specialized data-processing techniques. These have been developed by the Basin Analysis Group of the Alberta Geological Survey for geological, hydrogeological, thermal and hydrochemical parameters. The synthesized information can then be integrated into a descriptive steady-state model which forms the basis of the numerical simulation of various processes and phenomena related to present fluid, heat and mass flows in the subsurface. These techniques have been applied to studies of the baseline hydrogeology of the Cold Lake oil sands deposit area and the deep waste disposal site at Swan Hills, Alberta, Canada. They are also applicable to petroleum exploration, *in situ* oil recovery, gas storage, geothermal resource evaluation and mineral exploration. Future research directions are to focus on the inclusion of models for mass transport and on the determination of the palaeohydrogeological regimes in sedimentary basins.

Approach to basin analysis

Conventional basin analysis is concerned with the structural, stratigraphic and palaeogeographic study of large, regionally disturbed packages of sedimentary rocks. The last two or three decades have seen a major revolution in the techniques of basin analysis, which now occupies the attention of a wide range of specialists including stratigraphers, sedimentologists, palaeontologists, structural geologists, geophysicists and geochemists. Five fundamental changes have occurred over this period in the science of sedimentary geology:

1 development of the capability of explaining the origin of sedimentary rocks through facies studies and facies models;
2 application of a genetic approach to stratigraphy using the 'depositional systems' method;
3 major advances in the evolution of modern seismic–stratigraphic techniques to allow chronostratigraphic correlation and the mapping of fine architectural details of entire basins;
4 comprehensive development of plate tectonic theory;
5 refinements in chronostratigraphy, especially through developments in radiometric dating techniques and magnetic reversal stratigraphy.

All these recent advances in conventional basin analysis have been widely applied, especially by petroleum explorers. Indeed, it must be rightfully acknowledged that much of the basic information on which the recent advances are predicated has resulted from efforts to explore for energy and mineral resources in sedimentary basins.

Sedimentary basins comprise two entities: (1) the rock framework, the understanding of the history of which is the subject of conventional basin analysis, and (2) the contained fluids, which are dominantly formation waters with effectively minor amounts of hydrocarbons (natural gas, condensate, conventional crude oil, heavy oil and bitumen) (here, it must be understood that we are concerned with fluid, heat and mass transport). Development of our understanding of the origin, migration and accumulation of formation fluids has fallen mainly to the petroleum industry. During the past two or three decades, when major advances were being made in conventional basin analysis as well as in our understanding of formation fluids, a revolution in thought was under way among groundwater hydrogeologists, specifically with respect to developments in the modelling of groundwater flow processes. We believe that considerable advantages can be gained by integrating knowledge from conventional basin analysis, our understanding of formation fluids and the techniques of groundwater modelling. We have termed this approach 'dynamic basin analysis', and its objective is to synthesize and evaluate the present geological, hydrogeological, thermal and geochemical situation in sedimentary basins as well as their past history.

From GOFF, J. C. & WILLIAMS, B. P. J. (eds), 1987, *Fluid Flow in Sedimentary Basins and Aquifers*, Geological Society Special Publication No. 34, pp. 31–44.

Applications of dynamic basin analysis

Dynamic basin analysis is effectively the development of a comprehensive, qualitative and quantitative understanding of dynamic processes in sedimentary basins: specifically, sedimentary processes, fluid flow, heat transfer and mass transport. When properly integrated, both present and past dynamic processes can be evaluated. This can only be achieved through a multidisciplinary computer-oriented team comprising geologists, geochemists, geophysicists, hydrogeologists, reservoir engineers, numerical modellers and computer experts.

The maxim 'the present is the key to the past' applies equally well to dynamic basin analysis as it does to other branches of geology. Applications of an understanding of the *present* rock framework and dynamic fluid flow, heat transfer and mass transport processes include the following:

1 determination of the effects of *in situ* methods for the recovery of energy resources;
2 evaluation of the feasibility and effects on the underground environment of deep waste disposal;
3 evaluation of the sub-surface environment for the storage of valuable resources, such as natural gas, or the confinement of hazardous materials;
4 exploitation of geothermal resources.

Although an understanding of the rock framework and formation fluids in existing sedimentary basins is extremely important in the search for energy and mineral resources, it is clear from what we know at this time that neither the hydrocarbons nor the strata-bound ore deposits originated in place under the present hydrodynamic and geothermal regimes. Accordingly, determination of the *past* dynamics of sedimentary basins should lead to a better understanding of the origin, migration and accumulation of both hydrocarbons and strata-bound ore deposits, as well as the maintenance of geopressures and the origin of geothermal reservoirs.

Hydrogeology data base

Fundamental to dynamic basin analysis is the acquisition, interpretation, and evaluation of a very diverse suite of point data into synthesized form so that it can be used as input to a mathematical model of flow processes in the subsurface. This requires the development of a data-base management system and specialized data-processing techniques. These have been developed by the Basin Analysis Group of the Alberta Geological Survey for geological, hydrogeological, thermal and hydrochemical parameters. The main features of the computer-based methodology are presented in this paper together with some examples of its application to a study of the baseline hydrogeology of the Cold Lake oil sands deposit and an evaluation of the Swan Hills area for deep waste disposal. More complete details of the methodology can be found in Bachu *et al.* (1987).

The hydrogeology data-base software is written in FORTRAN 77 and is implemented on VAX/VMS. It is structured as a threaded tree, within which extensive use is made of variable-length records and nested variable-occurrences of sub-records and/or data elements. Point data are organized in two levels: level A contains only raw data, and level B contains data which require processing, e.g. hydraulic heads determined from drill-stem test reports. The spatial location of point data is characterized by the Dominion Land Survey (DLS) coordinates and by their depth. Currently, 18 record types have been defined in the data base (Fig. 1), although provisions have been made to accommodate up to 50. The data-base master record appears only once in the data base, and the site master, location, stratigraphic pick and four header records relating to deep wells appear once per site; the remaining records may appear many times per site.

The input sources for the hydrogeology data base are also shown in Fig. 1. Primary information on well location, stratigraphic picks, drill-stem tests and core analyses are obtained from magnetic tapes compiled by the Energy Resources Conservation Board (ERCB); all this information resides in level A. The right-hand side of Fig. 1 indicates information not available on the ERCB magnetic tape. Additional interpretation of well logs is generally also necessary in order to define specific aquifers and aquitards in the areas studied. Microfilms of drill-stem test charts have been interpreted to determine hydraulic heads and permeabilities, and hard copy of formation-water analyses has been entered and interpreted; both these types of information require interactive evaluation before the point data are of use in hydrogeological studies. Records for data from shallow wells and surface waters can be added to the hydrogeology data base as necessary.

The structure of the hydrogeology data base allows for easy entry, access, linking and processing of the point-related information in either the entire sedimentary basin or portions of it.

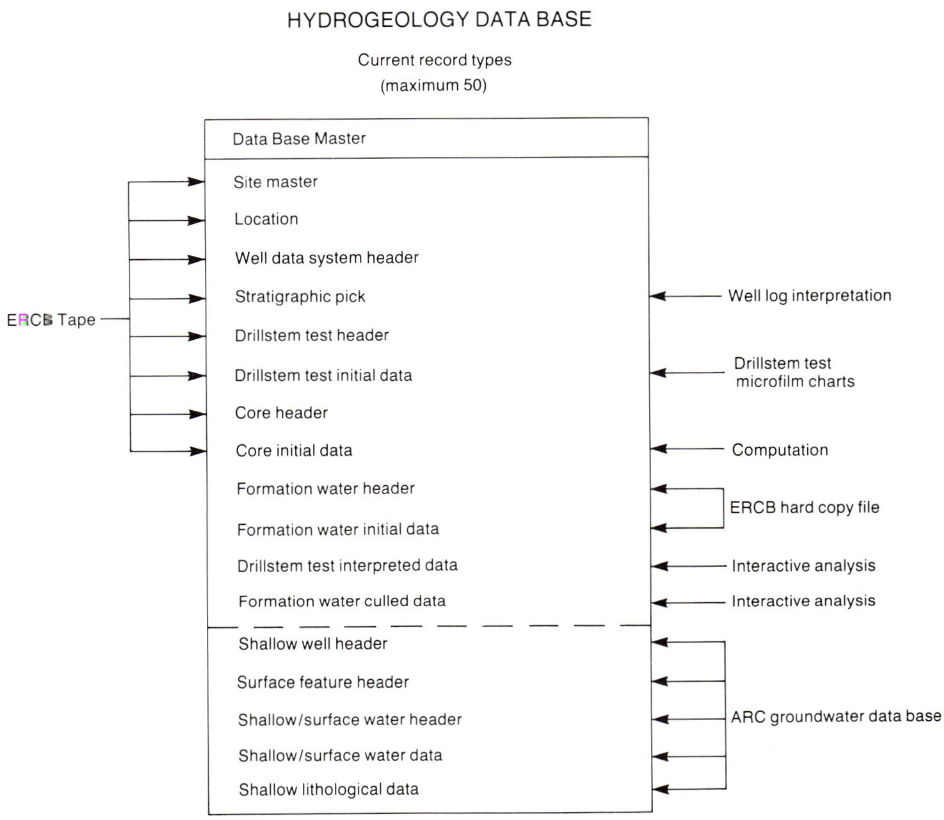

FIG. 1. Record types and input sources for the hydrogeology data base.

Data processing

General comments

The modelling of sedimentary basins requires the handling of immense amounts of point data distributed within a three-dimensional framework; even relatively small blocks of well-drilled sedimentary basins contain a wealth of information which can only be handled reasonably using computer-based technology. For example, our evaluation of the Swan Hills area, Alberta, for deep waste injection was carried out in a study area of 15 760 km^2 (about 40 000 km^3 of Phanerozoic sediments). The hydrogeological evaluation was based on the examination and interpretation of stratigraphic information from 3276 wells, 635 drill-stem tests, 3477 core analyses and 645 formation-water analyses. The specialized computer software described in this paper was used to evaluate the three-dimensional underground flow regime for seven major aquifers and seven intervening aquitards, and only 8.4 months of professional time, 4.0 months of technical support and 4.8 months of data-entry time were required for this procedure.

Because of the sheer size of the data sets it is necessary to forego considering each data element on an individual basis, or refining each relationship between elements in pursuit of exact solutions, in favour of synthesizing the point information to a more generalized form such as maps, statistical characteristics and/or mathematical relationships. Minimization of human intervention in the synthesis process is to be encouraged. This means that data points are seldom checked individually unless, for example, they create 'mountains' or 'black holes' on maps or 'tails' on frequency-distribution charts. The mountains, black holes and tails generally result from erroneous data and the individual point, or points, can easily be checked and corrected or removed from level B of the data base.

The stratigraphic data are processed first in order to build up the geometric framework of the stratigraphic units and, following processing of the flow data (fluid flow, mass and heat transport), the individual stratigraphic units are combined into aquifers, aquitards or aquicludes. The raw data from drill-stem reports are all interpreted in a consistent and fast manner using a specially designed interactive and graphic software package. Permeabilities and porosities from core analyses are weighted to obtain representative values of the cored intervals. The geometry and hydraulic characteristics of the hydrostratigraphic units are used as input into the three-dimensional flow model. Stratigraphic information is treated as a series of surfaces (elevation relative to sea-level datum) and a special package of cartographic software (GEOPLTR) transforms the data location from DLS to Cartesian coordinates, thus allowing for the posting of the data points. The irregular distribution of data points for each stratigraphic surface is then transformed into a regular grid of values and contoured using the SURFACE II software package (Kansas Geological Survey 1978). Human examination of the maps then allows the detection of possible errors and anomalies in the original data, which are either corrected or removed. Flow data are processed in a similar way, except that they are first assigned to individual stratigraphic units by an automatic interpolation procedure. After allocation, the flow data are synthesized and displayed as contour maps, cross-sections, distribution maps or a variety of statistical distribution plots which vary with the parameter being studied.

It is beyond the scope of this paper to describe in detail the wide variety of software developed for treating the stratigraphic data, thermal data and hydraulic data. Instead, the methods used to synthesize the hydrochemical data will be used as an example of the type of software developed.

Example: processing of formation waters

Information on the physical and chemical properties of formation waters is important in hydrogeological studies for several reasons. First, unlike crude oil and natural gas, formation waters comprise the aqueous continuum which exists throughout all sedimentary basins. Second, regional patterns of some components of formation waters reflect the flow path of that water and thereby can be used to assist in the interpretation of potentiometric surface patterns. Third, unlike crude oil and natural gas, formation waters may take part in water–rock reactions, and knowledge of these reactions can be useful in interpreting the history of the formation waters. Fourth, maps of variation in the density of the formation waters are essential for correcting hydraulic heads determined from drill-stem tests and for the production of accurate potentiometric surface maps.

The main source of information on formation waters in Alberta is the hard-copy file of the ERCB. Most of the analyses were obtained by the producing companies and are highly variable with respect to the techniques used for sample preservation in the field, the method of collection and the analytical techniques. Because of these facts and the very large size of the data base it was necessary to develop special software to cull, manipulate and otherwise sort this information before plotting on maps.

Culling

The main objective of all culling, graphical and statistical techniques for evaluating formation-water analyses is to obtain information which is not readily apparent from the analyses. The specific treatment technique depends to a great extent on the nature of the required information and the size of the data base. Culling criteria, several simple graphical procedures, the use of dummy values and data-transformation techniques, and four fairly sophisticated statistical techniques for the treatment of standard formation-water analyses, are presented by Hitchon (1985). The size of the ERCB data base, however, together with the rather simplistic nature of the chemical and physical data available, means that sophisticated statistical techniques for data analysis are not generally justified. Further, as noted by Hitchon (1985), most simple graphical methods have limited utility with large data bases, and this is particularly true when only the major ions have been determined. As a result of these limitations the approach to treating the vast number of data present in the ERCB formation-water data file has been to develop culling criteria for removing analyses which are not representative of the formation water in the underground environment and to find methods of treating the remaining data, which are still rather voluminous, to produce maps with reasonably smooth regional trends.

Most of the formation-water samples were collected from drill-stem tests or surface facilities such as well heads, treaters and separators. Some samples came from holding tanks or were produced by bailing, and are consequently of questionable quality. Formation waters from drill-stem tests or from surface facilities produced early in the history of the well may be contaminated with drilling fluid or mud-filtrate water. A poor (high) recovery position in the drill-stem test-fluid column is usually indicative of diluted formation water. Additional contamination may result from the use of KCl muds, acid washes or washes from cement jobs. In the case of water samples obtained from producing oilfields, contamination may also be due to water injection into the reservoir; this can probably be evaluated with knowledge of the reservoir history. Although poor collection procedures and inadequate sample preservation are criteria which would normally justify culling, the pertinent information is not usually available. When an equivocal analytical procedure is known to have been used, this would also justify removal of the sample from the data base. In addition to these limitations, generally only the common ions Ca^{2+}, Mg^{2+}, Cl^-, SO_4^{2-}, HCO_3^- and CO_3^{2-} are determined, with Na^+ being calculated stoichiometrically as the difference between the sum of the anions and the sum of the cations. Thus, although incomplete analyses and a poor ionic balance would normally be additional reasons for culling formation-water analyses, the latter is not a valid criterion where Na^+ has been determined by difference. Despite all these variables which may affect the composition of formation waters in the ERCB data base, it has been possible to develop culling criteria and data-manipulation techniques which result in reliable maps of the regional chemical and physical properties of the formation waters.

The flow chart for the interactive processing of formation-water analyses from point data to synthesized information is shown in Fig. 2. The formation-water header record (see Fig. 1) contains a list of pointers to all analyses of formation waters for each individual well, including location identification, the number of analyses and the depths sampled. Data relevant to a single analysed sample are contained in the formation-water initial data record; because the formation water may be produced from more than one well (e.g. from a batch separator) or from more than one stratigraphic interval or may be related to a drill-stem test already on file, this record may contain a number of pointers to other well locations. In addition to the analytical data (up to 73 determined components) this file also contains information on the method of production, the date sampled and the laboratory that performed the analysis.

Following manual entry of the formation-water analyses and ancillary information from the ERCB data file, the chemical information is examined by a culling routine, the objective of which is to remove incomplete and obviously erroneous analyses. Thus samples that are clearly contaminated with washes from cement jobs, acid washes and KCl muds are removed; also excluded are analyses where Ca^+ and Mg^+ have been determined as equivalent Ca^+ and those for

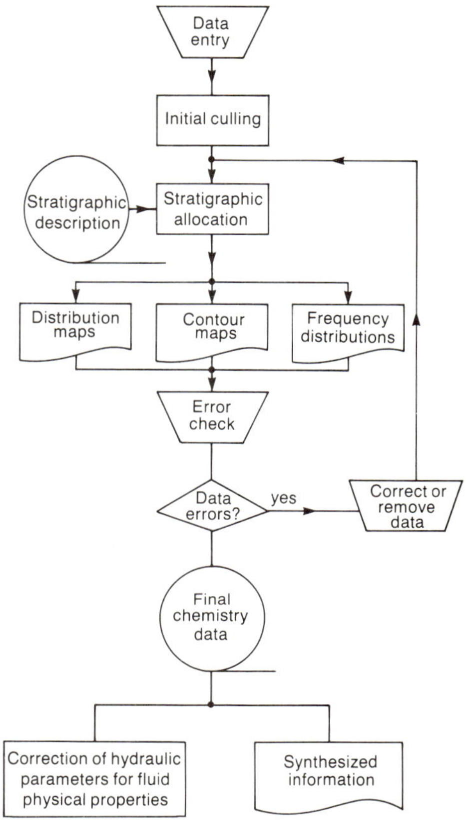

FIG. 2. Flow chart for interactive processing of formation-water analyses from point data to synthesized information.

which any of Ca^{2+}, Mg^{2+}, Cl^-, SO_4^{2-}, HCO_3^- and CO_3^{2-} are missing (a zero value for CO_3^{2-} can be accepted). For the few analyses in which Na^+ and K^+ have been determined separately, a check is made on the cation–anion balance. Na^+ is then calculated 'by difference' on the remaining analyses; even if Br^- and I^- are reported, which is rare, the Na^+(diff) value is calculated without the values for Br^- and I^-. Total dissolved solids (TDS) are then calculated. The need to recalculate Na^+(diff) and TDS (calculated) results from our observation that these numbers are sometimes in error in the hard-copy files. The resulting formation-water culled-data record contains the following information for further testing:

Na^+(diff), Ca^{2+}, Mg^{2+}
Cl^-, *Br^-*, *I^-*, HCO_3^-, CO_3^{2-}, SO_4^{2-}
TDS(calculated), *TDS(evap. 110 °C)*, *TDS(ignition)*
Density, resistivity

The values in italics are optional. The formation-water culled-data file is then separated into derivative files by stratigraphic or hydrostratigraphic units, and it is these derivative files from which the contour maps, distribution maps and frequency distributions are prepared (see Fig. 2).

Processing

Once the most obviously erroneous analyses have been removed, the remaining analyses can be examined using a variety of techniques. Because many geochemical data approximate a log-normal distribution, it is possible to use cumulative frequency plots of the logarithm of selected ions as a culling criterion. The most geochemically conservative ion reported in standard formation-water analyses is Cl^-, and accordingly this is the best one for constructing cumulative log frequency plots. A typical example is shown in Fig. 3(a). Computer printouts are then obtained of the individual analyses at the less than 10% and greater than 90% frequency intervals. Rapid manual examination of these analyses allows removal of all or part of the tails and after possibly several iterations the final satisfactory cumulative frequency profile is obtained (Fig. 3b). Experience shows that the majority of formation-water analyses removed by this technique were subject to either evaporation in holding tanks or contamination by drilling mud. Although other ions could be plotted using cumulative log frequency plots, only the use of Cl^- has proved to be satisfactory.

Contour maps of the regional distribution of selected chemical and physical properties of formation waters from the cleaned-up data base can then be plotted to identify the remaining anomalous samples. The most useful parameters for this purpose are TDS(calculated), Cl^-, SO_4^{2-} and density. More rigorous culling of the remaining anomalous samples could then be achieved using trend surface maps, although this technique has not been used so far. Many other culling methods can be devised, but it is doubtful whether the reliability of the information present in the standard formation-water analyses in the ERCB data file justifies them at this stage. As an example of the degree to which these procedures are effective in culling a formation-water data base, we can cite the results of our work in the Cold Lake area, Alberta. Out of more than 3100 wells in the study area, formation-water analyses were available from only 1894 wells (about 60%). A total of 3650 formation waters from these wells was entered into the data base, of which 2366 (65%) remained in the formation-water culled-data file. After examination of the cumulative log frequency plots and removal of anomalous data

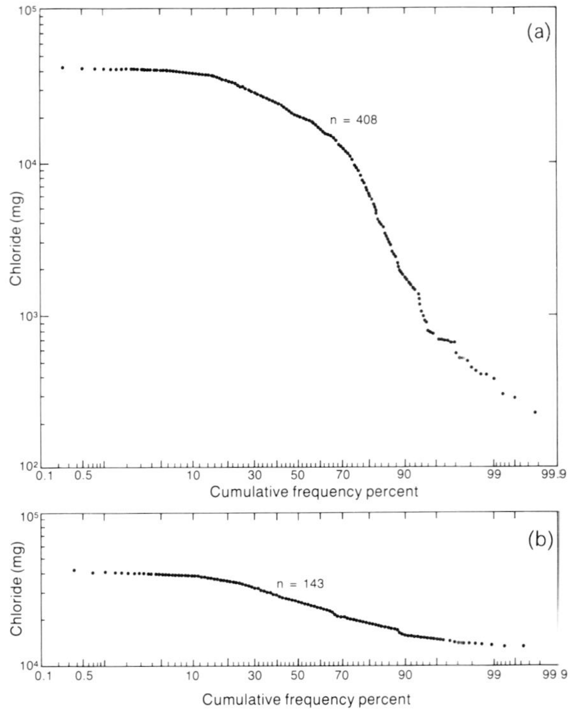

FIG. 3. Cumulative frequence plots of log Cl^- in formation waters from the Lower Cretaceous Viking Formation, Cold Lake study area, Alberta: (a) original data base; (b) after removal of the tails following manual examination of the individual analyses.

from the various contour maps, the final data base comprised 852 analyses which represented only 23% of the original data base. Although, perhaps, this final data base could be purged of additional analyses by using more rigorous culling techniques, we have considerable confidence in the general regional trends of the resulting formation-water composition maps. In addition to printed maps and diagrams, the grid of regional variations in the density of formation water in any specific stratigraphic unit is used as input to the drill-stem test records for correcting hydraulic-head values.

Hydrogeological synthesis

Synthesis and evaluation of the stratigraphic, hydraulic and hydrochemical data result in the identification of individual hydrostratigraphic units, i.e. the individual aquifers, aquitards and aquicludes. The first step in the hydrogeological synthesis is to use all the pertinent interpreted information to characterize regionally the flow within and between the identified aquifers. More detail about the various techniques used for this purpose can be found in Bachu *et al.* (1987). The final product is a three-dimensional conceptual and descriptive model of the sedimentary basin or block being studied. The model consists of regular grids of values defining the stratigraphy, the potentiometric surfaces of the flow in aquifers and the distributions of chemical components and temperatures in the sub-surface environment. Also the main flow parameters, hydraulic conductivities and storativities are statistically characterized by averages, range of variation, standard deviations etc. for each stratigraphic and/or hydrostratigraphic unit. On the basis of conservation principles, computation of the mass balance of the fluids entering and leaving the sedimentary block helps to calibrate the effective average hydraulic parameters characterizing each unit.

The conceptual/descriptive model of the sedimentary basin can be used in numerical simulations of flow phenomena. For the Cold Lake and Swan Hills areas presented subsequently, the model FE3DGW developed by Pacific Northwest Laboratory, U.S.A. (Gupta *et al.* 1979, 1984), was used to simulate the natural steady-state flow conditions and the hydraulic effects of deep waste injection. We shall now summarize the results of modeling fluid flow in the Cold Lake and Swan Hills areas, Alberta, in terms of simplified block diagrams of the model areas showing the input and output flow rates and the rates of cross-formational flow. Figure 4 shows the position of the study areas in relation to Alberta.

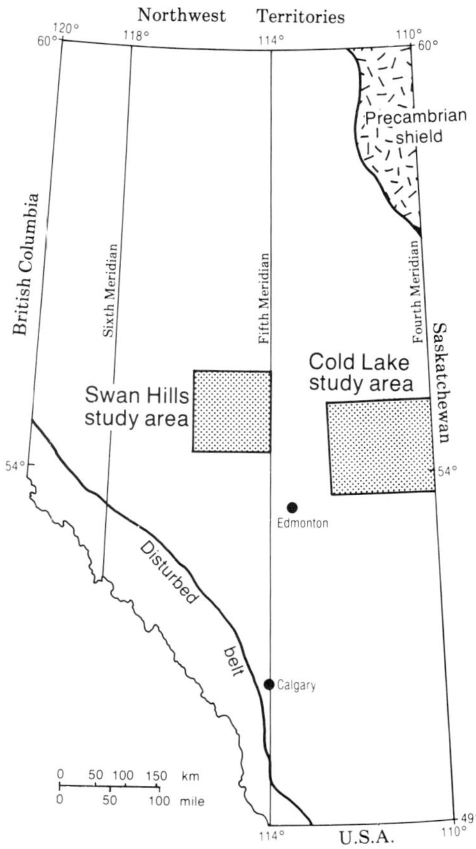

FIG. 4. Location of Cold Lake and Swan Hills study areas, Alberta.

Cold Lake study area

The Cold Lake study area is defined as Tp 55-69, R 1-17, W4 Mer, and covers approximately 23 800 km^2 in E central Alberta. The Cold Lake oil sands deposit occupies a significant portion of the eastern half of the study area. The objective of the study was to determine the baseline hydrogeological situation prior to development of commercial *in situ* recovery operations in the oil sands deposit because of the possible impact these might have on the natural flow system, specifically on the near-surface potable groundwater resources.

The regional Phanerozoic history was synthesized from the literature, together with stratigraphic information from more than 3100 wells. Nearly 1200 drill-stem tests and 3650 formation-water analyses as well as 1550 core analyses were used to describe the hydraulic parameters of the stratigraphic units and formation waters. This information allowed the identification of seven aquifers, eight aquitards and three aquicludes between the Precambrian basement and the base of the Quaternary deposits.

In hydrogeological studies it is important to determine the geometry of the aquitards and aquicludes because the absence of these water-retarding hydrostratigraphic units allows juxtaposition of the aquifers. Figure 5 shows the study area with the boundaries of some critical hydrostratigraphic units and a diagrammatic dip (NE–SW) cross-section through the study area. The generalized stratigraphy and hydrostratigraphy of the model area are shown in Fig. 6. For the purposes of numerical simulation, the Middle Devonian Prairie Formation halite (D) was designated as the basal aquiclude. Shale sequences which act as aquitards are the Upper Devonian Ireton Formation (F), the Cretaceous Clearwater Formation (K), the Joli Fou Formation (M) and the Colorado Group (O). With the exception of the Colorado Group, which extends across the entire study area, the three aquitards and the basal aquiclude (D) are discontinuous.

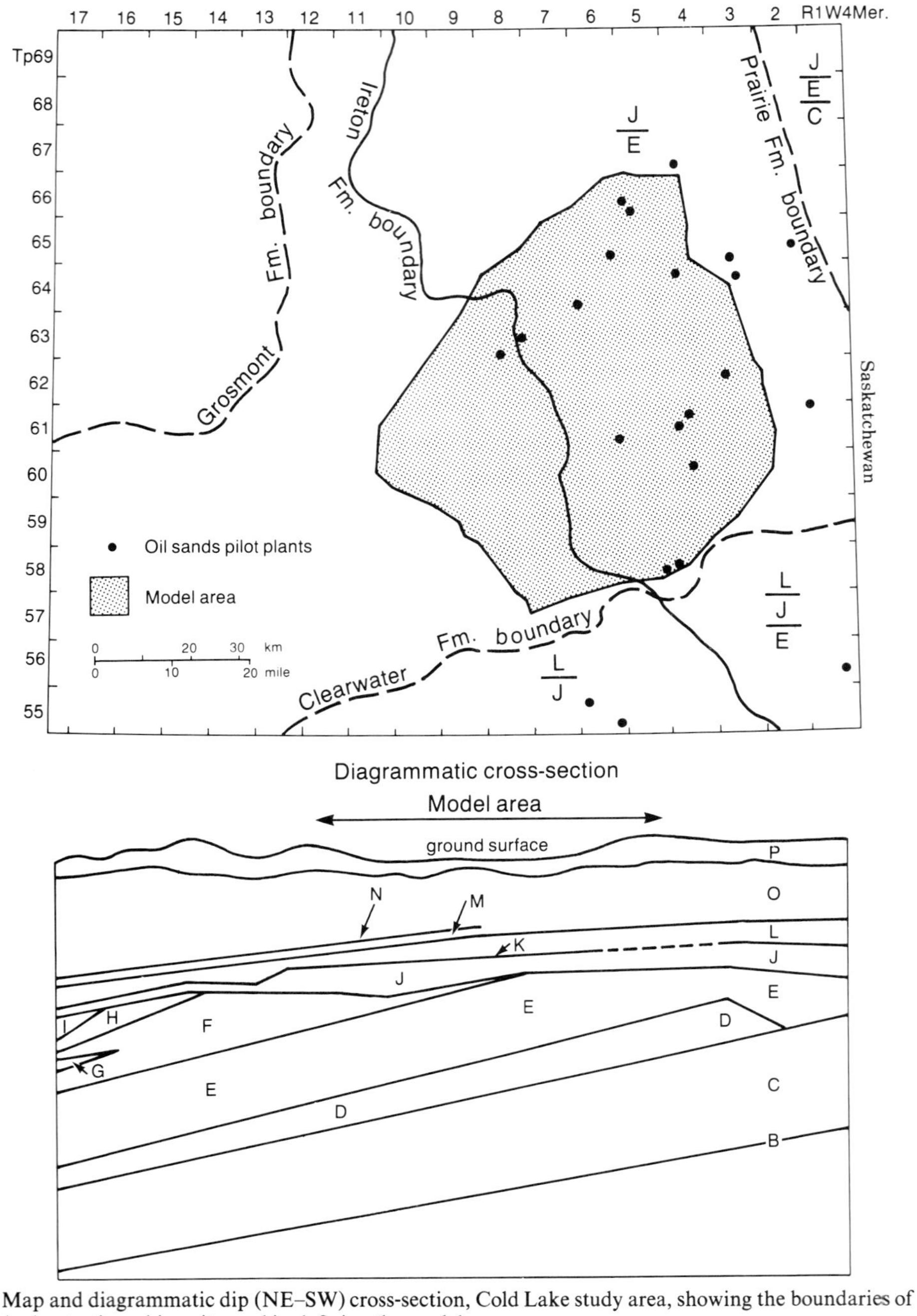

FIG. 5. Map and diagrammatic dip (NE–SW) cross-section, Cold Lake study area, showing the boundaries of critical hydrostratigraphic units used in defining the model area.

Absence of the Prairie Formation halite owing to solution and the presence of the pre-Cretaceous unconformity act in combination in the north-eastern corner of the study area to juxtapose the 'Lower' Mannville Group (J), the Beaverhill Lake and Watt Mountain Formations (E) and the Keg River Formation (C) aquifers. Down-dip from this NW corner of the study area, the absence of the Ireton Formation (F) allows juxtaposition of the 'Lower' Mannville Group (J) and the Cooking Lake and Beaverhill Lake Formations (E). In the SE corner of the study area, however, where both the Clearwater Formation (K) and the Ireton Formation (F) are

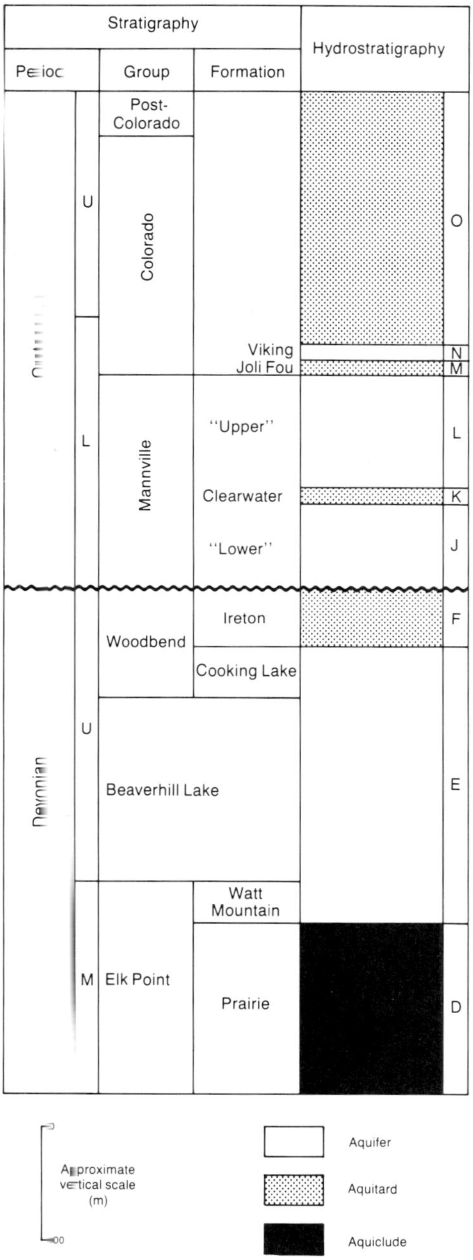

FIG. 6 Generalized stratigraphy and hydrostratigraphy of the model area (see Fig. 5 for location of the model area), Cold Lake study area, Alberta. Note: Hydrostratigraphic units F and N are not present throughout the model area (see Fig. 7).

absent, 'Upper' and 'Lower' Mannville Group rocks are in contact with the Beaverhill Lake Formation carbonates beneath the pre-Cretaceous unconformity. Another region in which strata of vastly different ages come in contact is in the NW part of the study area where the arenaceous 'Lower' Mannville Group is in direct contact with the porous dolomites of the Grosmont Formation. The final model area was selected to include the majority of the oil sands pilot plants and to be within the limits of the Clearwater Formation aquitard and outside the boundary of the Grosmont Formation which acts as a significant drain in the regional flow system. Following several tests, it was determined that the Quaternary hydrostratigraphic units could be decoupled from the pre-Quaternary units, and as a consequence the final sequence modelled included all hydrostratigraphic units between the bedrock (top of unit O) and the top of the Prairie Formation aquiclude (base of unit E).

The block diagram in Fig. 7 shows the Phanerozoic sequence modelled and the input and output flow rates as well as the rates of cross-formational flow.

Swan Hills study area

A detailed hydrogeological study was carried out in a region defined as Tp 62-74, R 1-13, W5 Mer, comprising an area of 15 760 km^2 effectively centred on the special-waste injection site of the Alberta Special Waste Management Corporation. The objective was the selection of aquifers for environmentally safe disposal of special wastes. This required (1) identification of the major hydrogeological and economic constraints, (2) an analysis of the natural flow system for all hydrostratigraphic units between the Lower Cretaceous Viking Sandstone aquifer and the Precambrian basement, (3) numerical simulation of the flow in effectively the entire Phanerozoic sequence to obtain conformance with the natural flow system and to calibrate the numerical model and (4) perturbation of the natural state at an injection rate of 225 m^3 day^{-1} in two potential injection aquifers to test the degree of confinement of the pressure build-up. The study was based on examination and interpretation of stratigraphic information from 3276 wells, 635 drill-stem tests, 3477 core analyses and 645 formation-water analyses. Although two different model areas were evaluated, we shall confine our attention in this paper to the smaller model area (4250 km^2) immediately around the special-waste injection site.

The stratigraphy and hydrostratigraphy of the Swan Hills area is considerably more complex than that of the Cold Lake area. Not only is the Phanerozoic section twice as thick, although there is effectively only the same number of hydrostratigraphic units, but there are two major unconformities which allow juxtaposition of

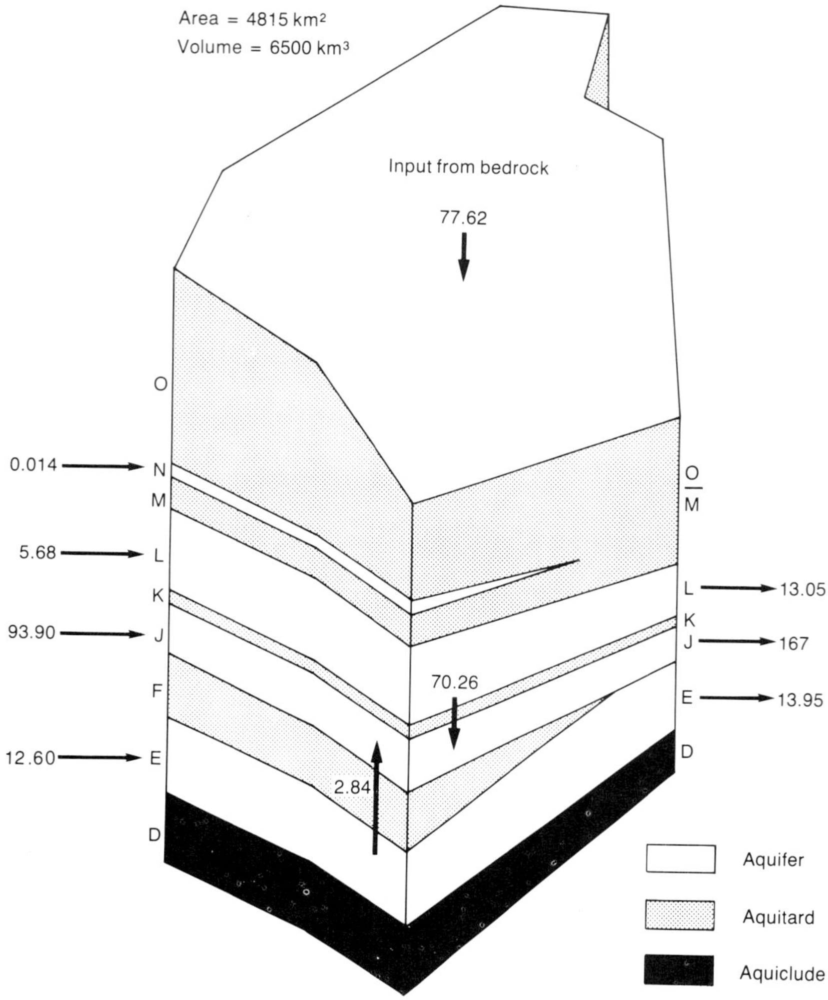

FIG. 7. Block diagram of the Phanerozoic sequence in the Cold Lake model area showing input and output flow rates and the rates of cross-formational flow (all flow rates in m^3 day^{-1}).

hydrostratigraphic units of contrasting hydraulic properties and which are associated with regoliths or karstic phenomena, both of which may short-circuit flow systems. Figure 8 shows the relation of the model area to the boundaries of critical hydrostratigraphic units at the pre-Devonian (Fig. 8a) and pre-Cretaceous (Fig. 8b) unconformities, as well as a diagrammatic dip (NE–SW) cross-section (Fig. 8c) through the injection site. The generalized stratigraphy and hydrostratigraphy at the injection site are shown in Fig. 9.

At the pre-Devonian unconformity (Fig. 8a) four Cambrian aquifers (Lynx, Pika, Eldon and Basal Sandstone) sub-crop against Middle Devonian strata; the intervening blank areas correspond to the sub-crop regions of the Sullivan, 'Upper Eldon' and Mount White–Cathedral–Stephen aquitards respectively. Immediately overlying the Cambrian throughout the central part of the study area and oriented in a NW–SE direction is a regolith of Granite Wash, ranging up to 70 m thick. This unit, shown stippled in Fig. 8(a), is a major short-circuit at the base of the Middle Devonian. The onlapping nature of the Middle Devonian Elk Point Group allows the sub-cropping Cambrian aquifers to be overlain by a succession (from E to W) of an aquiclude (Lotsberg–Basal Red Beds), aquifer (Ernestina Lake), aquitard (Red Beds), aquifer (Keg River–Contact Rapids) and finally, across the entire

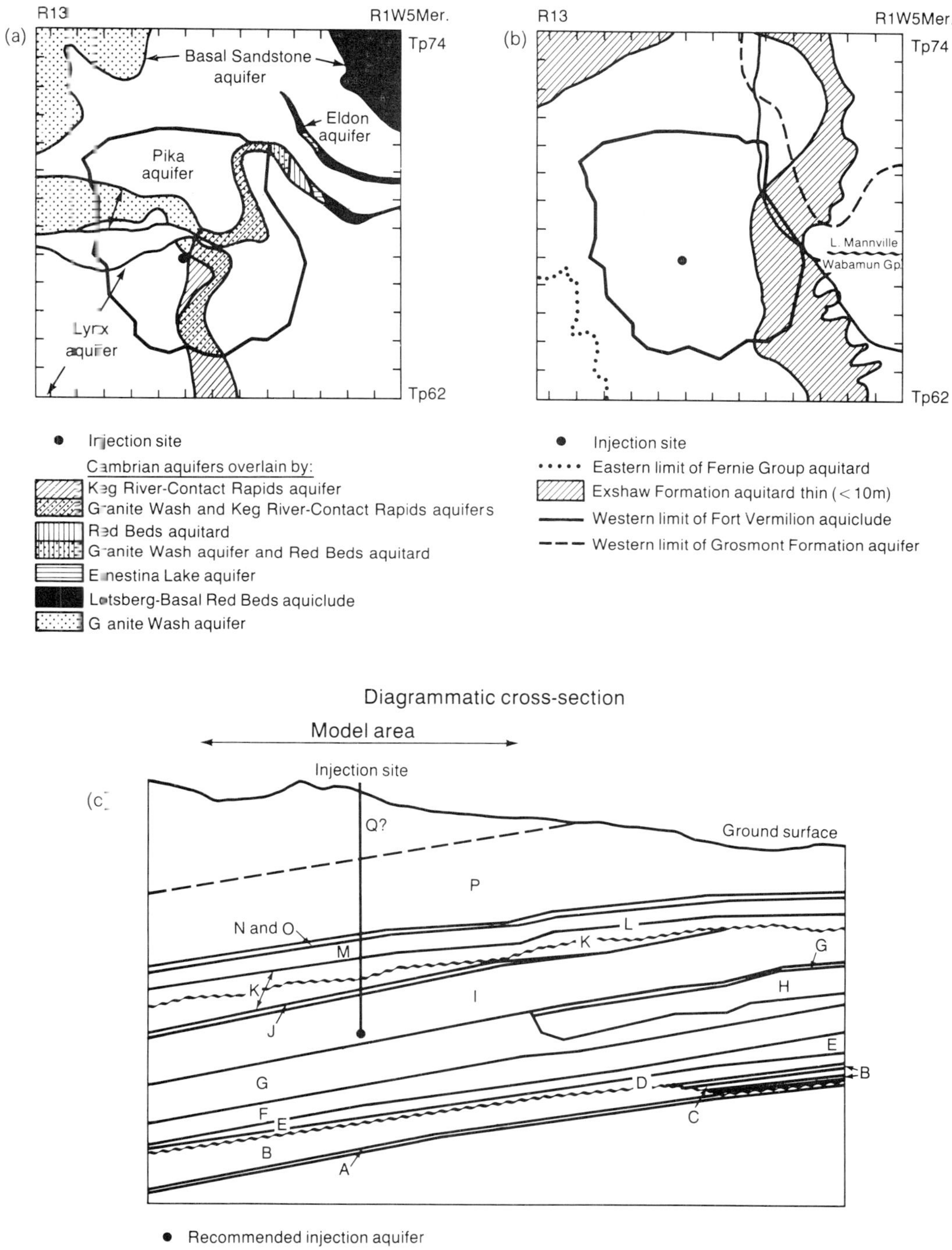

FIG. 8. Diagrammatic sketches of critical hydrostratigraphic situations, Swan Hills study area: (a) pre-Devonian unconformity; (b) pre-Cretaceous unconformity; (c) dip (NE–SW) cross-section through the injection site.

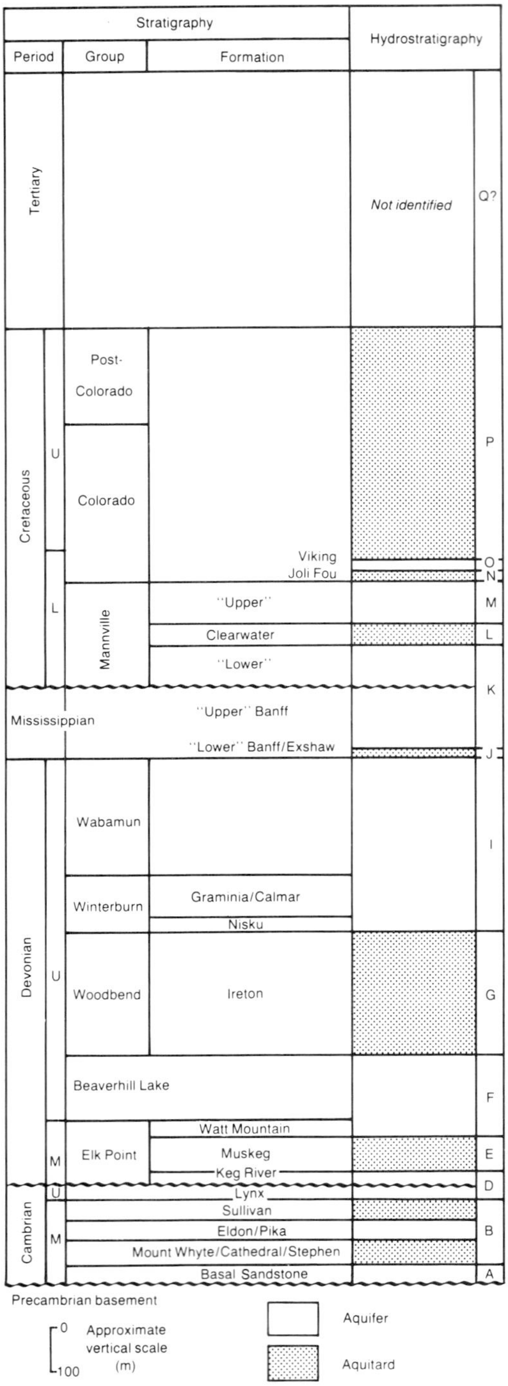

FIG. 9. Generalized stratigraphy and hydrostratigraphy at the injection site, Swan Hills study area, Alberta.

study area, the Muskeg aquitard (E). Thus, although the Cambrian aquifers within the study area may be short-circuited through the overlying Granite Wash and through both the Ernestina Lake and Keg River–Contact Rapids aquifers, with respect to the Middle Devonian, the entire study area is covered by the Muskeg aquitard which effectively separates the pre-Muskeg hydrostratigraphic units from the overlying regional Watt Mountain and Beaverhill Lake aquifers (F).

In ascending order the second critical area is that associated with the Grosmont Formation (Fig. 8b, hydrostratigraphic unit H). This unit ranges up to more than 200 m thick, but is overlain by only a thin sequence (maximum 20 m) of Upper Ireton Formation shales (G); in contrast, the Ireton Formation is about 300 m thick over most of the study area. Our work in the Cold Lake area shows that the Grosmont Formation acts as a major drain to the surrounding flow system and its presence in the eastern third of the study area results in significant cross-formational flow through the thin overlying Upper Ireton Formation shales.

The pre-Cretaceous unconformity presents the third potentially critical hydrostratigraphic situation. Along the eastern margin of the study area, the 'Lower' Mannville Group aquifer (K) lies directly on the carbonates of the Upper Devonian Wabamun Group (I) (see Fig. 8b). Immediately to the W of the Wabamun Group sub-crop is a narrow band, about 20 km wide, in which the combined 'Lower' Mannville–'Upper' Banff porous unit (K) is separated from the Wabamun Group carbonates by less than 10 m of Exshaw Formation shale (J). As a result, in effectively the entire eastern third of the study area there is hydraulic continuity between the 'Lower' Mannville aquifer and the underlying carbonates of the Upper Devonian Wabamun and Winterburn Groups (I), and the latter units are then separated from the underlying Grosmont Formation (H) in the same part of the study area by only the thin Upper Ireton Formation shales (G).

The extreme NW corner of the study area (see Fig. 8b) also has just a thin sequence of Exshaw Formation shales, but its position far from the injection site makes its presence less critical. For a similar reason, the thin (about 20 m) Fernie Group shales in the extreme SW of the study area are not likely to be critical with respect to the injection site.

The model area described in this paper was selected so that it was centred on the special waste injection site and lay between (i.e. outside) the boundaries of the Grosmont Formation and Fernie Group. The sequence modelled lay between the Lower Cretaceous Viking Sandstone

(O) and the Precambrian basement; this effectively provided an aquitard above the modelled sequence of about 550 m of Upper Cretaceous Colorado and post-Colorado Group shales (P), plus 600 m of generally argillaceous Tertiary strata.

The block diagram in Fig. 10 shows the NW half of the area modelled, and the input and output flow rates as well as the rates of cross-formational flow; note that these rates refer to the entire block modelled.

It is beyond the scope of this paper to describe the arguments and calculations carried out in reaching a recommendation for a preferred injection aquifer. It is sufficient to note that the aquifer recommended and subsequently successfully drilled was the Wabamun–Winterburn hydrostratigraphic unit (I).

Future research directions

We believe that the approach we have taken to dynamic basin analysis is sound, but we recognize that much development work is yet to be done if the end result is to be achieved. Included among the parameters to be integrated into the present data-base system are those concerned with crude

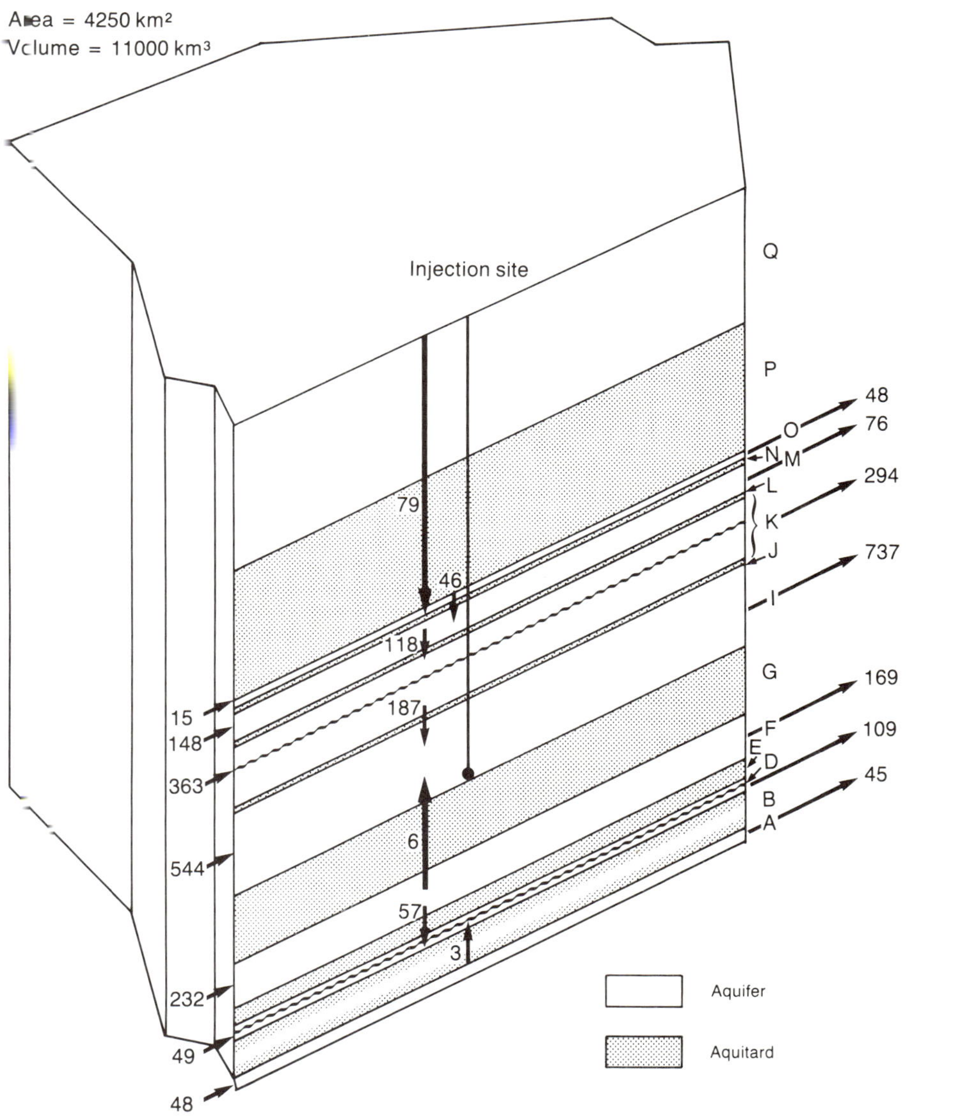

FIG. 10. Block diagram of the Phanerozoic sequence in the NW half of the Swan Hills model area showing input and output flow rates and the rates of cross-formational flow for the entire block (all flow rates in m^3 day^{-1}).

oil and natural gas composition, as well as properties of kerogen which relate to the thermal maturation history of the sedimentary rocks. Even this additional information will only allow modelling of the present rock framework and its contained fluids. To understand the dynamic history (*sensu lato*) of sedimentary basins, it is necessary to develop 'back-stripping' models which can be integrated with fluid flow, heat transport and mass transport models. With respect to integrating these various types of models in order to determine the palaeohydrodynamics of sedimentary basins, we recognize that one of the most difficult matters to be considered is the fixing of reasonable boundary conditions, no matter which model type is being considered. At this stage in development the end result is apparently intransigent. Despite this rather pessimistic viewpoint, the potential practical value justifies the effort to develop truly integrated models for dynamic basin analysis.

ACKNOWLEDGMENTS: Permission to include information obtained under contracts with Alberta Environment and the Alberta Special Waste Management Corporation is greatly appreciated. The authors especially acknowledge the technical programming assistance of M. Brulotte and the editorial and technical support of M. M. Madunicky.

References

BACHU, S., SAUVEPLANE, C. M., LYTVIAK, A. T. & HITCHON, B. 1987. Analysis of fluid and heat regimes in sedimentary basins: techniques for use with large data bases. *Amer. Assoc. Petroleum Geologists Bull.*, **71**, pp. 822–843.

GUPTA, S. K., COLE, C. R. & BOND, F. W. 1979. *Finite-element Three-dimensional Groundwater (FE3DGW) Flow Model Formulation, Program Listings and Users' Manual*, Pacific Northwest Laboratory, Richland, WA.

——, —— & PINDER, G. F. 1984. A finite-element three-dimensional groundwater (FE3DGW) model for a multiaquifer system. *Water Resour. Res.* **20**, 553–63.

HITCHON, B. 1985. Graphical and statistical treatment of standard formation water analyses. *In:* HITCHON, B. & WALLICK, E. I. (eds) *Practical Applications of Ground Water Geochemistry, Proc. 1st Canadian/American Conf. on Hydrogeology*, pp. 225–36, National Water Well Association.

KANSAS GEOLOGICAL SURVEY 1978. *SURFACE II Graphics System.*

BRIAN HITCHON, S. BACHU, C. M. SAUVEPLANE & A. T. LYTVIAK, Basin Analysis Group, Alberta Geological Survey, Alberta Research Council, PO Box 8330, Postal Station F, Edmonton, Alberta T6H 5X2, Canada.

Post-Palaeocene evolution of regional groundwater flow systems and their relation to petroleum accumulations, Taber Area, southern Alberta, Canada

J. Tóth & T. Corbet

SUMMARY: Three regionally extensive fluid dynamic systems have been mapped in a 23 500 km^2 portion of the Western Canada Sedimentary Basin in south-eastern Alberta. The Modern Land Surface System reaches depths of 300 m and is well adjusted to the modern relief. The subjacent Erosional Rebound System occupies most of the 600 m thick predominantly shale Colorado Aquitard including its often gas-bearing sand members. It is characterized by extensive regions of sub-hydrostatic pressures. Below the Colorado Aquitard is the Cypress Plain System which has strong upward-directed forces and mild but definite lateral components. The lateral components oppose the modern relief and are oriented from the centre of the study area toward its western and eastern boundaries where most of the known oil fields are also located. The three dynamic systems are interpreted to be transient stages of sequentially superimposed gravitational and dilatational force fields. The gravitational fields depend on topography and the dilatational field results from elastic rebound of the Colorado Aquitard in response to erosional removal of 700 m of overburden during Pliocene and Pleistocene times.

The flow pattern that existed during the time of the Cypress Plain was generated after Palaeocene (58 Ma) or Eocene (36 Ma) times by a palaeo-relief dominated by the NE–SW-trending Bow Island Topographic High. It was composed of and maintained as cross-formational gravity-flow systems throughout at least the Miocene (i.e. for more than 30 m.y.). Oil migrated to and was entrapped in the discharge regions of these systems. Subsequently, dilation of the Colorado Aquitard created the hydraulic sink in the Erosional Rebound System and induced gas to exsolve and accumulate in the sands.

Based on a one-dimensional solution of the diffusion equation, periods of 10 m.y. and 0.1 m.y. respectively were calculated to be required for pore pressures below and above the Colorado Aquitard to adjust to changes in the land surface. Compared with inferred ages and durations of palaeo-topographies, these times agree with the hypothesized evolutionary history of flow systems. There was sufficient time for flow in the Cypress Plain System to adjust to the post-Eocene relief but not enough time has yet elapsed for it to re-adjust to the Modern Land Surface which was initiated during middle Pleistocene (1 Ma) times. In contrast, the Modern Land Surface System has had 10 times the 0.1 m.y. required for it to adjust to the modern relief, and thus its apparent steady state is expected.

According to the hydraulic theory of petroleum migration (Tóth 1980) the principal agent in the transport and accumulation of hydrocarbons in geologically mature basins is gravity-induced cross-formational groundwater flow. Geological maturity as used in the present context stipulates a stable undeforming rock framework which is devoid of fluid-impelling forces caused by mechanical stresses such as those due to compaction, compression or dilation. From an exploration point of view one important prediction of the theory is that in mature basins hydrocarbons will accumulate preferentially in the discharge, or ascending-flow, regions of the flow systems. However, this spatial relation between water-flow patterns and petroleum accumulation may be modified by the presence of force fields that are due to deformation of the rock framework.

Furthermore, owing to resistance to flow and compressibility of the rocks, a change in the mechanical stress field or the configuration of the land surface entails a delay in the adjustment of the fields of both deformational and gravity flow, respectively (Neuzil & Pollock 1983; Tóth & Millar 1983). The time lag, or the period of transiency, may be sufficiently long to mask the cause-and-effect relations between regions of hydrocarbon emplacement and their generating flow systems. Therefore it is imperative that the evolutionary history of the formation-fluid flow fields is also understood. The objective of the study reported in this paper was to analyse the spatial and temporal relations between hydrocarbon accumulations and the formation-fluid flow generated by the combined forces of gravity and rock deformation.

The area of study provides an opportunity of analysing the effects of erosion and rock deformation on the sub-surface flow and, consequently,

From GOFF, J. C. & WILLIAMS, B. P. J. (eds), 1987, *Fluid Flow in Sedimentary Basins and Aquifers*, Geological Society Special Publication No. 34, pp. 45–77.

FIG. 1. Location of the Taber Area.

on hydrocarbon distributions. It covers approximately 23 500 km^2 in SE Alberta, Canada (Fig. 1). Owing to rapid uplifting and intensive erosion more than 700 m of sediments have been removed from the area since Palaeocene time, giving rise to major changes in the topography and to elastic rebound of the rock formations, particularly during the last 5 m.y. Also, owing to intensive exploration for and exploitation of oil and gas in the area, a sufficient data base exists. The study constitutes essentially a synthesis and interpretation of geological and hydrodynamic information produced by the petroleum industry and/or found in the literature. Because of the novelty and complexity of both the problem and the approach used it was deemed necessary to introduce the paper by a theoretical overview of groundwater flow conditions in a topographically modified and mechanically-deforming geological environment. Theoretical situations of flow, hydraulic head and pressure patterns, with which observed field patterns are compared for the purpose of interpretation, are also presented in this overview.

Principles of groundwater flow in evolving drainage basins

Flow equations and their application

Among possible factors that may generate groundwater flow in evolving drainage basins two important ones are the configuration of the water table and the mechanical stresses in the rock framework. Differences in the elevation of the water table induce gravitational flow, and changes in the mechanical stress field may result in compressional or dilatational flow owing to deformation of the rocks. During the evolution of a basin both the water-table relief and the stress field may undergo modifications causing a concomitant change in the pathways and intensities of groundwater flow. The shape of the water table may be modified, for instance, by erosional sculpturing of the land surface or changes in the climate. Changes in mechanical stresses may be effected by compression or dilation of the rock skeleton owing to erosional loading or unloading respectively. The relations between these environmental effects and groundwater flow can be expressed by an equation which is also used in the theory of consolidation (Neuzil & Pollock 1983):

$$\nabla^2 h + \frac{\alpha}{K}\frac{\partial \sigma}{\partial t} = \frac{\rho g(\alpha + n\beta)}{K}\frac{\partial h}{\partial t} \qquad (1)$$

where h is the hydraulic head, K is the hydraulic conductivity, n is the porosity, α is the compressibility of the rock, β is the compressibility of water, σ is the load, ρ is the relative density of water and g is the acceleration due to gravity.

The second term on the left-hand side of Equation 1 is related to the deformation of the rock skeleton caused by a change in loading and resulting in the deformational component of the flow. Under conditions of constant load $\partial\sigma/\partial t = 0$ and no deformational flow exists. Equation 1 then becomes what is known as the diffusion equation or the general equation of the gravity flow of groundwater:

$$\nabla^2 h = \frac{\rho g(\alpha + n\beta)}{K}\frac{\partial h}{\partial t} \qquad (2)$$

The term on the right-hand side of Equation 2 comprises the factors controlling those transient changes in the gravitational component of the flow field caused by the time changes in the field of hydraulic heads. These changes can be imposed on the boundaries of the flow region by, for instance, sediment accretion or erosional unloading. In the absence of such changes in h, i.e. when $\partial h/\partial t = 0$, Equation 2 reduces to the Laplace equation

$$\nabla^2 h = 0 \qquad (3)$$

Equation 3 describes the distribution of hydraulic heads in a conservative potential field and can be used to calculate the steady-state gravity component of regional groundwater flow.

The evolutionary state of the groundwater flow field in a given geographic region can be evaluated, in principle, by comparing observed and calculated patterns of hydraulic heads or pore pressures, or both. In practice, however, the application of the above equations to real field situations is usually rendered difficult by an insufficient knowledge of the physical parameters (K, n and α) of the rocks and their regional distributions, an insufficient knowledge of the geological history determining the initial and boundary conditions for the various evolutionary stages and, finally, computational difficulties resulting from the large contrasts in permeabilities, the large time steps due to the geological time-scale of the processes and the large numbers of nodal points necessitated by the three-dimensional nature of most real-life problems. Nevertheless, judicious application of the above equations to simple type situations allows the definition of diagnostic features of groundwater flow fields under various boundary conditions and at different stages of evolution. These features can then be used to characterize observed situations and to establish, at least conceptually, a probable sequence of stages of flow-field evolution.

Before the above principles can be applied to an actual field situation a few basic concepts must be reviewed.

Steady-state patterns of regional groundwater flow

Groundwater flow is said to be steady state if its direction and intensity do not change with time. Consequently, hydraulic heads and pore pressures in a steady-state flow field are also constant with respect to time. In nature, steady-state conditions are considered to exist in regions where the effects of water-table changes and rock deformation on the flow are negligible. These effects can be negligible either because of their small original magnitude or because a sufficiently long time lapse has allowed them to decrease to an insignificant level. For purposes of analysis steady-state patterns can be calculated using Equation 3.

The simplest regional flow pattern generated by gravity occurs in a unit groundwater basin (Fig. 2) which is a geometrically delimited three-dimensional block of the Earth's crust that is homogeneous with respect to hydraulic conductivity. Its upper boundary is an axially symmetrical topographic depression with water tables rising linearly from the valley bottom to the water divides, and it is bounded by an effectively impermeable horizontal stratum at the base and by vertical planes beneath the water divides through which no flow occurs because of laterally continuing symmetry of the land surface relief. The regional groundwater flow pattern which develops in a saturated unit basin with gravity being the sole driving force of water movement is a basic or unit pattern. In Fig. 2 such a flow pattern is represented by the flow lines **q** which are orthogonal to the lines of equal hydraulic head h calculated using the Laplace equation (Equation 3) for the given boundary conditions (Tóth 1962):

$$\left(\frac{\partial h}{\partial x}\right)_{x=0}=\left(\frac{\partial h}{\partial x}\right)_{x=s}=\left(\frac{\partial h}{\partial z}\right)_{z=0}=0 \qquad (4)$$

where $h=h(x)$.

Three fundamentally different groundwater flow regimes can be distinguished in a unit basin: recharge or inflow, transfer or throughflow and discharge or outflow, which are located in the areas of recharge, midline and discharge respectively (Fig. 2). Relative to the water table,

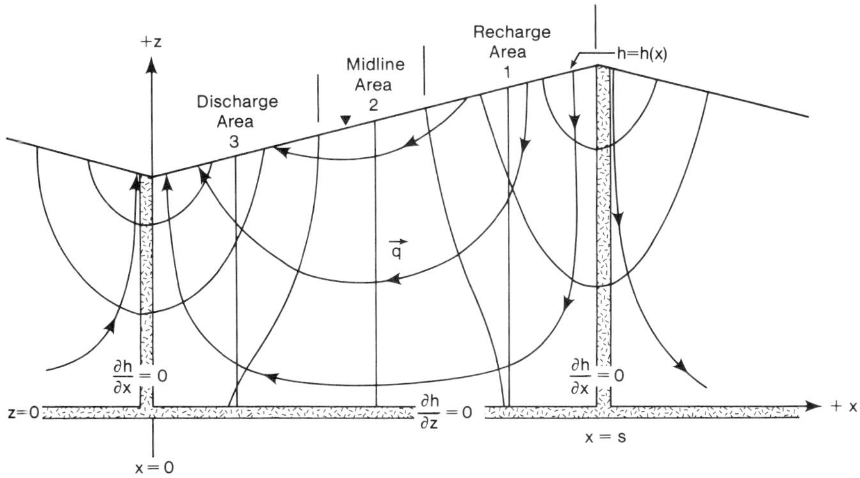

FIG. 2. Patterns of flow and hydraulic head in a unit groundwater basin.

groundwater flow is descending, lateral and ascending in these three regions. The intensity of movement decreases with increasing depth and away from the area of midline; in the lower corners of the flow field near-stagnant conditions exist. With increasing depth, water levels decline in the recharge area, remain constant in the vicinity of the midline and rise in regions downslope from it, possibly to elevations above the land surface.

The above conditions are also shown by the relations of pore pressures to depth (Fig. 3). For a given depth the actual pore pressures are lower than, equal to and higher than the nominal hydrostatic values calculated for the same depth for areas of recharge, midline and discharge respectively (Figs 2 and 3, lines 1, 2 and 3). Consequently, the angle that a measured $p(d)$ curve makes with the nominal hydrostatic line (Fig. 3) is indicative of the relative flow directions and thus of the nature of the flow regime in the area.

Departures from the basic flow pattern can be introduced by complexities of the topographic relief and heterogeneities of the permeability of the rock framework. A discussion of general flow patterns is beyond the scope of the present paper and can be found in various previous publications (Tóth 1963, 1980, 1985; Freeze & Witherspoon 1967; Freeze & Cherry 1979).

Flow patterns in the field can be evaluated from measurements of water levels and formation pressures. Numerous examples of measured field patterns can be found in the literature, and many of them have been reviewed by Tóth (1980, 1985).

Transient conditions of the gravitational component of regional flow

As was pointed out in connection with Equation 2, transient conditions of the gravitational component of regional groundwater flow can be induced by a change in boundary conditions. An example of such a change is the sculpturing and general lowering of the land surface by erosion. If the elevation of the land surface is lowered at a rate greater than the rate of hydraulic head dissipation as controlled by the physical properties of the rocks, pore pressures will be greater than hydrostatic throughout the basin (Bradley 1975). After erosion stops pressures will gradually adjust to hydrostatic values that correspond to the new elevation of the land surface. The time required for the pore pressures to equilibrate with the changed boundary conditions is termed the adjustment time (Tóth & Millar 1983). During the time of adjustment flow is in a transient state.

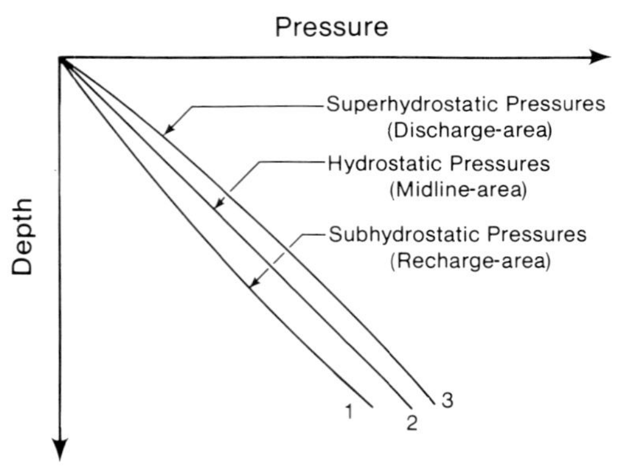

FIG. 3. Vertical distribution of pore pressure in a unit groundwater basin.

Transient water levels or pore pressures, with adjustment periods commensurate with the human time-scale, are commonly observed at relatively shallow depths or near the source of perturbation. Examples are responses in groundwater levels to precipitation or pumping. However, it is impossible to observe water-level adjustments on the geological time-scale or over basinal distances, and therefore it is necessary to resort to calculation for their estimation. Adjustment times on the geological time-scale can be evaluated quantitatively using Equation 2. These times have been calculated for several aquifers in the Taber Area and will be discussed in a later section.

Transient flow due to compaction or dilation of the rock framework

Transient flow and pressure conditions in deep basins with compacting or elastically expanding rock frameworks have been the subject of several theoretical analyses including those by Bredehoeft & Hanshaw (1968), Jacquin & Poulet (1970, 1973), Neuzil & Pollock (1983) and Bethke (1985). It appears from these studies that both positive and negative excess pressures can be generated and maintained for geologically significant lengths of time by realistic rates of sedimentary or erosional unloading. However, numerical results are very sensitive to the values of the physical constants of the rocks and to the rates of the geological processes involved and can therefore be used as guides at best. Nevertheless, these studies provide conceptual bases from which at least the nature, direction and relative magnitude of changes in the flow and pressure conditions, which occur in response to the deformational history of the basin's rock skeleton, can be postulated.

For example, erosional unloading results in a decrease in the weight supported by the rock framework and the pore fluids, and may cause a concomitant decrease in pore pressures. Reduced pressures will be maintained until the mass and pressure deficiency is satisfied by flow, initially from adjacent rock bodies and ultimately from the land surface or other source areas. At this final stage the aquitard is fully expanded and the decrease in weight is completely transferred to the rock framework. Pore pressures may even become negative or tensile depending on the rate and amount of unloading and on the compressibility, permeability and porosity of the rocks. Such a case was calculated by Neuzil & Pollock (1983) on the basis of Equation 1 and is illustrated in Fig. 4. In this example pressures are kept hydrostatic at both the upper and lower boundaries implying atmospheric conditions at the land surface and a highly permeable stratum at the base which is connected to the land surface. Careful examination of Fig. 4 shows that the deviation between the calculated actual pressures and the nominal hydrostatic pressures increases up to 3 m.y. after which it decreases, but vanishes only beyond 5 m.y.

Conceptualized evolutionary history of groundwater flow and fluid dynamic parameters in a topographically modified and mechanically deforming geological environment

On the basis of the previous considerations it is possible to conceptualize the evolutionary changes in regional groundwater flow and its

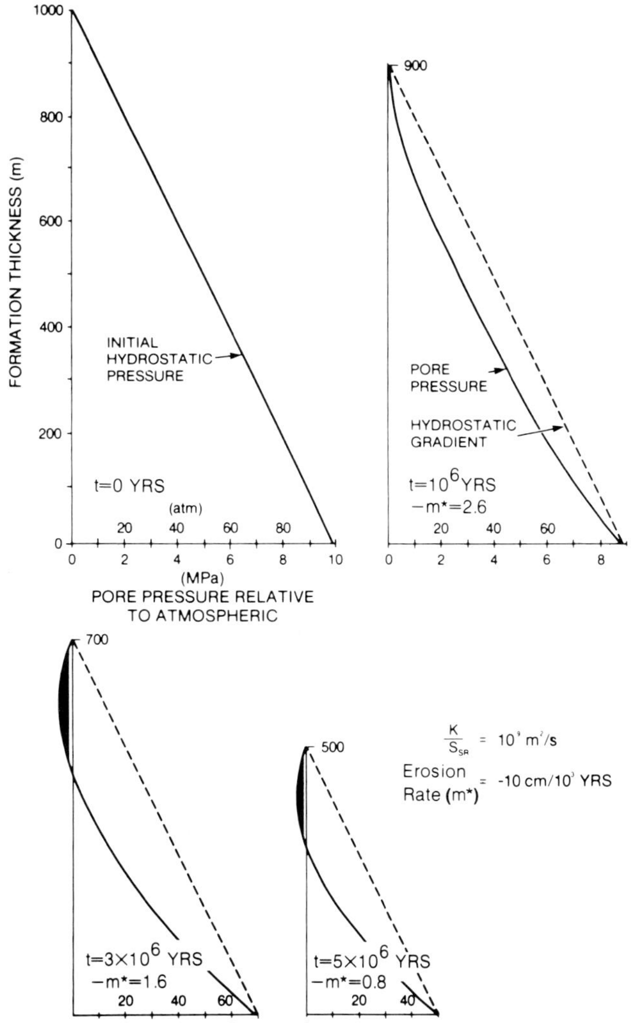

FIG. 4 Hypothetical progression in time of pore pressures in a unit 1000 m thick eroded to 500 m in 5 m.y. Negative pore pressure or tension is indicated by the shaded areas. (After Neuzil & Pollock 1983.)

associated fluid dynamic parameters in geological basins where the topography and mechanical stresses undergo modifications. Such a conceptualized evolutionary history for a simple basin is presented in Figs 5–13. The original upward-convex land surface S_1 is assumed to have been lowered and reversed by erosion to the new concave surface S_2 and the centrally positioned thick aquitard is assumed to have expanded in response to the decrease in loading. Although in this hypothetical evolutionary sequence the changes in topography and loading are assumed to have occurred instantaneously, the main stages of evolution are believed to represent a meaningful first approximation to the post-Palaeocene fluid-flow history in the Cretaceous and younger sediments of the Taber Area.

During stage $t_{-\infty}$, i.e. prior to the erosional event at $t = t_0$, the gravity component $\mathbf{q}_g = -K(\text{grad } h_g)$ of the total regional flow field $\mathbf{q}_T$ is adjusted to the prevailing topography and is in steady state (Fig. 5A). No deformational flow

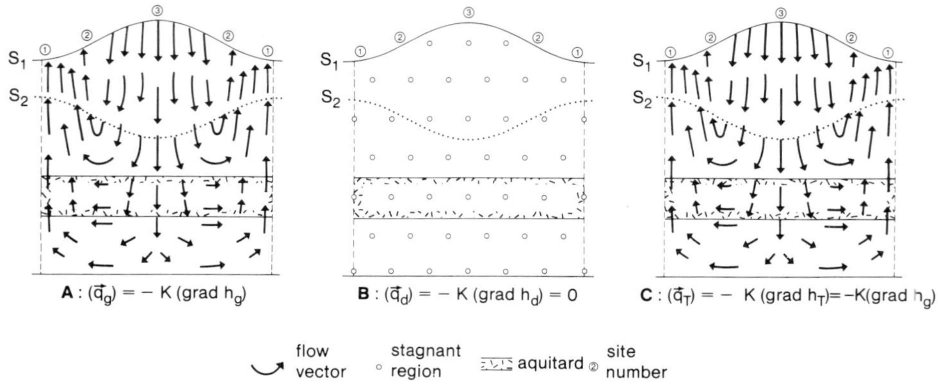

FIG. 5. Conceptual evolution of groundwater flow in a simple basin at time $t_{-\infty}$.

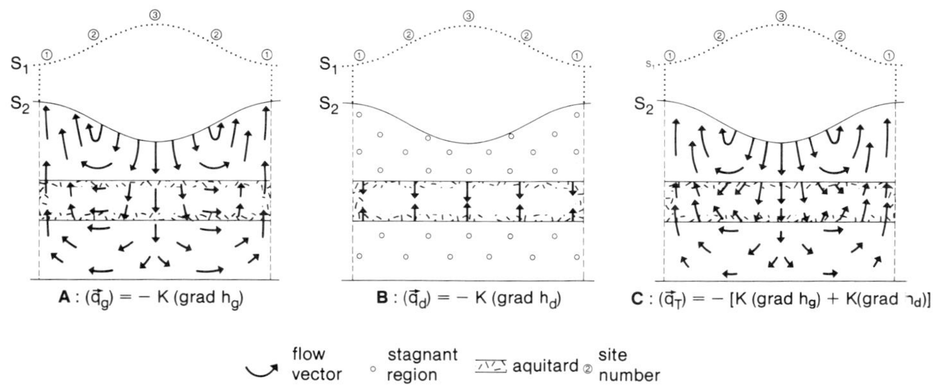

FIG. 6. Conceptual evolution of groundwater flow in a simple basin at time t_0.

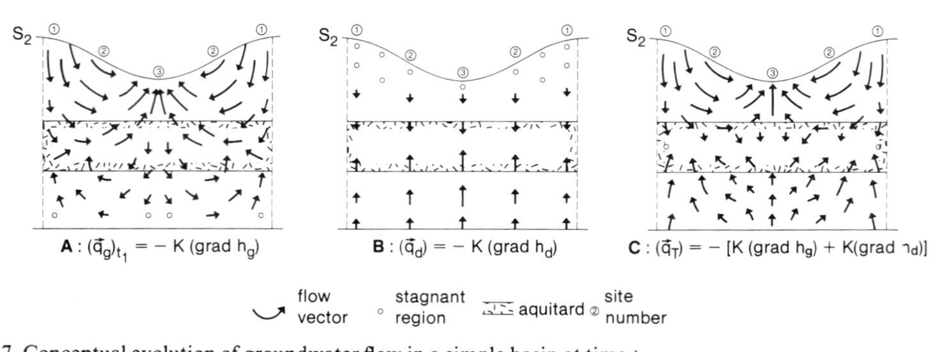

FIG. 7. Conceptual evolution of groundwater flow in a simple basin at time t_i.

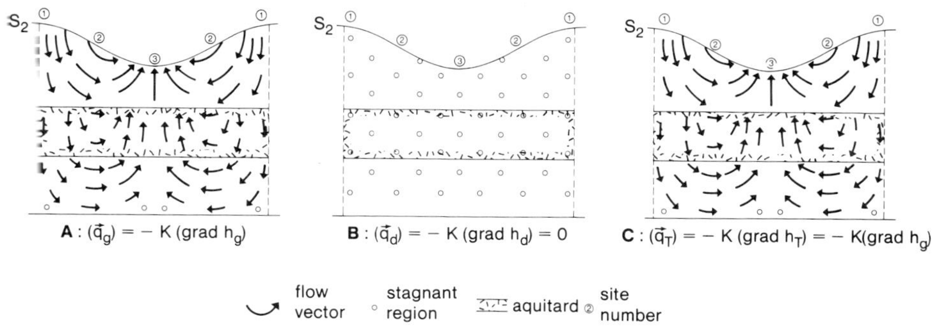

FIG. 8. Conceptual evolution of groundwater flow in a simple basin at time t_∞.

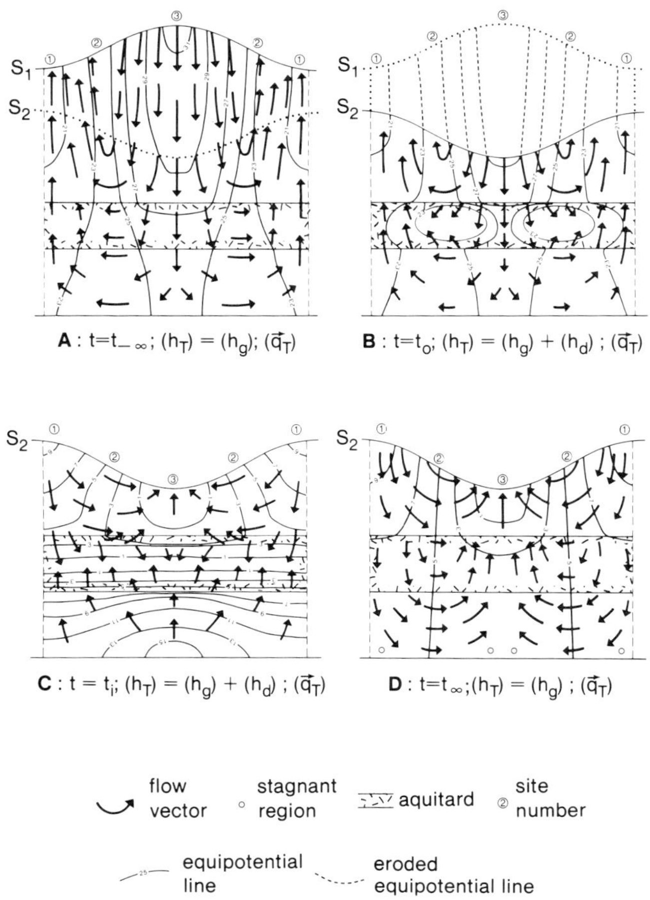

FIG. 9. Conceptual evolution of hydraulic head distributions in a simple basin at times (a) $t_{-\infty}$, (b) t_0, (c) t_i and (d) t_∞.

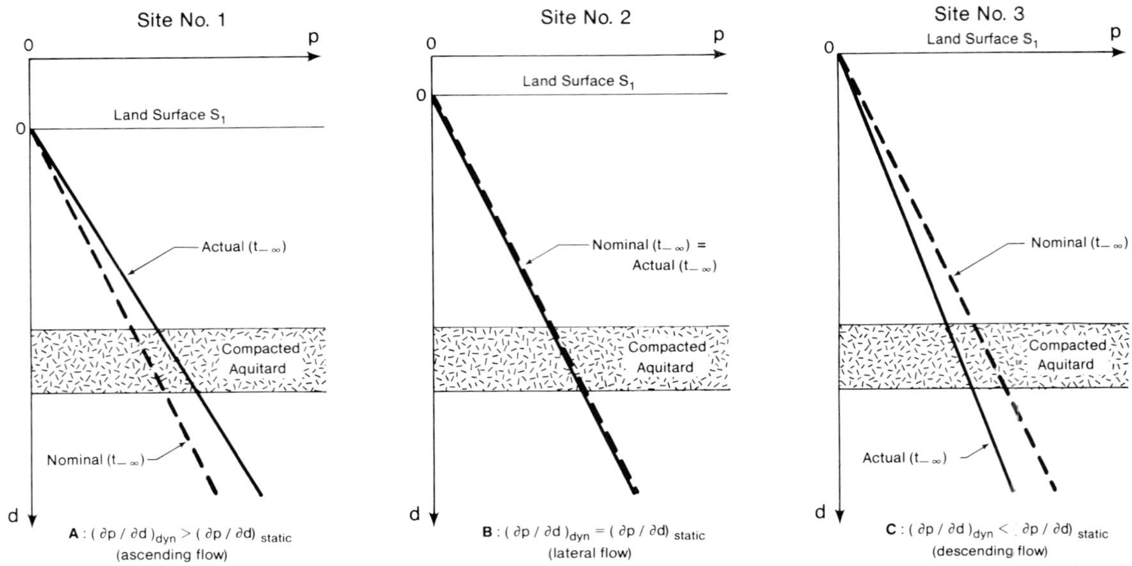

FIG. 10. Conceptual evolution of pressure versus depth in a simple basin at time $t_{-\infty}$.

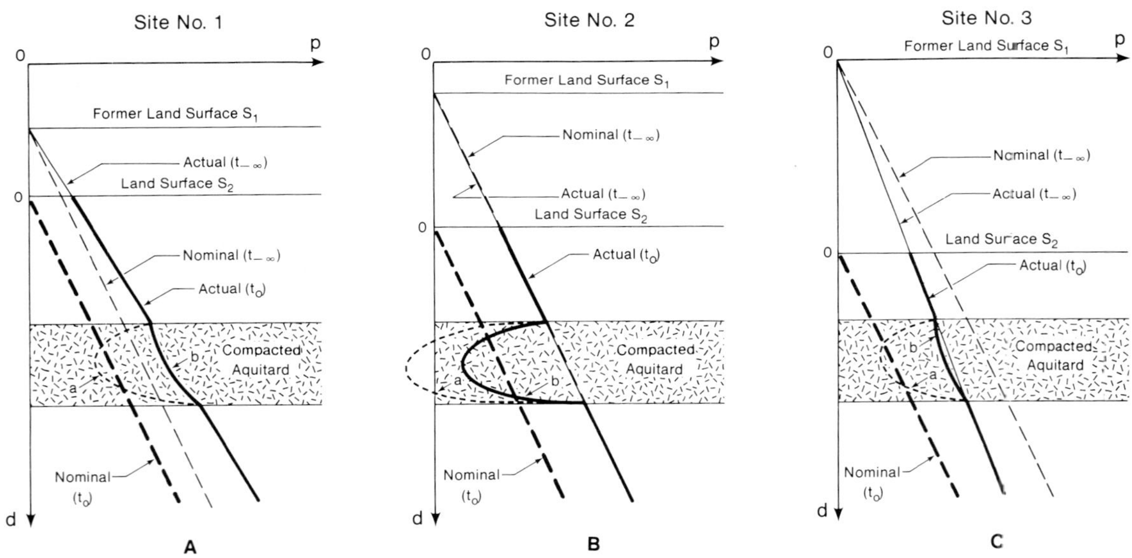

FIG. 11. Conceptual evolution of pressure versus depth in a simple basin at time t_0.

components exist, i.e. $\mathbf{q}_d = -K(\text{grad } h_d) = 0$, because the rock framework has been free from mechanical deformation, i.e. free from driving forces that are due to compression or dilation, for a time long enough for possible excess pressures to dissipate (Fig. 5B). Consequently, the resultant pattern of total flow (Fig. 5C) is identical with that of the gravitational component.

The distribution of hydraulic heads h_T associated with the flow field at stage $t_{-\infty}$ (Fig. 9A) is characterized by those features known from the unit basin (Fig. 2). The reference for the hydraulic-head values in Fig. 9 is the lowest point on the surface S_2 which is assigned an elevation of zero units. Water levels decline with increasing well depths in the recharge, remain constant in the throughflow and increase in the discharge areas, i.e. at sites 3, 2 and 1 respectively (Fig. 9A). The same conditions are indicated by rates of vertical pressure increase which are lower

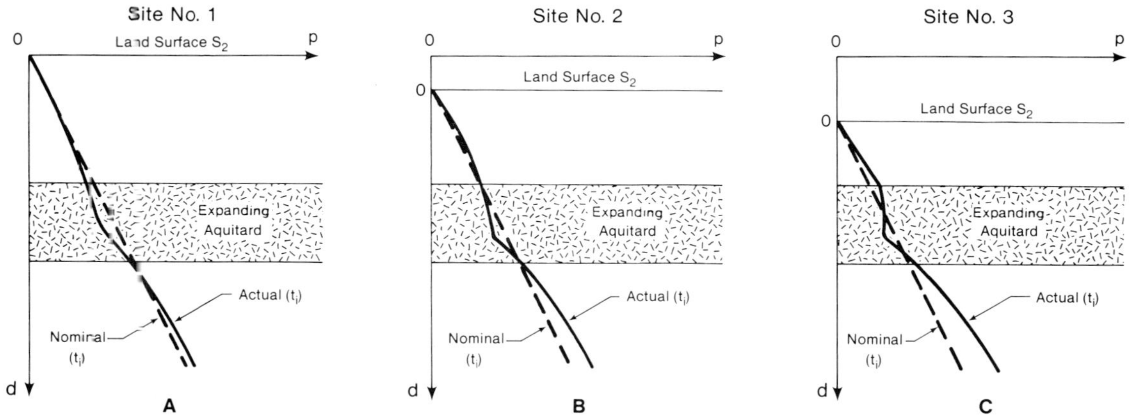

FIG. 12. Conceptual evolution of pressure versus depth in a simple basin at time t_i.

than, equal to and greater than hydrostatic when related to the topographic elevations of the three localities (Figs 10C, 10B and 10A respectively).

At time $t = t_0$ the land surface is assumed to change instantaneously from S_1 to S_2 by an erosional removal of the material from the zone encompassed between these two surfaces. The gravitational flow pattern $\mathbf{q}_g$ is suddenly out of adjustment with the new surface configuration (Fig. 6A), and the elastic rebound reduces pore pressures inside the aquitard generating inward-oriented dilatational flow $\mathbf{q}_d = -K(\text{grad } h_d)$ (Fig. 6B). Therefore the resultant flow field at the start of the evolutionary history is made up of the sum of the gravitational and deformational flow components (Fig. 6C): $\mathbf{q}_T = \mathbf{q}_g + \mathbf{q}_d$.

In constructing the flow field illustrated in Fig. 6C it was assumed that the vertical force components $\partial h_g/\partial z$ in the descending and ascending limbs of the gravitational field were greater than the dilatational force components $\partial h_d/\partial z$ at the corresponding locations in the aquitard. Consequently, the resultant flow field (Fig. 6C) retains the continuity of the original gravitational field across the aquitard, although with appropriate modifications in direction and intensity. However, an alternative scenario can also be envisaged by assuming deformational forces that are greater than the gravitational forces. In this case the gravitational flow is suppressed in areas of oppositely directed forces and the inward-acting tensional forces prevail: the continuity of the flow fields above and below the aquitard is disrupted and flow is oriented toward the aquitard's centre along both its upper and lower surfaces. The hydraulic-head distribution shown in Fig. 9B represents the first alternative outlined for stage t_0. It indicates a potentiometric minimum inside the aquitard. This feature is closed, however, and does not extend throughout the entire stratum.

The pressure–depth relations shown for stage t_0 in Fig. 11 illustrate three characteristic features of the initial pressure conditions: (1) reduced pressures in the aquitard caused by its erosional rebound; (2) maintenance of the original senses of the vertical flow components in the aquifers under the former inflow and outflow regions; (3) pressures in excess of hydrostatic with respect to the current surface S_2 caused by a lowering of the previous surface S_1.

As mentioned earlier, a pressure reduction which is sufficiently large to intersect vertical flow completely could develop in the aquitard. In this case vertical pressure gradients are less than hydrostatic in the upper part of the aquitard and greater than hydrostatic in the lower part, and induce descending and ascending flow (segments a in Figs 11A, 11B and 11C respectively). If, however, conditions remain as illustrated in Fig. 6C, downward and upward flow are maintained under the former recharge and discharge areas (sites 3 and 1 respectively) and the rates of vertical pressure changes remain less than hydrostatic under site 3 and greater than hydrostatic under site 1 (segments b in Figs 11C and 11A respectively). However, flow is convergent in the aquitard from above and below under the midline regions (segment b in Fig. 11B).

The continuation of the pre-erosional senses of the flow in the aquifers is indicated by the preservation of the original slope relations between the hydrodynamic and hydrostatic $p(d)$ curves both above and below the aquitard (Figs 11A, 11B and 11C).

Finally, the positive nature of the excess pressures in the aquifers at stage t_0 is shown by the fact that the position of the actual $p(d)$ curves is to the right relative to the current nominal values.

The patterns of flow and dynamic parameters during stages t_i develop gradually by diffusion, i.e. relaxation, from the initial conditions and under the effects of the new boundary surface S_2. The gravity-flow component at a time $t=t_i$ is characterized by a field whose upper and lower regions have been generated by two different topographic surfaces, namely S_2 and S_1 respectively, which belong to two different geological periods. Flow in the upper aquifer and upper portions of the slowly diffusive aquitard is nearly or completely adjusted to S_2 and can be considered to be steady state (Fig. 7A). In the lower aquifer, however, flow is still maintained in the original pre-t_0 direction by fluid potentials generated by surface S_1 but, not being reinforced, it is in a transient state and gradually losing intensity (Fig. 7A). Owing to the different times of generation of its different portions this type of flow pattern can be appropriately termed heterochronous.

Because of a relatively easy communication with a free upper boundary, and thus with recharge, through the highly diffusive upper aquifer the dilatational stresses in the upper reaches of the flow field tend to be reduced by time $t=t_i$ (Fig. 7B). Pressure deficiencies, however, are not easily eliminated in the lower portions of the aquitard: a lack of flux (because $\partial h/\partial z=0$) across the basal boundary results in continuing strong upward-directed driving forces in the lower aquifer.

The resultant flow field (Fig. 7C) comprises three diagnostically different dynamic systems. The system in the upper aquifer is nearly or completely adjusted to the modern topography and is in a quasi-steady state. Its portions penetrating the aquitard are oriented from areas of current recharge to areas of current discharge. The middle, or aquitard, system is characterized by forces with vertical components oriented toward the interior from both the upper and lower regions of this slowly diffusive unit and with horizontal components oriented in the general lateral directions of the dynamic systems in the upper and lower aquifers respectively. The third dynamic system is found in the lower aquifer. Its principal direction is vertically upward into the lower portions of the aquitard with, however, a noticeable horizontal component in the direction of the pre-t_0 stage steady-state regional flow field which was generated by the former topographic surface S_1. The lateral orientation of the third dynamic system is in agreement with the previous topographic slope and is opposite to the modern slope.

The hydraulic-head distribution of stage t_i reflects the flow conditions outlined above (Fig. 9C). Its principal features are (1) a topographically adjusted steady-state pattern in the upper aquifer, (2) a potentiometric sink in the aquitard and (3) a slowly decaying transient field in the lower aquifer with dominant upward directions and a lateral potentiometric slope from the areas of modern discharge (former recharge, site 3) to modern recharge (former discharge, site 1) (Fig. 9C).

During the time period between t_i and t_0 pore pressures adjust considerably to the changed boundary conditions. The results of this process of relaxation are illustrated by the $p(d)$ curves constructed for stage t_i (Fig. 12). (Details of $p(d)$ relations found at outer and interior boundaries of the flow region have been omitted for simplicity.)

In the upper aquifer previous superhydrostatic pressures (generated in site 1 at t_0 by the dual factors of land-surface lowering and upward flow) have dissipated and been reduced by negative dynamic pressure differences of descending flows in the modern (t_i) recharge area (Fig. 12A, site 1). At site 2 or the midline region (Fig. 12B) the hydrostatic excess pressures of stage t_0 have decreased to normal hydrostatic values by t_i, and in the discharge area (Fig. 12C, site 3) dynamic superhydrostatic pressures are associated with ascending flow.

Pressure relaxation is slower in the aquitard than in the aquifers owing to the aquitard's lower hydraulic diffusivity. Nevertheless, original pressure deficiencies have been reduced and more evenly spread in this central unit. Vertical pressure gradients are now uniformly below hydrostatic values in the upper portion of the aquitard and above hydrostatic values in the lower portion, indicating the convergent nature of flow into the dilated stratum from all adjacent rocks (Figs 12A, 12B and 12C).

Superhydrostatic pressures still prevail everywhere in the lower aquifer but their magnitude has noticeably decreased compared with the initial values at t_0. The magnitude of the remaining anomaly depends on the shift in the nominal hydrostatic curves from the original ($t_{-\infty}$) to the current (t_i) position and on the nature of the new hydrodynamic regime (recharge, midline or discharge). Thus the extant anomaly is least in the modern recharge area (Fig. 12A) where a minimum change in topographic elevation has caused a minimum shift in the hydrostatic curve (Fig. 11A) and where the switch from a discharge to a recharge position results in large pressure gradients with respect to time, thereby accelerating the process of equilibration. However, a large drop in the original land surface has resulted in a large difference between the former and current values of the nominal hydrostatic pressures in the previous recharge area (Figs 11C and 12C). The dissipation of this already large pressure difference also tends to be slowed down by the positive dynamic pressures associated with the ultimate discharge nature of site 3.

The final stage of a full cycle of flow-system evolution at $t = t_\infty$ is characterized by the complete adjustment to the topography (and thus by a steady state) of the gravitational component of the flow field throughout the entire depth of the basin and by a complete disappearance of the deformational flow field (Figs 8A and 8B respectively). At this stage the basin is referred to as mature; the resultant flow field is identical with the gravitational component (Fig. 8C).

The hydraulic-head distribution indicates these conditions by exhibiting continuously decreasing values from site 1 through site 2 to site 3 (Fig. 9D). The zone of minimum potentials of stage 1 has vanished. Flow in the lower aquifer is reversed relative to the previously existing and now extinct regional system.

The vertical pressure gradients reflect normal unit basin conditions (Figs 3 and 13), i.e. hydrostatic values in the midline regions (Figs 3 and 13E, site 2) and negative and positive dynamic pressure differences in the recharge and discharge areas (Figs 13A and 13B, sites 1 and 3 respectively).

The evolutionary patterns of flow, heads and pressures outlined above may be disrupted and/or distorted by a superposition of multiple cycles of basin evolution, by differing history in different parts of the initial basin and by a heterogeneous permeability distribution. Nevertheless, the general principles, concepts and features of the evolutionary stages of regional groundwater flow outlined above may be useful in analysing the flow-system history of a given area. These considerations will be applied to the Taber Area in the following sections.

Geology

The most important aspect of the geology of the Taber Area for the present study is its control of the spatial distribution of rock permeability, porosity and compressibility. Because these hydraulic properties are closely represented by the rock lithologies, this section is primarily confined to a description of the distribution of gross lithology. Structural features that may have affected post-Palaeocene erosional patterns are also discussed. The lithological descriptions given below are based on a volume published by the Alberta Society of Petroleum Geologists (1966).

The sedimentary cover is a relatively thin (1800–2600 m thick) flat sequence of carbonates, evaporites and clastics resting on the crystalline Precambrian surface. The sequence can be roughly divided into upper and lower halves, the lower half consisting of Palaeozoic carbonates and clastics and the upper half of Mesozoic clastic rocks of predominantly Cretaceous age. A thin layer of Mississippian rocks is included with the Mesozoic section.

The spatial distribution of the hydraulic properties is described by dividing the rock sequence into hydrogeological units (Fig. 14). The first-order unit is the hydrogeological formation, which is designated as an aquifer or aquitard and usually comprises a geological formation or assemblage of formations. The second-order unit is the hydrogeological group, the hydraulic properties of which can be considered to be homogeneous on a regional scale. The seven designated groups are shown in cross-section in Fig. 15, the location of which is shown in Fig. 16. Fluid dynamic systems comprise groundwater flow

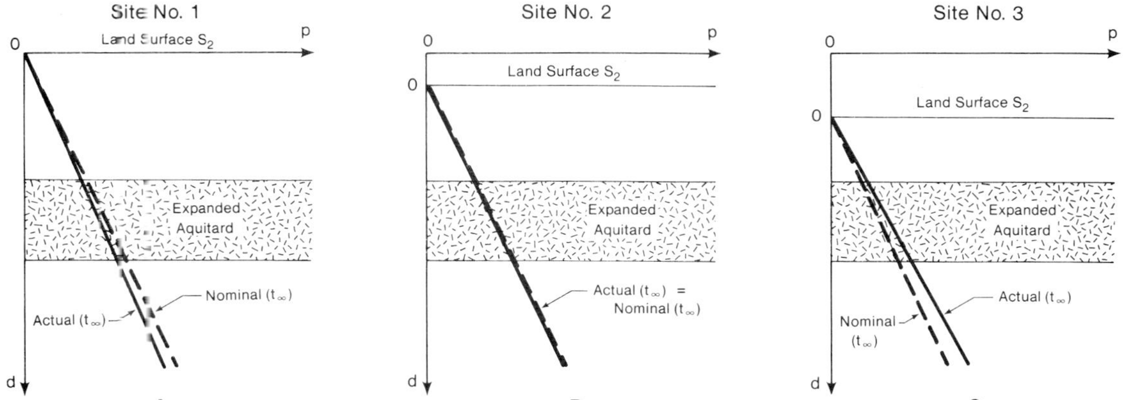

FIG. 13. Conceptual evolution of pressure versus depth in a simple basin at time t_∞.

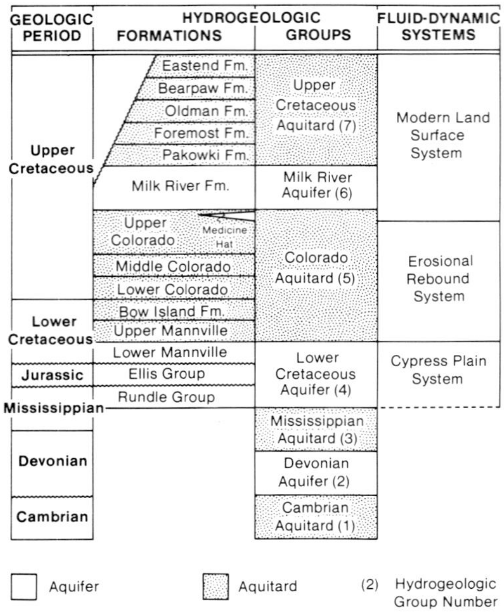

FIG. 14. Table of hydrogeological units.

systems of common history, age and controlling factors, and will be discussed in the section on current groundwater flow patterns. All seven hydrogeological groups are required for the calculation of times of adjustment of fluid pressures to changes in the land surface. However, flow patterns will be presented only for the upper four groups (Mesozoic section). The Mesozoic groups will therefore be described in more detail than the Palaeozoic groups.

The values of the hydraulic properties assigned to the hydrogeological groups are shown in Table 1 (see also Fig. 14). The hydraulic conductivity values for the Lower Cretaceous and Milk River Aquifers were determined from drill-stem tests and from values of rock-core permeability obtained from the provincial Energy Resources Conservation Board. Other hydraulic conductivity values and the specific storage values were estimated using values measured for rocks of similar lithologies.

The lowest of the three Palaeozoic hydrogeological groups is the 250–425 m thick Cambrian Aquitard which is composed of shale and some carbonates. The 350–500 m thick Devonian Aquifer consists of vuggy dolomite and bioclastic limestone. At the top of the Palaeozoic sequence is the Mississippian Aquitard comprising 200–300 m of crinoidal limestone, shale and anhydrite.

The Lower Cretaceous Aquifer, which contains most of the important oil reservoirs in the study area, comprises three hydrogeological formations (Fig. 14). The bottom boundary of this group is the first base of significant porosity in the Mississippian bioclastic carbonates and does not coincide with a geological formation boundary. The thickness of Mississippian rocks included in this hydrogeological group ranges from zero to 150 m, and averages 60 m. The Ellis Group is a 30–125 m thick shale and sandstone sequence of Jurassic age which is separated from rocks above and below by major unconformities. The Lower Mannville unit consists of 30–60 m of non-marine sandstones and shales which were deposited as basal fill on the sub-Cretaceous unconformity.

The 500–700 m thick Colorado Aquitard, which comprises five hydrogeological formations, is the most effective aquitard in the Mesozoic sequence. The Upper Colorado hydrogeological formation is a massive marine shale which contains a thin sandstone (the Medicine Hat sandstone) near its top in the eastern half of the Taber Area. The three units below the Upper Colorado unit (Middle Colorado, Lower Colorado and Bow Island) are primarily shale but contain numerous thin, although laterally continuous, sandstone lenses. The degree of sandstone development within these formations varies considerably over the Taber Area. Because the sandstone lenses often contain commercial deposits of natural gas, considerable information about their hydraulic properties and fluid pressures is available. As will be discussed in more detail below, the fluids within the sandstones characteristically have lower pressures than fluids in the deeper Lower Cretaceous and shallower Milk River Aquifers. Non-marine shale and siltstone are the

TABLE 1.

Hydrogeologic Group and Number	Thickness (m)	Hydraulic Conductivity K (m/s)	Specific Storage S_o (cm^{-1})	K/S_o (cm^2/s)
Upper Cretaceous Aquitard (7)	140	10^{-11}	5×10^{-7}	2×10^{-3}
Milk River Aquifer (6)	110 ⊙ P_1	5×10^{-7}	5×10^{-6}	10
Colorado Aquitard (5)	370	10^{-12}	5×10^{-7}	2×10^{-4}
Lower Cretaceous Aquifer (4)	340 ⊙ P_2	4×10^{-7}	5×10^{-6}	8
Mississippian Aquitard (3)	300	10^{-9}	5×10^{-7}	2×10^{-1}
Devonian Aquifer (2)	410	3×10^{-8}	10^{-6}	3
Cambrian Aquitard (1)	320	10^{-10}	10^{-7}	1×10^{-1}

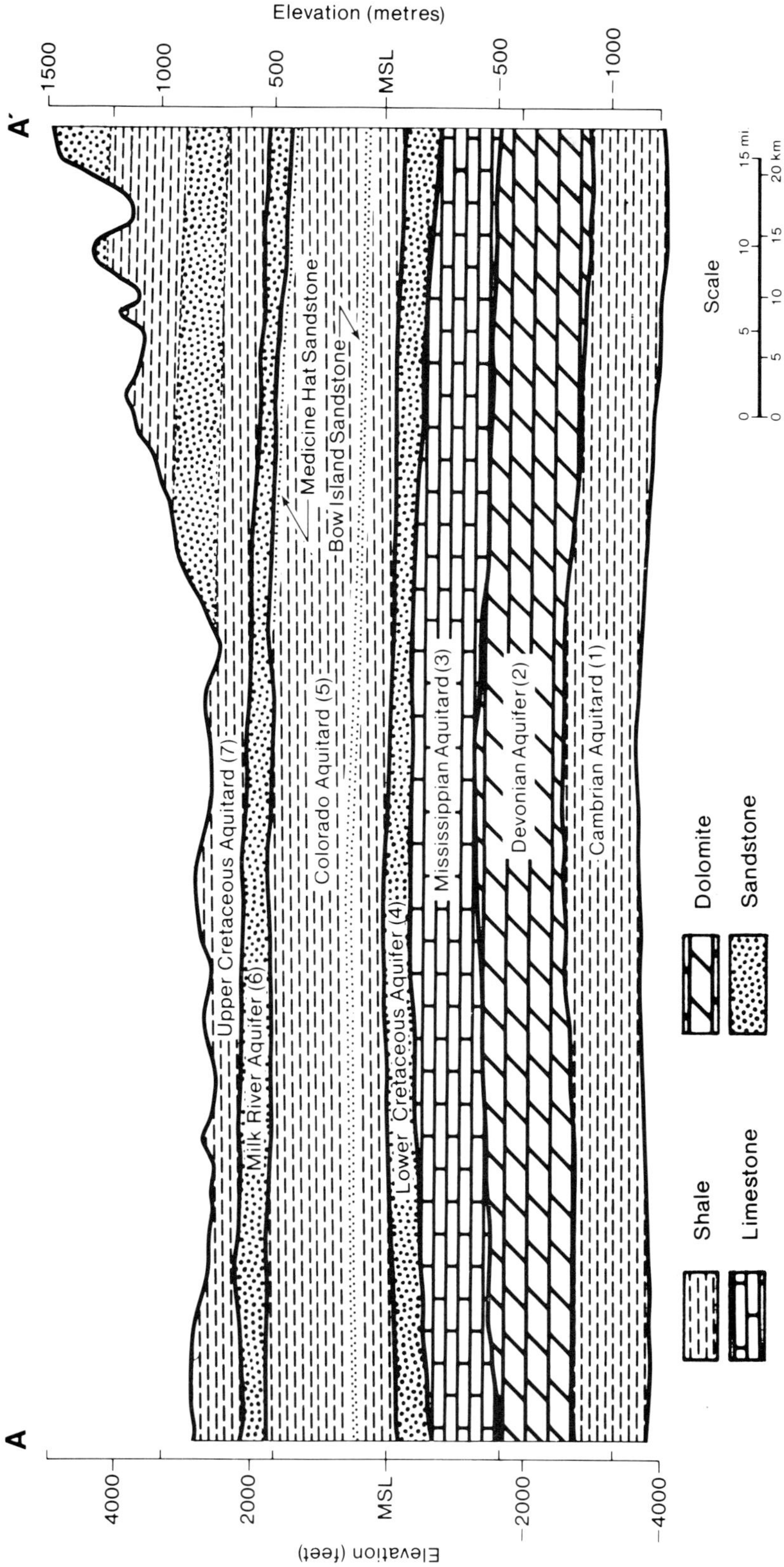

FIG. 15. Cross-section AA′ of hydrogeological groups (see Fig. 16 for location).

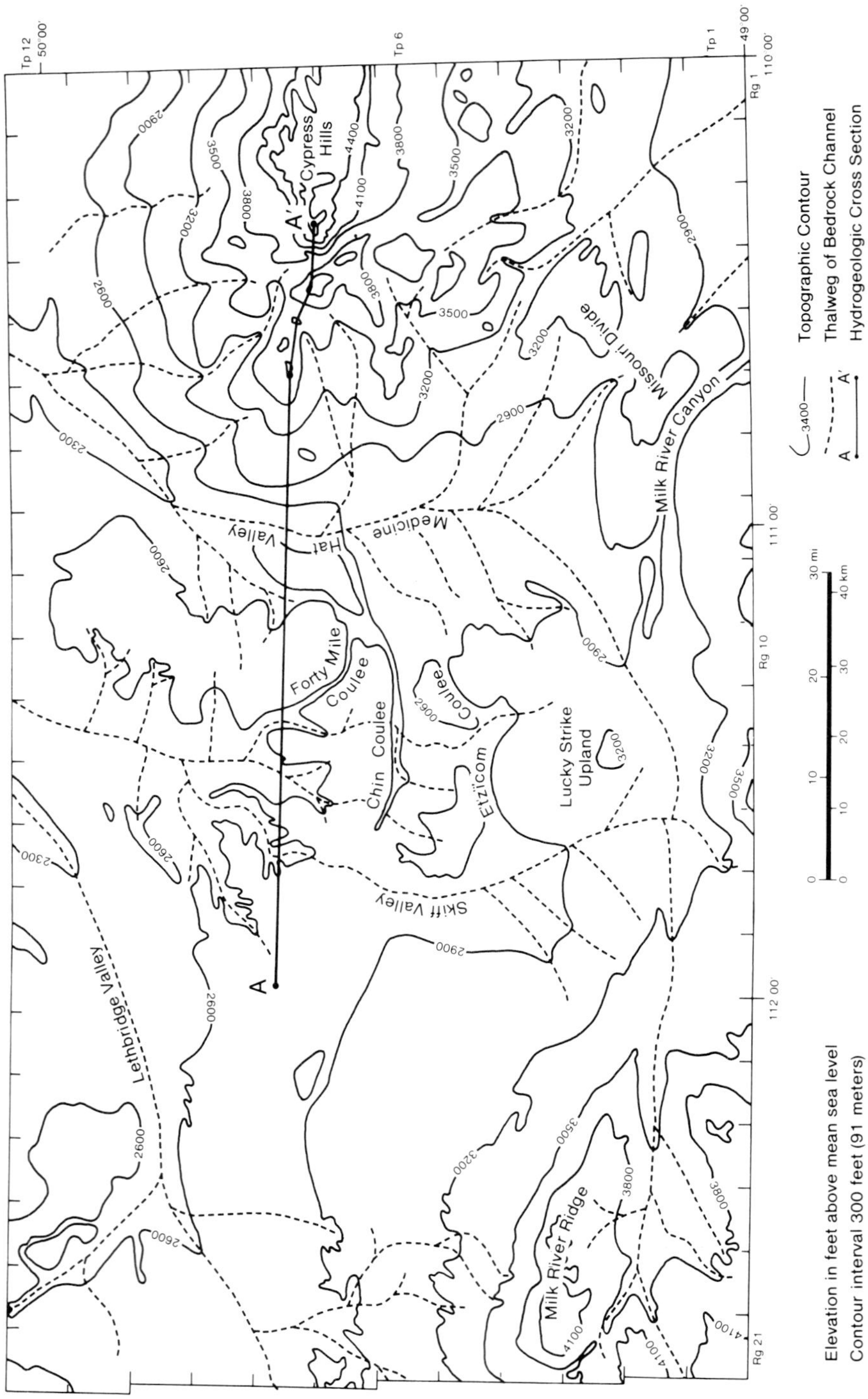

FIG. 16. Bedrock topography (after Westgate 1968).

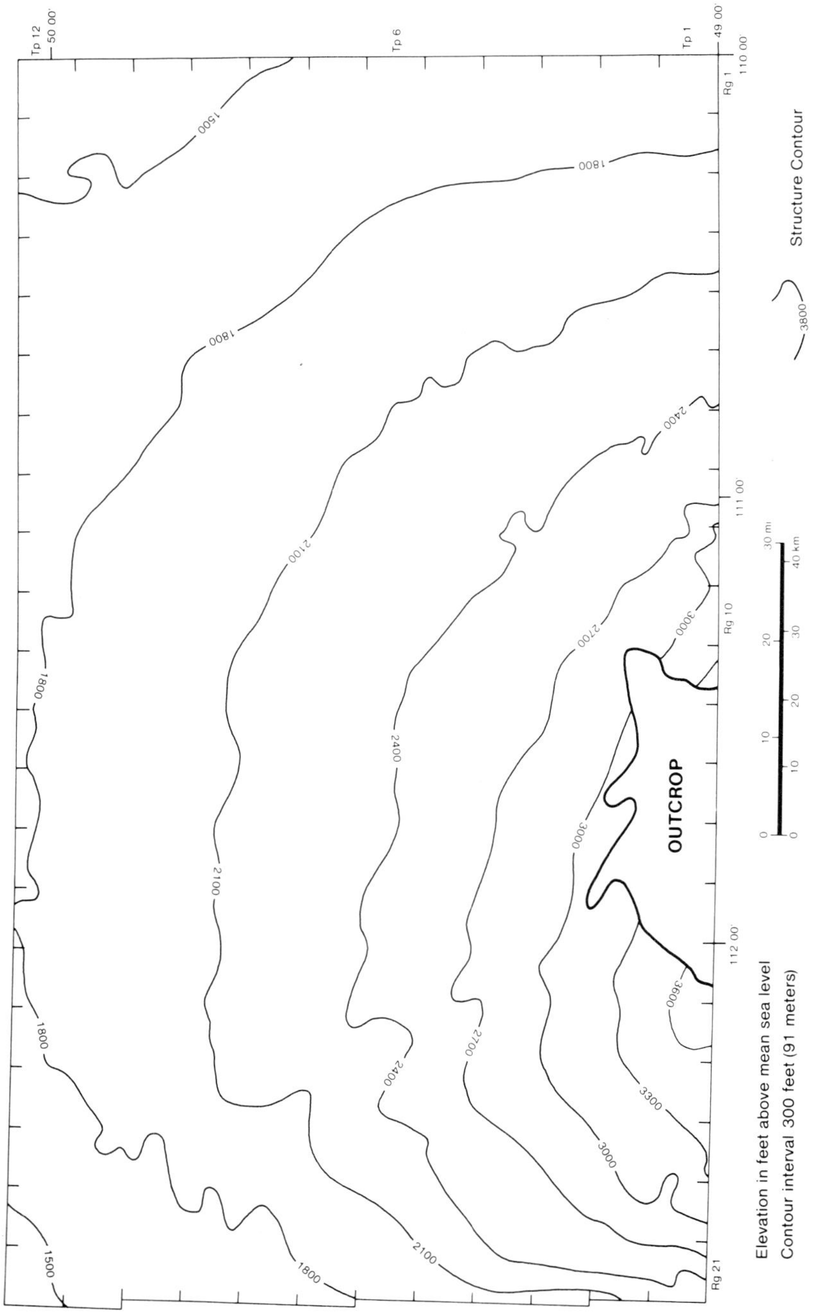

FIG. 17. Structure of the Milk River Formation (after Alberta Energy Resources Conservation Board 1970).

primary lithologies in the 60–150 m thick Upper Mannville unit.

The 125 m thick Milk River Aquifer is a sandstone, siltstone and sandy shale sequence which outcrops or is covered by a thin veneer of glacial deposits in the S central part of the study area. Its relatively high permeability renders it the most important source of groundwater in the region. This formation plunges in a fan pattern away from the outcrop area (Fig. 17), reaching depths of 200 m below the land surface to the N and over 450 m along the E and W borders of the Taber Area. The sandstones pinch out N of the area.

The Upper Cretaceous Aquitard is exposed at the land surface over all parts of the Taber Area except on the Cypress Hills (Fig. 16) and in a small area on the flank of the Sweetgrass Hills where the Milk River Aquifer outcrops (Fig. 17). Because the formations comprising this unit have been extensively eroded in post-Palaeocene time, its thickness varies from zero in the S central part of the Taber Area to 700 m under the Cypress Hills. Away from the Cypress Hills the Upper Cretaceous Aquitard consists of the Pakowki marine shale and argillaceous sandstones, siltstones and shales of the Foremost and Oldman Formations. Under the Cypress Hills, this unit also includes the 300 m thick marine shale of the Bearpaw Formation.

The structure of the area is well represented by the contours on top of the Milk River Formation (Fig. 17). Tovell (1958) suggested that this pattern is the result of the merging of three independent structures which are all most distinctly developed outside the Taber Area: the Kevin–Sunburst Dome, the Bow Island Arch and the uplift associated with the intrusion of the Sweetgrass Hills (Fig. 18). The resulting composite structure is often referred to as the Sweetgrass Arch. However, this name is also applied to an ancient structural feature which follows the trend of the Bow Island Arch and its southward extension. Therefore, in order to avoid confusion, the term Bow Island Arch has been adopted in the present paper to denote the NE–SW-trending component of the composite structure and to distinguish it from the older structure. The Kevin–Sunburst Dome and the Sweetgrass Hills intrusions formed at the same time as the thrusting of the front ranges of the Rocky Mountains, but the history of the Bow Island Arch and its relation to the Sweetgrass Arch is uncertain (Tovell 1958). The Bow Island Arch is probably related to the development of the Alberta syncline.

Faults locally affect the structure in one part of the Taber Area. Westgate (1968) describes faults with vertical displacements of over 900 m in the vicinity of Tps 8 and 9, Rgs 4 and 5. These faults may be caused by the emplacement of an igneous body at depth and seem to be an isolated occurrence.

Topographic evolution

The evolution of fluid-flow patterns was presented previously for a hypothetical basin whose topography is changed by erosion. To apply these concepts to the study area it is necessary to trace the evolution of its actual surface during post-Palaeocene time. The Cypress Hills (Fig. 16) are a striking indication of the amount of erosion that has occurred in this area since Miocene time. The plateau top of these hills, which is currently elevated 630 m (2070 ft) above the surrounding plains, is capped with gravels deposited by a river that once flowed from a site in the Rocky Mountains 300 km SW of the Cypress Hills. Reconstruction of the gradient of this river shows that up to 730 m of rock has been eroded from the central portion of the Taber Area in the last 10–20 m.y.

Any reconstruction of the evolution of palaeo-land surfaces necessarily involves some speculation. However, in this case enough information is available to determine a probable erosional history. The work of several researchers has focused on or included study of the evolution of the land surface in or near the Taber Area. Studies concerning the southern Alberta plains and adjacent areas include those of McConnell (1886), Collier & Thom (1918), Williams (1929), Alden (1932), Howard (1960) and Westgate (1968). Studies conducted in the Rocky Mountains W of the Taber Area include those of Pardee (1950) and Balley *et al.* (1966). Much of the work of these researchers involved the identification and correlation of topographic highlands that are erosional remnants of previous land surfaces. Several highlands in and near the Taber Area have been assigned to particular surfaces and times, and thus help to identify the general elevation of the land surface at various times in the past. The following synthesis is based on the work of these researchers in general and on their identification of remnant highlands in the Taber Area in particular.

The interior plains have experienced long periods of stability alternating with short periods of uplift resulting in rapid down-cutting by rivers since mid-Eocene time. Extensive peneplains, of which only a few erosional remnants remain, formed on the plains and in the mountain regions during the stable periods. Following Alden (1932), the terms Cypress Plain, No. 1 Bench and

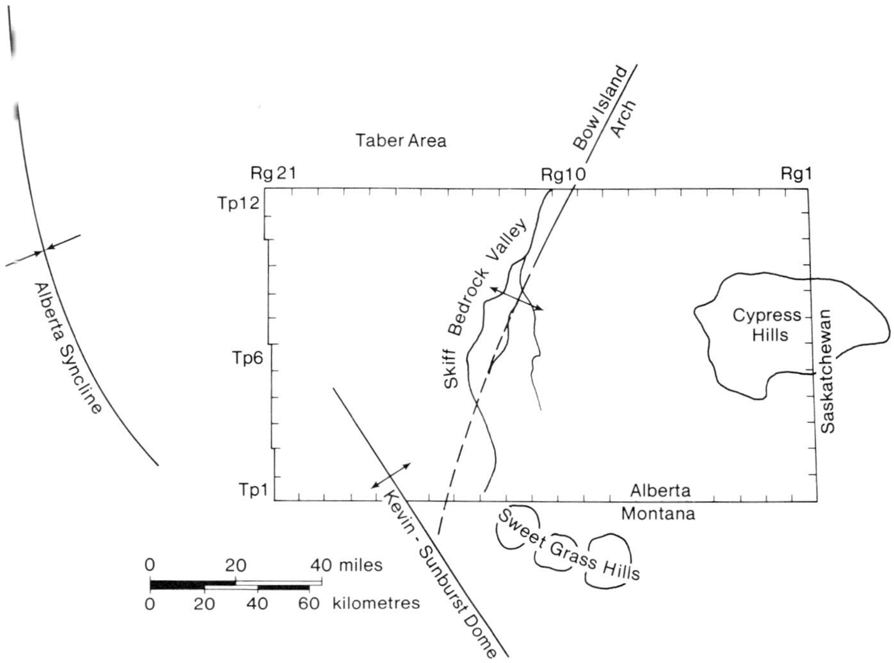

FIG. 18. The three structural components of the Sweet Grass Arch (i.e. the Kevin–Sunburst Dome, the Bow Island Arch and the Sweetgrass Hills) and location of the Alberta syncline.

No 2 Bench are used in this paper to mean three palaeo-land surfaces represented by erosional remnants. The term Modern Land Surface will be used for Alden's No. 3 Bench. The above terms thus refer to mature land surfaces, with hundreds of metres of topographic relief, which formed during the long intervals when rapid uplift was not occurring. The Cypress Plain and the Modern Land Surface are represented by surfaces S_1 and S_2 respectively in the conceptualized evolution of groundwater flow (Figs 5–8).

The erosional history is summarized in Fig. 19 which shows the general elevation and probable duration of existence of major land surfaces since Eocene time. The Cypress Plain may have existed for as long as 30 m.y. in contrast with the three younger land surfaces which all existed within the past 5 m.y. Approximately 700 m of rock has been eroded from the surface of the Taber Area since late Miocene time, with about 500 m of that rock eroded during early Pleistocene time.

The Modern Land Surface, which is the most recent surface considered, formed in Middle Pleistocene time. It differs from the current land surface only by minor changes that occurred during Pleistocene glaciation. Although the Taber Area was possibly glaciated as early as 700 000 to 800 000 years BP, the identifiable changes attributed to glaciation, i.e. deposition of significant thicknesses of drift and cutting of currently existing proglacial meltwater channels, occurred during the past 24 000 years (Westgate 1968). Therefore the Modern Land Surface was very similar to the current bedrock topography, except that it did not include the meltwater channels. These are the prominent E–W-trending channels with steep walls, the most important of which are the Milk River Canyon E of Rg 9, and the Etzikom, Chin and Forty-Mile Coulees (Fig. 16). The Modern Land Surface had a relief of approximately 790 m (2600 ft). Primary features were the Cypress Hills, the Milk River Ridge and the Lethbridge, Skiff and Medicine Hat Valleys. S of the Taber Area were the Sweetgrass Hills (Fig. 18).

The No. 2 Bench is of early Pleistocene age. Westgate (1968) correlated the divide between the Lethbridge and Medicine Hat drainages and the 'saddle' linking the Cypress and Sweetgrass Hills, called here the Missouri Divide (Fig. 16), with the No. 2 Bench. These regions have a current elevation of approximately 975 m (3200 ft).

The No. 1 Bench had formed by the end of the Miocene or early Pliocene, i.e. approximately 5 Ma, and was not greatly eroded until the Pleistocene, i.e. less than 2 Ma. The ancestral Cypress Hills, formed when streams cut down to

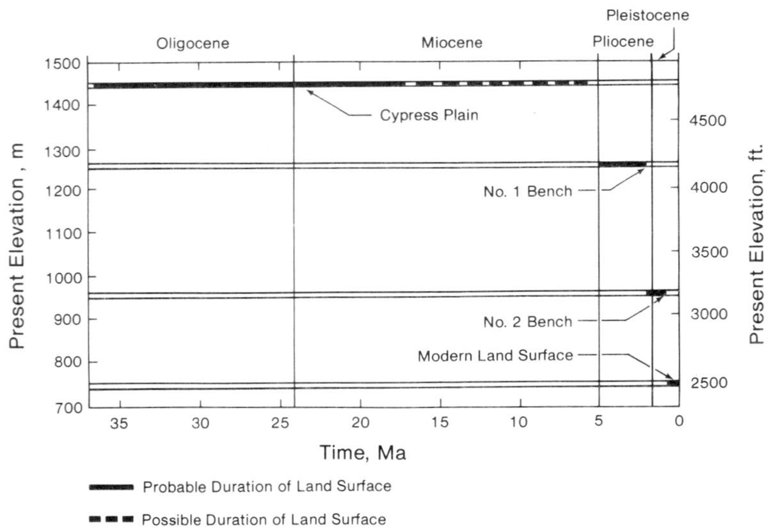

FIG. 19. Duration and elevation of post-Palaeocene land surfaces.

the level of the No. 1 Bench, stood about 180 m above the surrounding area. The buttes just W of the Cypress Hills (Fig. 16), e.g. in Tp 6, Rgs 3–4 and Tps 7–8, Rg 4, with tops at about 1280 m (4200 ft) are probably remnants of the No. 1 Bench (Westgate 1968). S of the Taber Area, the Sweetgrass Hills also stood above the base level of the No. 1 Bench. Alden (1932) suggested that the gravel-capped ridge with an elevation of about 1280 m (4200 ft) in Tp 1, Rgs 21–22, is a remnant of the No. 1 Bench, and that the Milk River Ridge may also correlate with this level but has been 'cut down to a somewhat lower level'.

The Cypress Plain, which represents the first recorded major break in uplift since the Rocky Mountains formed, probably developed by late Eocene or Oligocene time and existed at least until early Miocene time, and possibly until a major uplift occurred near the end of Miocene time roughly 5 Ma. Except for the Cypress Hills themselves, the Cypress Plain has been eroded from the Taber Area. It was probably a mature surface with hundreds of metres of relief and supporting rivers large enough to transport gravel containing pebbles over 15 cm in diameter to the Cypress Hills area from the Rocky Mountains. The general elevation of this surface is represented by the top of the Cypress Hills (present elevation 1460 m (4800 ft)).

We postulate that the Cypress Plain was similar to the Modern Land Surface except that a general drainage divide existed between the Medicine Hat and Lethbridge Valleys, located in the region of the Skiff Valley (Fig. 16). This pre-Pliocene divide, called here the Bow Island Topographic High, coincided with the axis of the Bow Island Arch (Fig. 18). Stelck (1975) showed that the region along this axis has a long history as an erosional high dating back to Silurian time and that this topographic feature affected sedimentation patterns as recently as Campanian time when the Oldman Formation was deposited. It is not unlikely that this region was also an erosional high on the Cypress Plain. The Skiff Valley is thus perceived to be a case of a reversal in topography that occurred when the core of the Bow Island Arch was eroded. The morphologies of the bedrock valleys support this concept, i.e. that the Skiff Valley is a feature more recent than the Medicine Hat and Lethbridge Valleys; the drainage area of the Skiff Valley is smaller than that of the other two and the longitudinal gradient of its thalweg is steeper. Headward erosion in the Skiff Valley seems to have captured waters of the Medicine Hat Valley. The Missouri Divide may be a remnant of another topographic high on the Cypress Plain. This region has probably acted as a divide between the ancestral Missouri River to the SE and rivers flowing NE to the Hudson Bay during Cypress Plain time much as it does today.

Based on the ranks of Cretaceous coals, Hitchon (1984) constructed a palaeotopographical map which agrees well with the general features of the Cypress Plain postulated above. The coal ranks were used to determine the thickness of sediments removed since the Rocky Mountain orogeny. The thicknesses were used in turn to construct the palaeotopographical map.

The map can be interpreted to represent primarily the Cypress Plain, because this surface lasted for a much longer time than other post-Palaeocene surfaces and thus was able to exert a dominant influence on the distribution of coal rank. Hitchon noted that the 'general drainage directions and the number of significant drainages' are similar to the pre-glacial drainage pattern, implying that the fundamental features have not changed since Eocene time. Hitchon's map (Hitchon 1984, Fig. 22) shows a major valley coinciding with the Medicine Hat Valley and a drainage divide located between the Skiff and Lethbridge Valleys. Another notable feature is the topographically high area coinciding with the Missouri Divide.

Figure 20 shows the postulated evolution of the land surface along line AA′ (see Fig. 16 for location). This line of section is analogous to the section of the conceptual development (Figs 5–7) in which the rapid erosion, i.e. approximately 700 m in less than 10 m.y., of the Cypress Plain to the Modern Land Surface is represented by the erosion of surface S_1 to surface S_2. The topographic high on the Cypress Plain in Fig. 20 represents the Bow Island Topographic High.

Calculated pressure adjustment times in the Lower Cretaceous and Milk River Aquifers

As discussed previously, time is required for pore pressures, and consequently hydraulic heads, to adjust to changes in the land surface. Adjustment times have been calculated for points P_1 and P_2 (Table 1) at the base of the Milk River and Lower Cretaceous Aquifers respectively. A one-dimensional solution of Equation 2 for a layered system (Tóth & Millar 1983) with the hydraulic properties given in Table 1 was used for the calculations. To simulate the effect of erosion an instantaneous drop in the land surface was assumed. These times can be used as an indication of whether or not hydraulic-head patterns in the Lower Cretaceous and Milk River Aquifers can be expected to be adjusted to the current surface or a palaeo-land surface.

Adjustment times at points P_1 and P_2 are approximately proportional to the hydraulic conductivities of the aquitard(s) overlying them (Tóth & Corbet 1985). Although there are no measured values of hydraulic conductivity available for shales in the Colorado and Upper Cretaceous Aquitards, values of 10^{-12} m s^{-1} and 10^{-11} m s^{-1} respectively based on literature data (Brace 1980; Neglia 1979) appear to be representative. Adjustment times calculated using these values are therefore considered accurate to within an order of magnitude.

Adjustment times have been calculated as functions of the relative head change, i.e. the ratio, R_{an}, of the head change actually accomplished at any given time during adjustment to the ultimate or total head change (Fig. 21). At point P_1, the times obtained for R_{an} values of 10%, 50% and 90% of hydraulic-head adjustment are 3500 years, 20 000 years and 78 000 years respectively. However, at point P_2 in the Lower Cretaceous Aquifer, i.e. below the Colorado Aquitard, the same degrees of head adjustment take 0.42 m.y., 2.1 m.y. and 7.0 m.y. Comparison of these times with the probable duration of palaeo-land surfaces (Fig. 19) provides a basis for relating existing head distributions to particular generating palaeo-topographies.

It appears from these comparisons that, with a duration of approximately 1 m.y., the Modern Land Surface existed for 13 times as long as is needed for a 90% head adjustment in the Milk River Aquifer. This head distribution has only been reinforced by the present topography since the end of the last glaciation, approximately 24 000 years ago. Attributing the hydraulic-head pattern in the Milk River Aquifer to the Modern Land Surface is therefore justifiable. However, only the Cypress Plain existed long enough (approximately 30 m.y.) to allow a complete head adjustment in the Lower Cretaceous Aquifer (7 m.y. required for 90% adjustment). Because the Cypress Plain has probably been maintained, albeit in a subdued form, to the end of the No. 1 Bench (approximately 2 Ma), it can be assumed that up to 50% of a potentiometric distribution related to the Cypress Plain may still be present in the Lower Cretaceous Aquifer.

Construction of maps and cross-sections of fluid potential

The current fluid-potential pattern was evaluated by producing potentiometric surface maps for four hydrogeological formations and one vertical hydraulic cross-section. Hydrogeological formations are the thinnest units for which it is practical to construct potentiometric surface maps. Hydraulic head values were calculated from fluid pressures determined during drill-stem tests or from water levels measured in water wells.

Construction of potentiometric surface maps for individual rock units is not meant to imply that flow is restricted to these units. On the contrary, these maps represent two-dimensional vertical projections of a continuous three-dimensional potential field, and cross-formational flow

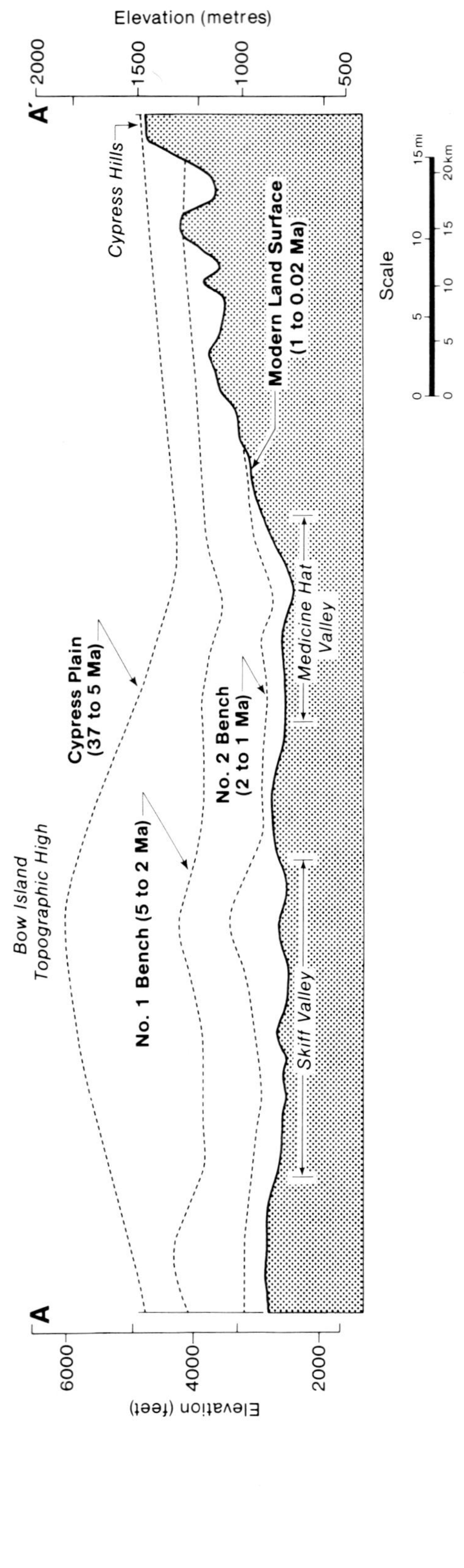

FIG. 20. Postulated evolution of the topography in post-Palaeocene time along cross-section AA′ (see Fig. 16 for location).

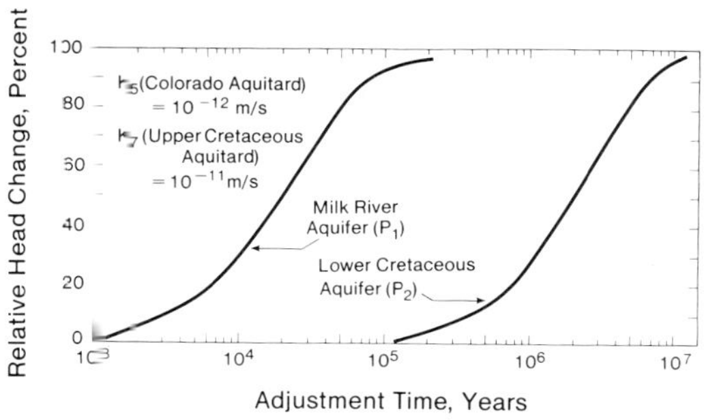

FIG. 2 . Adjustment times for the Milk River and Lower Cretaceous Aquifers.

is considered to be an important aspect of the flow pattern.

The fluid-potential distribution of the Milk River Aquifer was determined by re-interpreting original measurements of water levels in water wells (Meyboom 1960; Meyboom, personal communication and original field data).

Hydraulic-head values for all units deeper than the Milk River Aquifer were calculated from fluid pressures measured during drill-stem tests conducted in wells drilled by the petroleum industry. Stabilized pressures were determined using the Horner method (Johnson Testers 1964). Except for the Middle Colorado units, only drill-stem tests that recovered water were used; nearly all drill-stem tests within the Middle Colorado unit recover natural gas. In all cases the tests used met quality criteria that ensure that the measured pressures are sufficiently accurate (Tóth & Corbet 1985). All potentiometric data were calculated using fresh-water density. This is justified because the salinity of water in all units is predominantly in the 3000–9000 ppm range, with very few measured values greater than 15 000 ppm.

Considerable effort was made to identify hydraulic-head values that may not represent natural flow conditions. Misleading head values may occur for two primary reasons: interference from nearby producing wells, and different head values in hydrogeological formations open to the same test interval. The effect of fluid production in one well on pressures measured in nearby wells depends on production rates and schedules, fluid characteristics, the distance between production and observation, and local geology. However, most of this information is unavailable. Therefore it is assumed that the two available factors, time of production and distance to production, will give an indication of the relative effect of fluid withdrawal on individual pressure measurements. Specifically, it is assumed that, with other factors being equal and by analogy with water-well-testing principles (Bruin & Hudson 1961), the effect of production is directly proportional to an interference index $\log(t/r^2)$, where t is the length of time in years since production started prior to the pressure measurement and r is the distance in miles between the production well and the well in which the pressure was measured. The index was calculated for pressure values measured within 3 lateral miles (4.8 km) of and within the same hydrogeological group as a producing well. Hydraulic heads calculated using pressures with interference indices greater than 0.7 are indicated on the maps. This value is exceeded for half the pressure values for which interference indices were calculated. Long open-test intervals can introduce excessive vertical components of head gradient into the potentiometric surface maps. In some tests pressures were measured at several depths within the same hydrogeological formation; in these cases the heads were averaged.

Although some of the local features on the potentiometric surfaces are undoubtedly caused by the above deficiencies, it is felt that the basic features of these maps represent actual conditions. The questionable features have not been removed from the maps (Figs 25–27 below); a knowledge of the positions of possibly inaccurate head values allows the reader to identify these features. Some contour lines were arbitrarily made bolder to emphasize what the authors feel are the important regional features on the potentiometric surfaces.

Current groundwater flow pattern

The current flow pattern in the Mesozoic rocks of the Taber Area is heterochronous and is analogous to the pattern presented for time t_i in the conceptual development. It can best be described in terms of three fluid dynamic systems, which are delineated on the basis of the coherencies of their flow and pressure patterns, their age of generation and the nature of the generating forces instead of rock parameters (Tóth 1978). These three systems are the Modern Land Surface System, the Erosional Rebound System and the Cypress Plain System, representing fluid dynamic conditions mainly above, within and below the Colorado Aquitard respectively (Fig. 14). For simplicity these will be referred to as the Modern, Rebound and Cypress Plain Systems.

Modern Land Surface System

Flow patterns of the Modern System, which probably includes the uppermost portion of the

Colorado Aquitard, can be analysed using the potential distribution in the Milk River Formation (Fig. 14). Hydraulic-head values are available only for the central portion of the Taber Area where this formation is within 300 m of the land surface. The potentiometric surface (Fig. 22) has a total relief of about 275 m (900 ft) with most of the surface having an elevation in the range 820–910 m (2700–3000 ft). The characteristic feature of the Milk River potentiometric surface is a close correspondence to the bedrock surface (Fig. 16), confirming its adjustment to the Modern Land Surface. Major topographic highs and the thalwegs of bedrock valleys are overlaid on the Milk River potentiometric surface for comparison in Fig. 23. The fluid potential increases rapidly S of the Milk River Canyon (Tp 1, Rg 6–15), and W of Rg 15 the equipotential lines follow the trend of the Milk River Ridge. The Medicine Hat and Skiff Valleys correlate with depressions in the potentiometric surface, and the Lucky Strike upland (Fig. 16) coincides with a closed potentiometric high (Fig. 23). Nearly all wells in regions of fluid potential less than 910 m (3000 ft) flow currently or flowed previously, indicating an upward component of flow in this region.

It appears that the complete trajectories of flow in rocks above the Colorado Aquitard are contained within the Taber Area. Although the Milk River Aquifer is generally considered to be an example of a classical outcropping ‘confined’ aquifer, from the present study it appears that water also enters or leaves the formation through overlying or underlying shale beds. Recharge occurs in topographically high regions along the S, SW and E margins of the area: along the outcrop of the aquifer of the Milk River, along the Milk River Ridge and its northern flank, on the Cypress Hills and on the Lucky Strike Highland (Fig. 16). Discharge, or upward flow as indicated by flowing wells, occurs in the central portion of the Taber Area and is concentrated along the Lethbridge, Medicine Hat and Skiff Valleys (Figs 16 and 22). Flow paths are of the order of 50 km long. These flow paths are shown in the upper portion of the hydraulic cross-section AA′ (Fig. 24) which corresponds to Figs 7 and 9C of the conceptual development.

Erosional Rebound System

Hydraulic-head distributions in the Colorado Aquitard (Figs 24 and 25) are interpreted to indicate that this unit behaves as an energy sink owing to an elastic rebound of its shales in response to erosional unloading.

Figure 25 shows the potentiometric surface in sandstones of the Bow Island Formation which is included in the Colorado Aquitard (Fig. 14) and is located near the base of the Rebound System. This surface has a total relief of 550 m (1800 ft), with most of the values between 700 and 800 m (2300 and 2800 ft), and is everywhere below the land surface. Its main features are a ridge of high fluid potential in the S central portion of the area (the region above the 2600 ft (790 m) contour), major closed potentiometric depressions in the area of Tps 2–5, Rgs 5–8, and a steep flexure of the potentiometric surface towards the Alberta syncline in the western portion of the area. The closed depressions are probably due to natural gas production.

Information about the fluid potential distribution in the Middle Colorado unit (Fig. 14) is limited. Nearly all drill-stem tests in this zone recover only natural gas or drilling mud. Hydraulic heads are everywhere 30–90 m (100–300 ft) greater than those in the Bow Island Formation and 30–120 m (100–400 ft) less than those of the Milk River. However, comparison with the Milk River Aquifer is difficult because most head values available for the Middle Colorado unit are in the region where few data are available for the Milk River Aquifer.

Figure 24 shows that groundwater flow is into the Bow Island unit from both above and below, i.e. from the portion of the Rebound System above the Bow Island unit and from the Lower Cretaceous Aquifer.

Cypress Plain System

Groundwater flow within the Cypress Plain System is primarily upward into the Rebound System (Fig. 24). There are horizontal components of flow, however (Figs 26 and 27), which can be considered as relicts of a flow pattern that was adjusted to the land surface of the Cypress Plain.

The Mannville potentiometric surface (Fig 26) is relatively smooth with a relief of 150 m (500 ft). The most prominent feature is a region of fluid potential above 945 m (3100 ft) trending N–S through the central part of the Taber Area. Other important features are a potentiometric low trending NE from Tp 3, Rg 7, an increase in fluid potential toward the SE corner and a region of low fluid potential centred on Tp 2, Rg 15. The latter feature and an area of irregular fluid potential in the NW portion of the area were probably influenced by oil production.

The Ellis potentiometric surface (Fig 27) generally has hydraulic heads 15 m (50 ft) greater than the Mannville surface (Fig. 26). However, the difference increases to 90 m (300 ft) in the SE

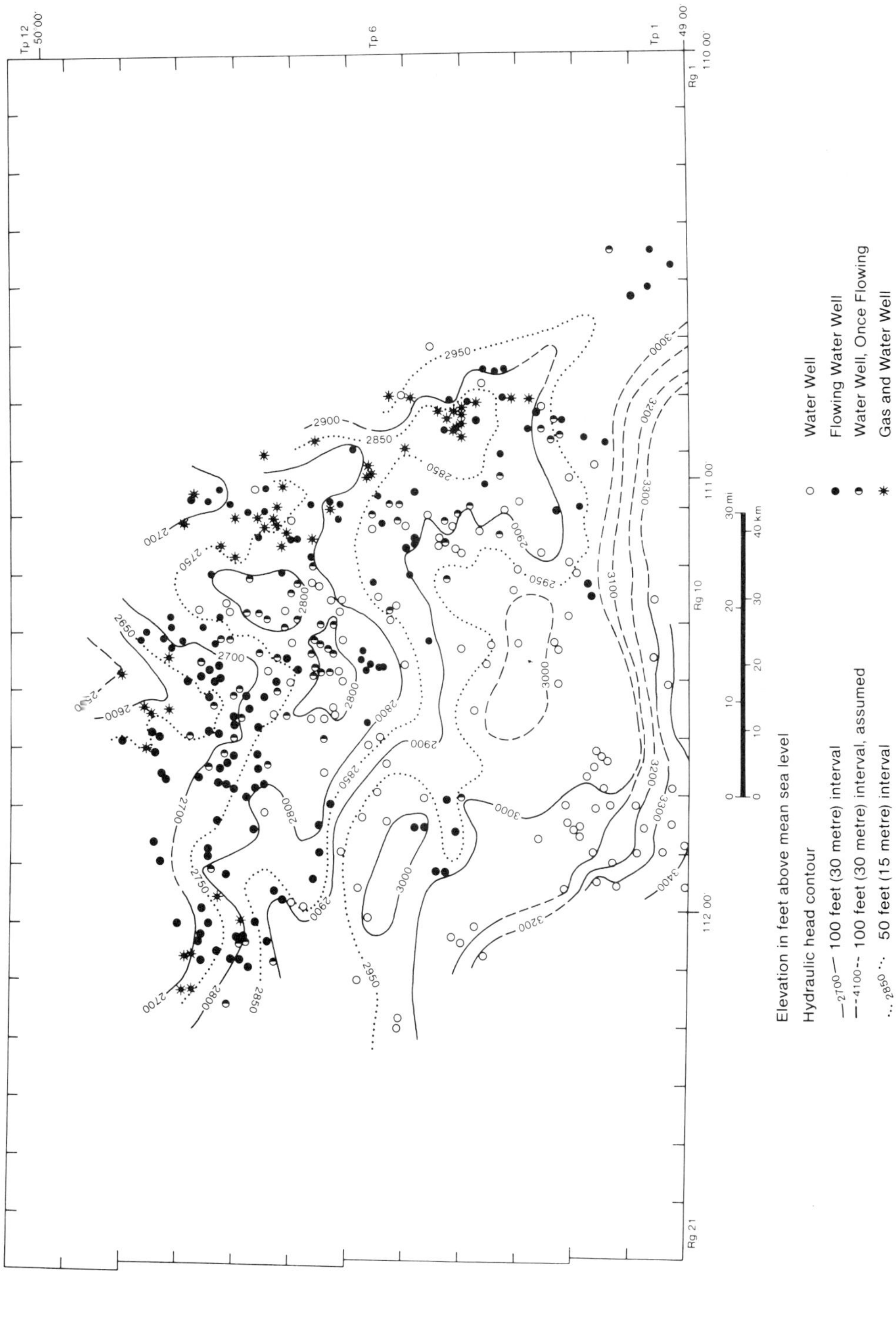

FIG. 22. Potentiometric surface of the Milk River Aquifer.

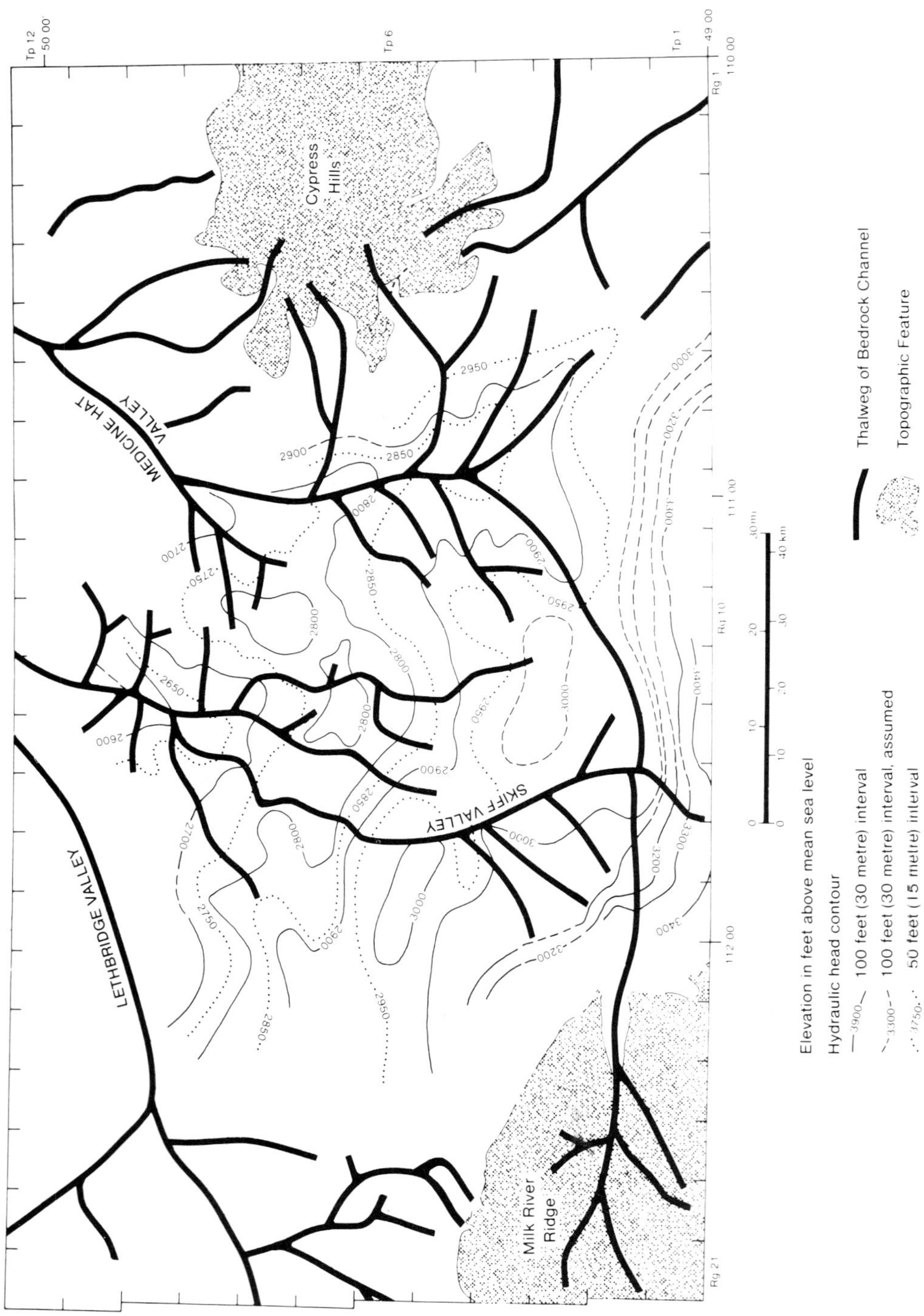

FIG. 23. Comparison of the Milk River Aquifer potentiometric surface with the bedrock topography.

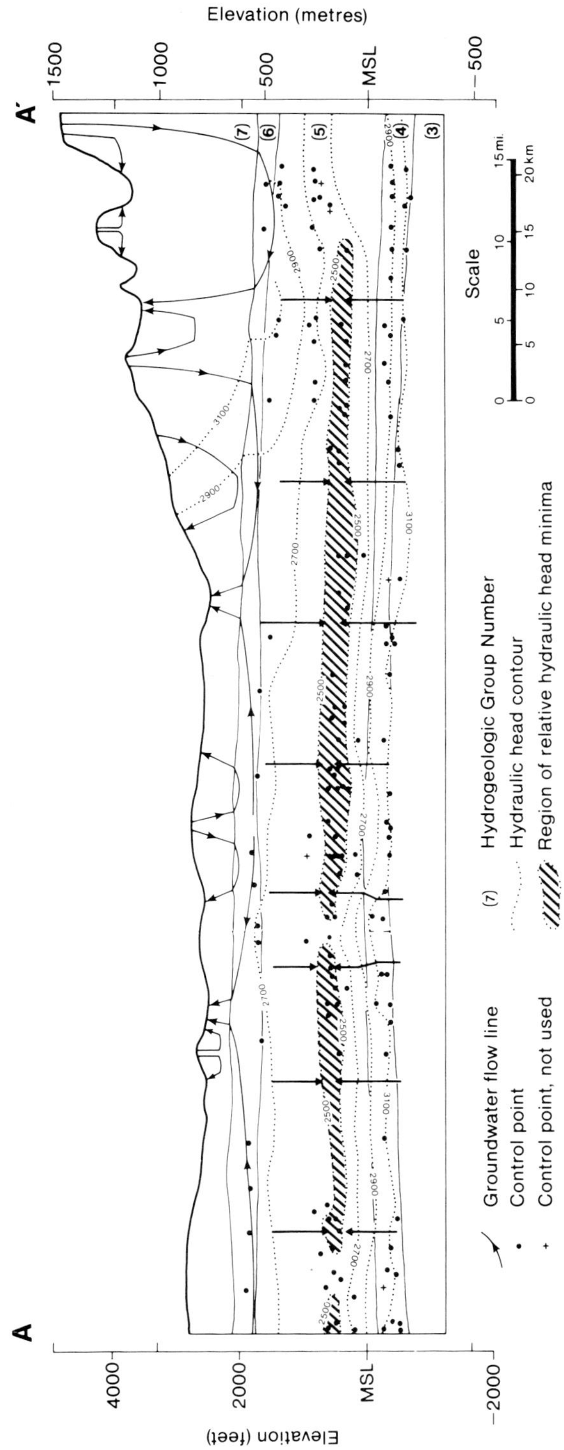

FIG. 24. Hydraulic cross-section AA′.

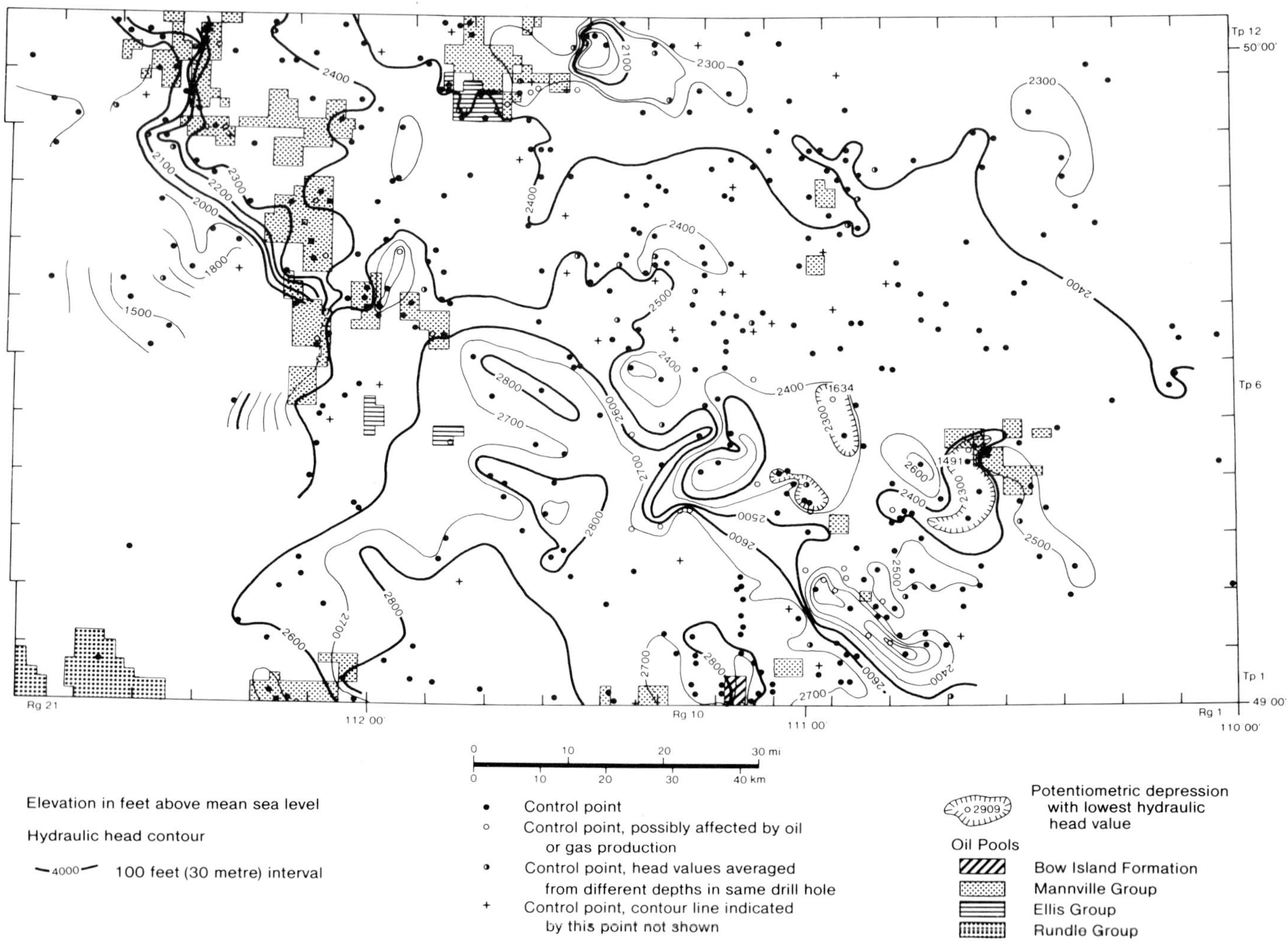

FIG. 25. Potentiometric surface of the Bow Island hydrogeological formation.

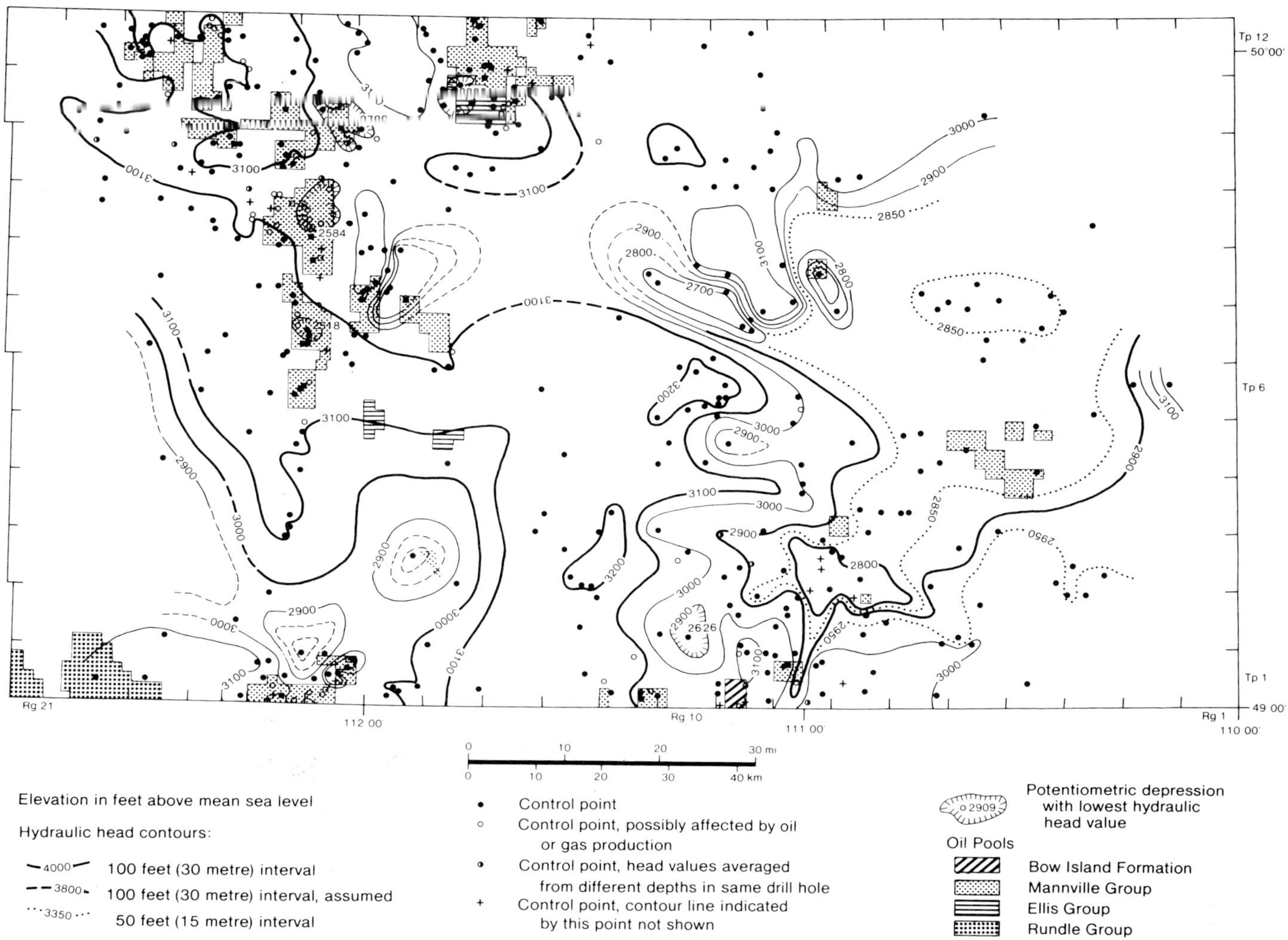

FIG. 26. Potentiometric surface of the Mannville hydrogeological formation.

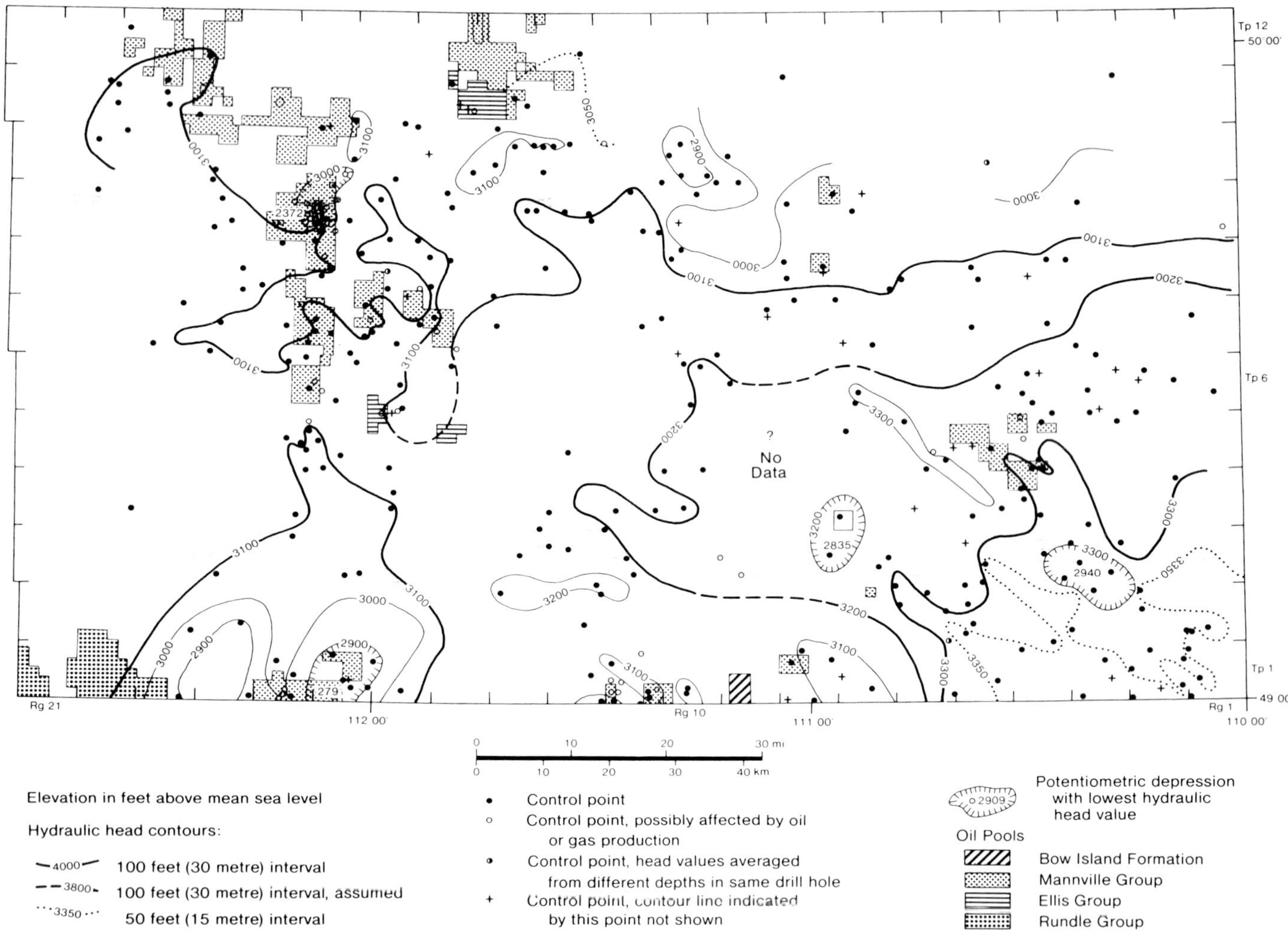

FIG. 27. Potentiometric surface of the Ellis hydrogeological formation.

corner of the Taber Area (Tps 1–7, Rgs 1–5). The primary feature of the head distribution is an increase in head to the SE corner with a general decline towards the N. W of Rg 13 the surface is fairly smooth with an elevation of about 945 m (3100 ft). The relief of the surface is 180 m (600 ft) with most of the surface between 915 and 1000 m (3000 and 3300 ft). In contrast with the Mannville unit there is no clearly defined channel below the Medicine Hat Valley in the potentiometric surface of the Ellis unit, although the paucity of data points makes this conclusion somewhat tenuous.

Age of water

The evolution of groundwater flow patterns as presented above has implications concerning the age of formation waters in the Taber Area. The potentiometric sink in the Colorado Aquitard isolates the Lower Cretaceous Aquifer hydraulically from the surface and holds water in place within the aquitard. Water in and below the Colorado Aquitard will not be exchanged with more recent meteoric water until the fluid potentials in these units adjust to some land surface. However, there has not been sufficient time for fluid potentials to adjust to any of the three relatively stable topographic surfaces that have existed since Cypress Plain time. Water currently present in the Rebound and Cypress Plain Systems must therefore have infiltrated during Cypress Plain time and is of pre-Pliocene age. Flow in the Modern System, in contrast, is adjusted to the Modern Land Surface. Since the flowthrough time of this system is probably of the order of tens or hundreds of thousands of years, some water in this system must have been recharged in Pleistocene time.

The stable isotopic compositions of formation waters can be interpreted to support these inferences about the age of waters in the Taber Area. Schwartz *et al.* (1981) found that the $^{18}O/^{16}O$ and $^{2}H/^{1}H$ ratios of water in the Milk River Formation (Modern System) are clearly different from those of the Bow Island and Sunburst Formations (Rebound System and Cypress Plain System respectively). The isotopic ratios of the Bow Island and Sunburst waters indicate a climate 12–13°C warmer than at present. Dorf (1960) showed that the climate of N America has changed considerably since early Oligocene time, and the climate that he indicates for the Taber Area during pre-Pliocene time is compatible with these ratios. Isotopic ratios in the Milk River Formation, however, change systematically from those in the S, which are expected for the present climate, to those in the N, which are the same as ratios in the Bow Island and Sunburst Formations. The systematic change in composition of the Milk River waters is best explained by megascopic (inter-formational) dispersion as suggested by Schwartz & Muehlenbachs (1979). By this mechanism, water recently recharged into the Milk River Aquifer may be mixing with upward-moving water from the Colorado Aquitard. The regional pattern of isotopic ratios in the Milk River Aquifer probably indicates an increase in the relative amount of pre-Pliocene water from the Colorado Aquitard along the flow path. However, a more recent interpretation (Schwartz *et al.* 1981) precludes upward flow from the Colorado Aquitard into the Milk River Aquifer, and thus is not in agreement with the flow pattern presented in this paper which includes the uppermost portion of the Colorado Aquitard as part of the Modern System.

Flow patterns during the time of the Cypress Plain and their relation to oil and gas accumulations

According to the hydraulic theory of petroleum migration (Tóth 1980), certain types of petroleum accumulations are formed by topographically induced cross-formational groundwater flow whose patterns are stable long enough for multiple pore volumes of water to carry and deliver hydrocarbons to given traps and reservoirs. The most recent flow conditions meeting these requirements appear to have existed during the time of the Cypress Plain. It is therefore reasonable to hypothesize that this flow has influenced the present locations of petroleum accumulations in the Taber Area.

It is impossible to reconstruct the details of the flow pattern that existed during Cypress Plain time because the pattern has been erased in the rock units above the Bow Island Formation and its remains in the deeper units have been modified by the fluid potential sink in the Colorado Aquitard. It is possible, however, to make general conclusions about the flow pattern of that time by relating available information about the topographic evolution and the presently observed potential distribution to evolutionary patterns of groundwater flow established theoretically.

Because the Cypress Plain land surface existed for at least 20 m.y., i.e. longer than the time required for flow patterns at the depth of study to adjust to a land surface, it is concluded that flow was steady state, cross-formational and gravity induced; it was analogous to that of time $t_{-\infty}$ of the conceptual development (Figs 5 and 9A). The strong upward components of force in the Lower Cretaceous Aquifer are due to low fluid potentials in the Colorado Aquitard, but the horizontal components are remnants of the flow pattern that existed during the time of the Cypress Plain. It is felt that the potentiometric surface of the Mannville unit (Fig. 26) contains the most complete record of the palaeo-flow pattern. The general pattern of lateral flow is away from two regions of high fluid potential: the SE corner of the study area, and a N–S-trending potentiometric rise centred along Rg 12. The increase in fluid potential in the SE corner is also clearly present on the Ellis potentiometric surface (Fig. 27), and the N–S feature can be observed on the Bow Island surface (Fig. 25). These two areas of high fluid potential coincide with the postulated topographic highs, the Bow Island Topographic High and the Missouri Divide, on the Cypress Plain.

In Cypress Plain time groundwater flow systems at least as deep as the Lower Cretaceous Aquifer were adjusted to the Cypress Plain topography. A schematic pattern of Cypress Plain flow is shown in Fig. 28. This figure corresponds to Figs 5 and 9A of the conceptual development. Meteoric water infiltrated into the system in topographically high regions, along the axis of the Bow Island Topographic High and along the Missouri Divide. Water movement was downward and away from these regions. Multiple pore volumes of water flowed cross-formationally through the Colorado Aquitard into the Lower Cretaceous Aquifer, turned laterally and then ascended to the land surface in topographically low areas. Upward flow occurred E and W of the Bow Island Topographic High with flow directed SE from this region converging with north-westerly flow in the region of low fluid potential in the Mannville unit near Tp 4, Rg 6 (Fig. 26). The length of flow systems was comparable with that of flow systems identified as currently existing in the Milk River Aquifer.

Nearly all oil production in the Taber Area is from the Lower Cretaceous Aquifer. The positions of producing oil fields in this unit are shown superimposed on the potentiometric surface maps (Figs 25–27) and on Fig. 28. These fields tend to occur in areas away from areas of high fluid potential, i.e. where flow in the postulated Cypress Plain pattern is upward. Hitchon (1984, Fig. 12) noted the presence of a groundwater divide extending from Montana to the Canadian Shield, the southern portion of which coincides with the Bow Island Topographic High. He suggested that this divide exerted a major influence on the position of major petroleum accumulations by hydraulically confining petroleum generated from source rocks in western Alberta to reservoirs W of the divide.

Commercially developed natural gas deposits are present in all major sandstone intervals in the Colorado Aquitard. Rice & Claypool (1981) present biochemical evidence that these deposits are of biogenic origin. They suggest that the gas came out of solution in response to erosional unloading during the Tertiary, and subsequently moved by capillary-pressure differentials from the shales to the sandstones. Nearly all drill-stem tests in the Colorado Aquitard above the Bow Island Formation recover natural gas or drilling mud; the sandstone layers contain very little water and are saturated with gas. Gas deposits generated by this mechanism would have been emplaced in post-Cypress Plain time and their positions would not necessarily be related to any groundwater flow pattern. Instead, gas fields would tend to occur in sandstone lenses surrounded by elastic shale in regions where thick layers of sediments have been eroded from the surface.

Summarizing remarks

The conceptualized evolution of groundwater flow systems presented in this paper incorporates and requires information from a variety of sources, including observed hydraulic-head distributions, structure and stratigraphy, a hypothesized development of the land surface, calculated pore-pressure adjustment times, isotopic compositions of formation waters and locations of oil and gas deposits. The strength of the conceptual model is its compatibility with various observed phenomena and theoretical considerations. Its importance is twofold: it provides a basis for understanding the present pattern of formation-fluid movement, and it provides insight into the groundwater flow systems and hydrological processes that influenced the current positions of oil and gas deposits.

Erosion of more than 700 m of rock from the surface of the Taber Area during the last 10–20 m.y. greatly influenced present groundwater and palaeo-groundwater flow patterns. Four palaeo-land surfaces have been identified. The most important, with respect to generation of flow patterns, are the Cypress Plain which existed in Oligocene and Miocene time, and the Modern Land Surface which formed in the Middle Pleistocene and lasted until approximately 24 000 years ago.

The current groundwater flow pattern is heterochronous: its upper portion was generated by

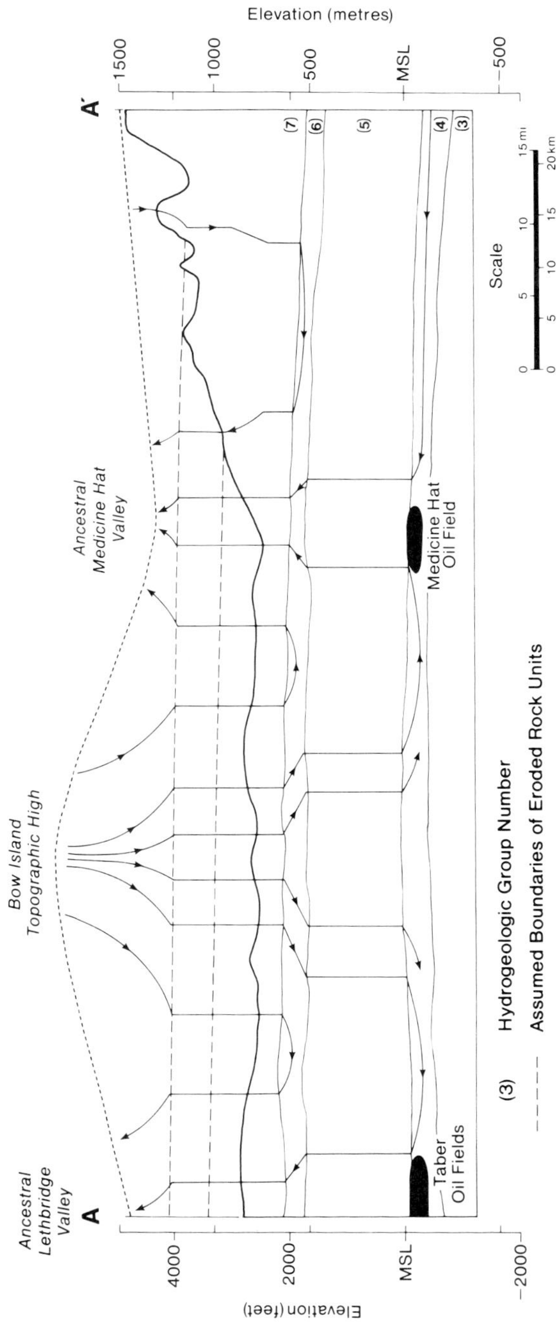

FIG. 28. Schematic flow pattern along the cross-section AA′ during Cypress Plain time.

the Modern Land Surface, but its lower portion contains relict components of a flow system generated by the Cypress Plain. The flow pattern is best described in terms of three regionally coherent fluid dynamic systems: the Modern Land Surface System, the Erosional Rebound System and the Cypress Plain System. The three dynamic systems are interpreted to be transient stages of sequentially superimposed gravitational and dilatational force fields. The gravitational fields depend on topography, and the dilatational field results from elastic rebound of the Colorado Aquitard in response to erosional removal of overburden during Pliocene and Pleistocene times.

The Modern Land Surface System reaches depths of 300 m and is well adjusted to the modern relief as is indicated by the close correspondence of the potentiometric surface of its primary aquifer (the Milk River Aquifer) with the bedrock topography. It appears that the complete trajectories of flow in the Modern Land Surface System are contained within the Taber Area. Recharge occurs in topographically high regions along the S, SW and E margins of the Taber Area, while discharge occurs in the central portion of the Taber Area where major bedrock channels are located.

The subjacent Erosional Rebound System occupies most of the 600 m thick predominantly shale Colorado Aquitard including its often gas-bearing sand members. It is characterized by extensive regions of sub-hydrostatic pressure. Groundwater currently flows into the lower portion of the Colorado Aquitard (the Bow Island Formation) from above and below. This flow distribution is interpreted to indicate that the Colorado Aquitard behaves as an energy sink owing to an elastic rebound of its shale in response to erosion during Pleistocene time.

Below the Colorado Aquitard is the Cypress Plain System which has strong upward directed forces and mild but definite lateral components. The lateral components oppose the modern relief and are oriented from the centre of the study area toward its western and eastern boundaries where most of the known oil fields are also located. The upward component of flow is towards the potentiometric sink in the Bow Island Formation, while the lateral components can be considered as relicts of the flow pattern that existed during the time of the Cypress Plain.

The Cypress Plain flow pattern was generated after Palaeocene (58 Ma) or Eocene (36 Ma) times by a palaeo-relief dominated by the NE–SW-trending Bow Island Topographic High. It was composed of and maintained as cross-formational gravity-flow systems throughout at least the Miocene (i.e. for more than 30 m.y.). Oil migrated to and was entrapped in the discharge regions of these systems. Subsequently, dilation of the Colorado Aquitard created the hydraulic sink in the Erosional Rebound System and induced gas to exsolve and accumulate in the sands.

ACKNOWLEDGMENTS: This project was made possible by financial support through the Alberta/Canada Energy Resources Research Fund, a joint programme of the Federal and Alberta Governments administered by Alberta Energy and Natural Resources, and a contribution from Canadian Superior Oil Ltd which has helped to defray computing costs; both of these are gratefully acknowledged.

References

ALBERTA ENERGY RESOURCES CONSERVATION BOARD 1970. *Structure Contour Map of the Milk River Formation*, Map Area 1, Scale 1:250 000.

ALBERTA SOCIETY OF PETROLEUM GEOLOGISTS 1966. *Geologic History of Western Canada*, 2nd edn, Alberta Society of Petroleum Geologists, Calgary.

ALDEN, W. C. 1932. Physiography and glacial geology of Eastern Montana and adjacent areas. *U.S. geol. Surv. prof. Pap. No. 174*.

BALLY, A. W., GORDY, P. L. & STEWART, G. A. 1966. Structure, seismic data, and orogenic evolution of southern Canadian Rocky Mountains. *Bull. Can. petrol. Geol.* **14**, 337–81.

BETHKE, C. M. 1985. A numerical model of compaction-driven groundwater flow and heat transfer and its application to the paleohydrology of intracratonic sedimentary basins. *J. geophys. Res.*, **90**, no. B8, pp. 6817–28.

BRACE, W. F. 1980. Permeability of crystalline and argillaceous rocks. *Int. J. rock Mech. min. Sci., Geomech. Abstr.* **17**, 241–51.

BRADLEY, J. S. 1975. Abnormal formation pressures. *Am. Assoc. petrol. Geol. Bull.* **59**, 957–73.

BREDEHOEFT, J. D. & HANSHAW, B. B. 1968. On the maintenance of anomalous fluid pressures, I. Thick sedimentary sequences. *Geol. Soc. Am. Bull.* **79**, 1097–106.

BRUIN, J. & HUDSON, H. E., Jr. 1961. Selected methods for pumping test analysis. *Ill. State Water Surv. Rep. Invest. No. 25*.

COLLIER, A. J. & THOM, W. T. 1918. The Flaxville gravel and its relation to other terrace gravels of the Northern Great Plains. *U.S. geol. Surv. prof. Pap. No. 108-J*, pp. 125–84.

DORF, E. 1960. Climatic changes of the past and present. *Am. Sci.* **48**, 341.

FREEZE, R. A. & CHERRY, J. A. 1979. *Groundwater*, Prentice-Hall, Englewood Cliffs, NJ.

—— & WITHERSPOON, P. A. 1967. Theoretical analysis of regional groundwater flow, 2, Effect of water-table configuration and subsurface permeability variation. *Water Resour. Res.* **3** (2), 623–34.

HITCHON, B. 1984. Geothermal gradients, hydrodynamics and hydrocarbon occurrences, Alberta, Canada. *Am. Assoc. petrol. Geol. Bull.* **68**, 713–43.

HOWARD, D. A. 1960. Cenozoic history of northwestern Montana and northeastern North Dakota with emphasis on the Pleistocene. *U.S. geol. Surv. prof. Pap. No. 326.*

JACQUIN, C. & POULET, M. J. 1970. Study of the hydrodynamic pattern in a sedimentary basin subject to subsidence. *45th Annu. Fall Meet. of the Society of Petroleum Engineers of AIME, Houston, TX, October 4–7, 1970*, Preprint, Paper SPE 2988.

—— & —— 1973. Essai de réstitution des conditions hydrodynamiques régnant dans un bassin sédimentaire au cours de son evolution. *Rev. Inst. fr. Pétr.* **27** (3), 269–97.

JOHNSTON TESTERS 1964. *Review of Basic Formation Evaluation*, Johnson Testers, Houston, TX.

MCCONNELL, R. G. 1886. Report on the Cypress Hills Wood Mountain and adjacent country. *Geol. nat. hist. Surv. Can. Annu. Rep.* **1**, 1C–85C.

MEYBOOM, P. 1960. Geology and groundwater resources of the Milk River Sandstone in southern Alberta. *Mem. 2*, Research Council of Alberta.

NEGLIA, S. 1979. Migration of fluids in sedimentary basins. *Am. Assoc. petrol. Geol. Bull.* **63**, 573–97.

NEUZIL, C. E. & POLLOCK, D. W. 1983. Erosional unloading and fluid pressures in hydraulically "tight" rocks. *J. Geol.* **91** (2), 179–93.

PARDEE, J. T. 1950. Late Cenozoic block faulting in western Montana. *Geol. Soc. Am. Bull.* **61**, 359–406.

RICE, D. D. & CLAYPOOL, G. E. 1981. Generation, accumulation, and resource potential of biogenic gas. *Am. Assoc. petrol. Geol. Bull.* **65**, 5–25.

SCHWARTZ, F. W. & MUEHLENBACHS, K. 1979. Isotope and ion geochemistry of groundwaters in the Milk River aquifer, Alberta. *Water Resou. Res.* **15**, 259–68.

——, —— & CHORLEY, D. W. 1981. Flow-system controls of the chemical evolution of groundwater. *J. Hydrol.* **54**, 225–43.

STELCK, C. R. 1975. Basement control of Cretaceous sand sequences in Western Canada. *Geol. Assoc. Can. Spec. Pap. No. 13*, pp. 428–39.

TÓTH, J. 1962. A theory of groundwater motion in small basins in central Alberta, Canada. *J. geophys. Res.* **67** (11), 4375–87.

—— 1963. A theoretical analysis of groundwater flow in small drainage basins. *J. geophys. Res.* **68** (16), 4795–812.

—— 1978. Gravity-induced cross-formational flow of formation fluids, Red Earth region, Alberta, Canada: analysis, patterns, and evolution. *Water Resour. Res.* **14**, 805–43.

—— 1980. Cross-formational gravity-flow of groundwater: a mechanism of the transport and accumulation of petroleum (the generalized hydraulic theory of petroleum migration). *AAPG Studies in Geology, No. 10, Problems of Petroleum Migration*, pp. 121–67.

—— 1985. The role of regional gravity flow in the chemical and thermal evolution of groundwater. *In:* HITCHON, B. & WALLICK, E. I. (eds) *Practical Applications of Ground Water Geochemistry, 1st Canadian/American Conf. on Hydrogeology, Banff, Alberta, June 22–26, 1984*, NWWA, Worthington, OH.

—— & CORBET, T. 1985. Investigations of the relations between formation fluid dynamics and petroleum occurrences, Taber Area, Alberta, *Final Rep., Alberta/Canada Energy Resources Research Fund, Contract No. U81–3R.*

—— & MILLAR, R. F. 1983. Possible effects of erosional changes of the topographic relief on pore pressures at depth. *Water Resour. Res.* **19** (6), 1585–97.

TOVELL, W. M. 1958. The development of the Sweetgrass Arch, southern Alberta. *Proc. geol. Assoc. Can.* **10**, 19–30.

WESTGATE, J. A. 1968. Surficial geology of the Foremost–Cypress Hills region, Alberta. *Res. Counc. Alberta Bull. No. 22.*

WILLIAMS, M. Y. 1929. The physiography of the southwestern plains of Canada. *Proc. Trans. R. Soc. Can.* **23**, 61–79.

J. TÓTH & T. CORBET,* Department of Geology, The University of Alberta, Edmonton, Alberta T6G 2E3, Canada.

*Present address: Department of Geology, 245 Natural History Building, 1301 West Green Street, Urbana, IL 61801, U.S.A.

Some aspects of the thermal regime and hydrodynamics of the western Canadian sedimentary basin

F. W. Jones & J. A. Majorowicz

SUMMARY: Approximately 65 000 bottom-hole temperatures from about 36 000 wells in the Prairies Basin of western Canada have been used to investigate the thermal regime there. Both lateral and depth variations in temperature gradients and heat-flow densities occur, and these are related to the hydrodynamics which is controlled by the topography. The vertical heat-flow density in the Mesozoic–Cenozoic formations is less than in the Palaeozoic formations in recharge areas, but is greater in discharge areas. A zone in which heat flow is approximately constant with depth exists in the central part of the basin between recharge and discharge areas. It is clear that basin-wide redistribution of heat by groundwater motion takes place. The relationship between geothermal gradients and hydrocarbon occurrences is studied for southern Alberta and it is found that the mean geothermal gradient in the Mesozoic and Cenozoic sediments is greater for areas where Mesozoic gas and oil pools exist than for areas where they do not occur.

A large number (approximately 65 000) of bottom-hole temperature (BHT) values from about 36 000 petroleum exploration wells in the Prairies basin of western Canada have been used to study the temperature regime there. The study area is shown on a map of the major physiographic and structural subdivisions of the basin in Fig. 1, and includes most of Alberta, the southern half of Saskatchewan and the south-western part of Manitoba.

The temperature field

The area was divided into 28.8 km × 28.8 km areas (3 × 3 township/range areas on the Dominion Land Survey System grid) and the temperature data were plotted as a function of depth for each area. Gradients were determined for two depth intervals, the Mesozoic–Cenozoic formations and the Palaeozoic formations. These two intervals were chosen because of the compositional change from mainly clastic rocks above the Palaeozoic erosional surface to the mainly carbonate–evaporite unit below. Least-squares straight lines were fitted to the data above and below the Palaeozoic erosional surface. The slope of the least-squares-fitted straight line below that surface gave the gradient for that interval (Grad 2). The gradient values above the Palaeozoic erosional surface (Grad 1) were estimated on the basis of the mean annual ground-surface temperature distribution in the area (Judge 1973) and the average Palaeozoic surface temperature obtained from the intercepts of the least-squares-fitted lines in the two intervals at the Palaeozoic erosional surface. An example of the calculations for one of the areas is given in Fig. 2.

Thermal conductivities were estimated from net rock analysis data from 158 wells uniformly distributed throughout the area, and calculations of heat-flow density have been carried out for two intervals: Mesozoic–Cenozoic (Q_1), and Palaeozoic (Q_2). It is clear from maps of Q_1 and Q_2 (Figs 3 and 4) that the basin heat is redistributed by groundwater motion which is related to topography. The results are described in detail in five papers (Majorowicz *et al.* 1984, 1985a, b, 1986; Jones *et al.* 1986) and are summarized in the schematic diagram of Fig. 5 which illustrates the relationship between the hydrodynamics of the basin and the temperature and heat flow. Downward groundwater motion in recharge areas (generally to the S and W) draws the isotherms downwards and produces a situation in which both temperature gradient and heat-flow density increase with depth. In discharge areas the upward motion draws the isotherms upwards towards the surface which results in decreases of both geothermal gradient and heat flow density with depth. The overall regional effect is clearly shown in a basin-wide plot of the quantity ΔQ_i which is the difference between the heat-flow densities in the Mesozoic–Cenozoic sediments and in the Palaeozoic sediments (i.e. $\Delta Q_i = Q_{1i} - Q_{2i}$) which is given in Fig. 6. Negative ΔQ values occur in the S and W (recharge) and positive ΔQ values lie to the N and E (discharge). A zone where $\Delta Q \approx 0$ lies through the centre of the basin. This is a zone between the recharge and discharge areas where the groundwater motion is approximately parallel to the surface. It is clear that the temperature regime is distributed by the water motion, and that both upward and downward components of heat transport exist in the sediments.

From GOFF, J. C. & WILLIAMS, B. P. J. (eds), 1987, *Fluid Flow in Sedimentary Basins and Aquifers*, Geological Society Special Publication No. 34, pp. 79–85.

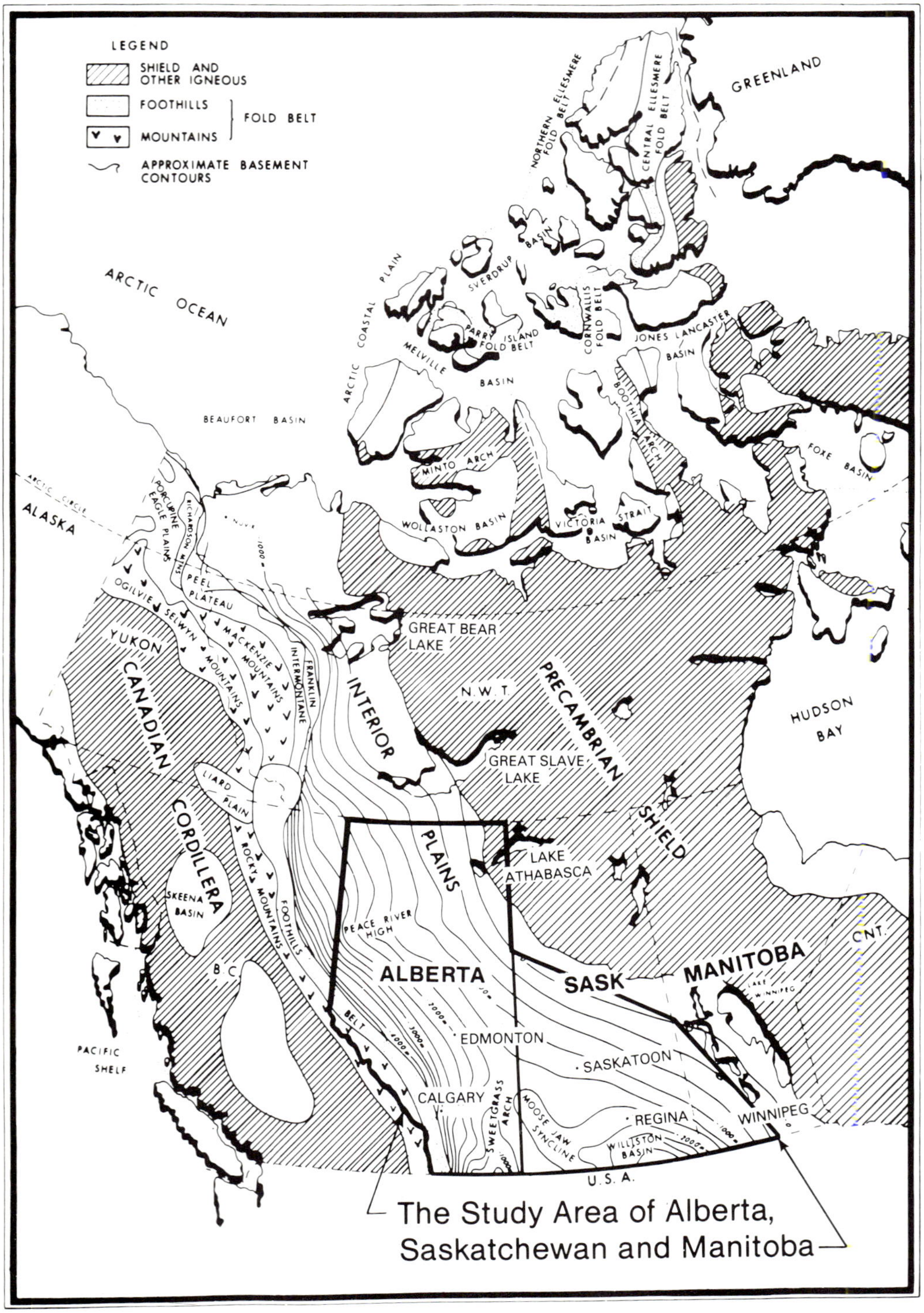

FIG. 1. The study area of Alberta, Saskatchewan and Manitoba shown on the map of the major physiographic and structural subdivisions and the sedimentary basin in western Canada (after Hay & Robertson 1984).

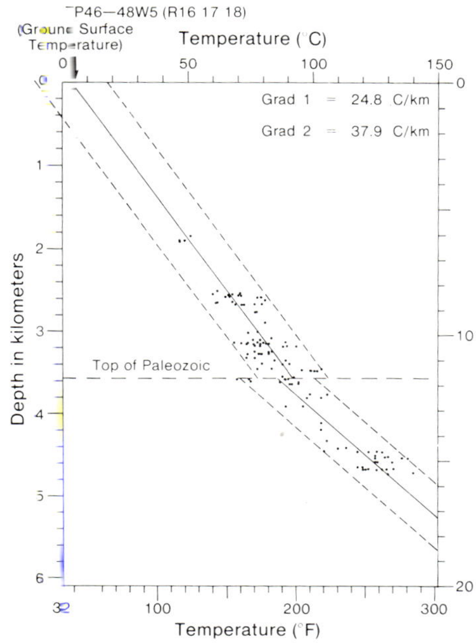

FIG. 2. Temperature as a function of depth for one 28.8 km × 28.8 km area (from Majorowicz *et al.* 1985a).

The relationship between the temperature field and hydrocarbon occurrences

Tóth (1980) described a relationship between the groundwater flow pattern and hydrocarbon accumulation. Descending water movement in recharge areas penetrates deeply into the sediments and causes geothermal lows. Such areas are unfavourable for oil and gas accumulation. However, upward water movement brings hydrocarbons into existing traps until trap capacity is reached. Simultaneously, the upward water motion brings heat toward the surface and produces both geothermal gradient 'highs' and heat-flow anomalies. The regional 'highs' in geothermal gradient and heat-flow density as seen in the Canadian Prairies basin correspond to the general direction of fluid migration eastward and upward as described by Hitchon (1984).

Within this framework, the differences between the geothermal gradient values where oil and gas pools occur compared with areas where they do not occur can be examined to investigate

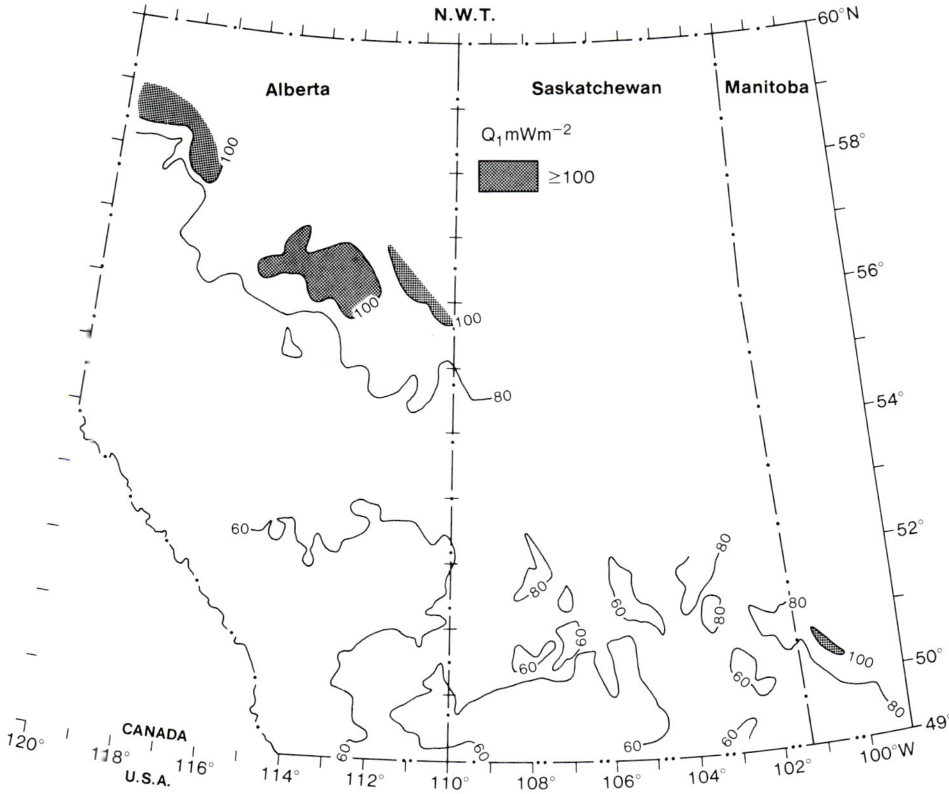

FIG. 3. Heat-flow map for the prairies basin of Alberta, Saskatchewan and Manitoba based on estimated heat-flow values Q_1 (in mW m^{-2}) above the Palaeozoic erosional surface (Majorowicz *et al.* 1985a, 1986).

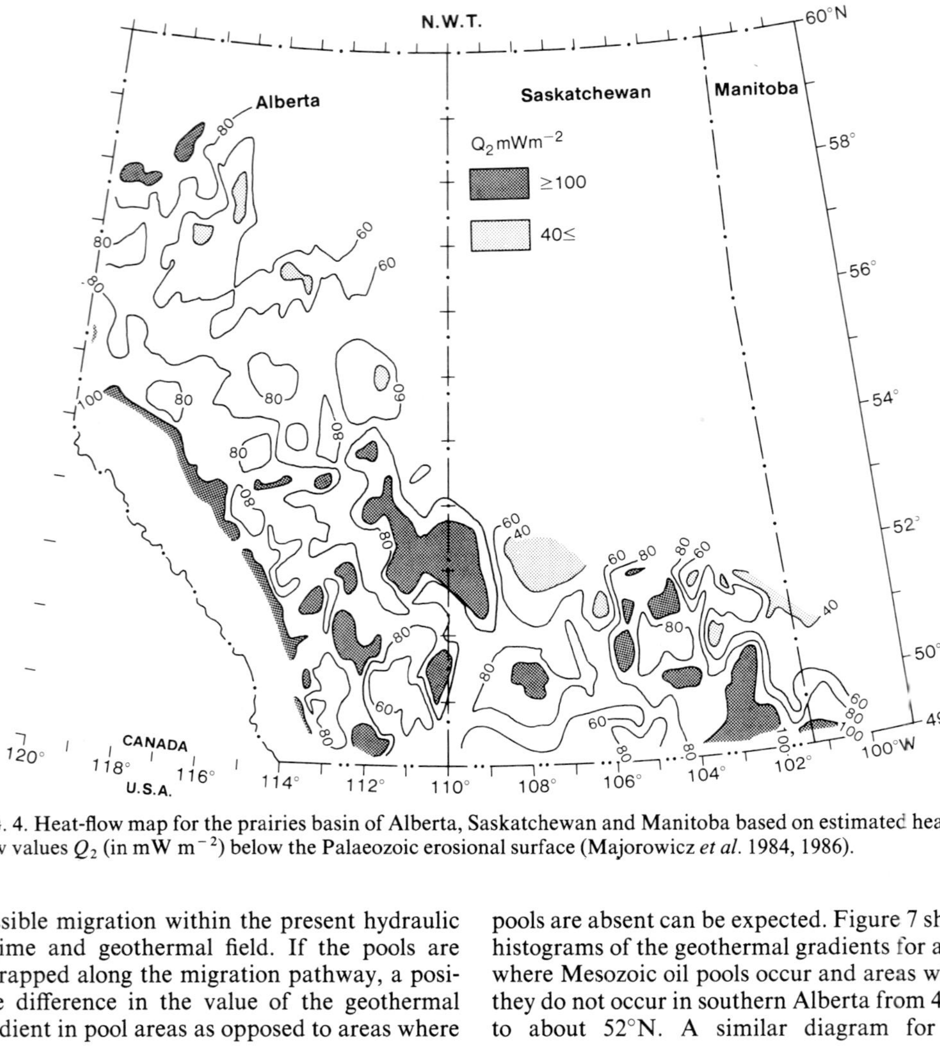

FIG. 4. Heat-flow map for the prairies basin of Alberta, Saskatchewan and Manitoba based on estimated heat-flow values Q_2 (in mW m^{-2}) below the Palaeozoic erosional surface (Majorowicz *et al.* 1984, 1986).

possible migration within the present hydraulic regime and geothermal field. If the pools are entrapped along the migration pathway, a positive difference in the value of the geothermal gradient in pool areas as opposed to areas where pools are absent can be expected. Figure 7 shows histograms of the geothermal gradients for areas where Mesozoic oil pools occur and areas where they do not occur in southern Alberta from 49°N to about 52°N. A similar diagram for the

Table 1. *Summary of geothermal gradient values for areas with and without hydrocarbon accumulations*

Age	Gradient mean (no. of observations)							
	Oil pools		Outside oil pools		Gas pools		Outside gas pools	
	Grad 1 (mK m^{-1})	(No. obs.)	Grad 1 (mK m^{-1})	(No. obs.)	Grad 1 (mK m^{-1})	(No. obs.)	Grad 1 (mK m^{-1})	(No. obs.)
Mesozoic	26.9	(145)	25.7	(826)	27.1	(431)	24.9	(540)
Post-Mannville	24.5	(29)	25.9	(942)	27.1	(390)	24.9	(581)
Mannville	27.2	(103)	25.7	(868)	27.2	(133)	25.7	(838)
Jurassic	29.7	(13)	25.8	(958)	27.8	(111)	25.6	(860)
Mean for the study area of southern Alberta	25.9	(971)						

After Jones *et al.* 1986.

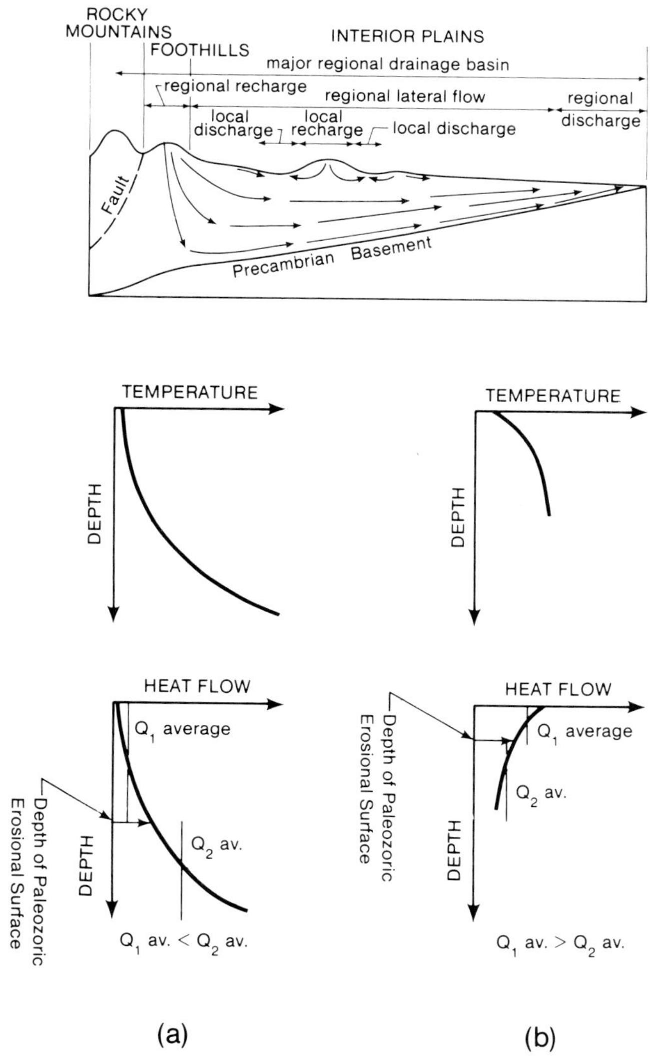

FIG. 5. Schematic model of the relationship between the hydrodynamic situation in the basin (after Hitchon 1984) and the temperature and heat flow (from Majorowicz *et al.* 1985a).

Mesozoic gas pools is shown in Fig. 8. In both cases the mean values of the gradients where pools exist are higher than those where they do not. The association between water motion and geothermal gradient variations and water motion and pool occurrences results in a relationship between pool occurrences and geothermal gradient highs.

Results from similar histograms for pools of various ages in this area of southern Alberta are summarized in Table 1. It is seen from this table that the relationship holds for all cases except Post-Mannville oil pools. This may indicate that the Post-Mannville pools have not undergone secondary migration, as previously suggested by Hacquebard (1977). Also, it is seen that the geothermal gradients in the pool areas are higher than the mean gradient for the whole area for all except the Post-Mannville oil pools. Further discussion with respect to particular pools is given by Jones *et al.* (1986).

ACKNOWLEDGMENTS: We would like to acknowledge the computing help of the Institute of Sedimentary and Petroleum Geology, Calgary, in preparing the heat-flow maps presented in Figs 3 and 4.

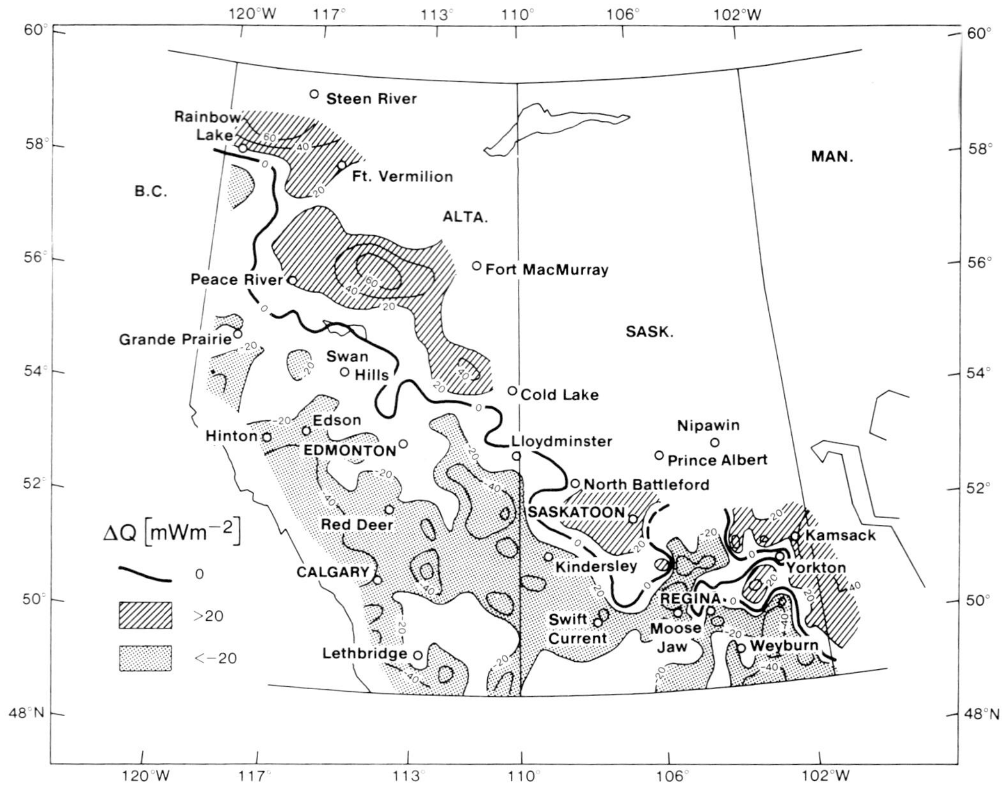

FIG. 6. Contour plot of the differences $\Delta Q_i = Q_{1i} - Q_{2i}$ (in mW m^{-2}) based on estimated values by Majorowicz *et al.* (1985b, 1986) for the prairie basin of Alberta, Saskatchewan and Manitoba (from Jones *et al.* 1985).

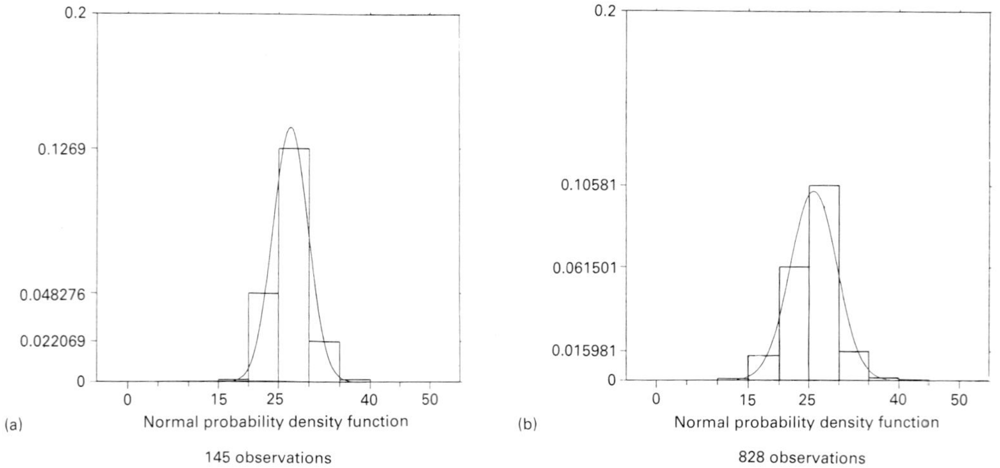

FIG. 7. Histogram of geothermal gradient values from (a) areas where Mesozoic oil pools occur (mean, 26.945; variance, 8.2983) and (b) areas where such pools do not occur in southern Alberta 49°N–52°N (mean, 25.754; variance, 15.059) (from Jones *et al.* 1986).

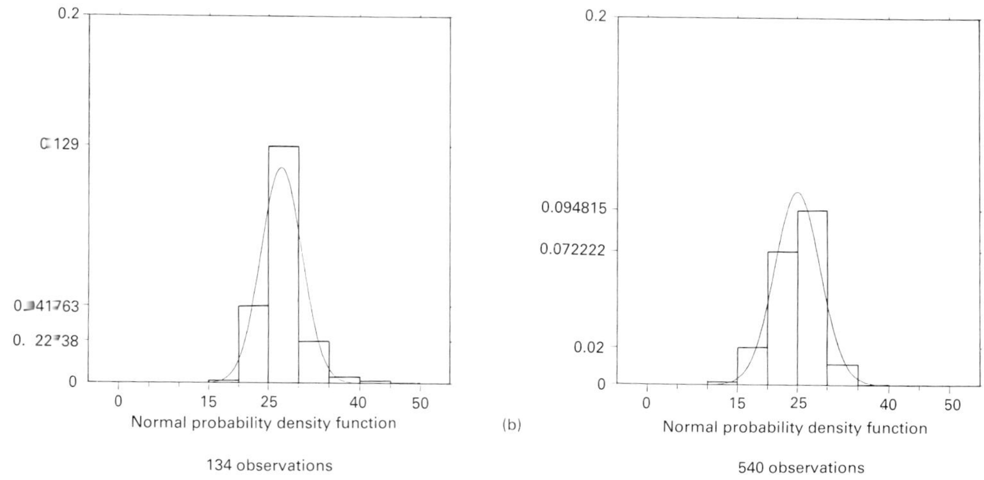

FIG. 8. Histogram of geothermal gradient values from (a) areas where Mesozoic gas pools occur (mean, 27.128; variance, 11.496) and (b) areas where such pools do not occur in southern Alberta 49°N–52°N (mean, 24.964; variance, 14.465) (from Jones *et al.* 1986).

References

HACQUEBARD, P. A. 1977. Rank of coals as an index of organic metamorphism for oil and gas in Alberta. *Geol. Surv. Can. Bull.* **262**, 11–23.

HAY, P. W. & ROBERTSON, D. C. 1984. Oil and gas development in western Canada. *Bull. Can. Petrol. Geol.* **32**, 74–82.

HITCHON, B. 1984. Geothermal gradients, hydrodynamics and hydrocarbon occurrences, Alberta, Canada. *Bull. Am. Assoc. petrol. Geol.* **68**, 713–43.

JONES, F. W., MAJOROWICZ, J. A. & LAM, H.-L. 1985. The variation of heat flow density with depth in the Prairies Basin of western Canada. *Tectonophysics*, in press.

——, ——, LINVILLE, A. & OSADETZ, K. G. 1986. The relationship between hydrocarbon occurrences and geothermal gradients in Mesozoic formations of southern Alberta. *Bull. Can. Petrol. Geol.* **34**, 226–39.

JUDGE, A. S. 1973. Deep temperature observations in the Canadian North. *2nd Int. Conf. on Permafrost*, pp. 35–40, National Academy of Sciences, Washington, DC.

MAJOROWICZ, J. A., JONES, F. W., LAM, H.-L. & JESSOP, A. M. 1984. The variability of heat flow both regional and with depth in southern Alberta, Canada: effect of groundwater flow? *Tectonophysics*, **106**, 1–29.

——, ——, —— & —— 1985a. Terrestrial heat flow and geothermal gradients in relation to hydrodynamics in the Alberta Basin, Canada. *J. Geodynam*, **4**, 265–83.

——, ——, —— & —— 1985b. Regional variations of heat flow differences with depth in Alberta, Canada, *Geophys. J.R. astr. Soc.* **81**, 479–87.

——, —— & JESSOP, A. M. 1986. Geothermics of the Williston Basin in Canada in relation to hydrodynamics and hydrocarbon occurrences. *Geophysics*, **51**, 767–79.

TÓTH, J. 1980. Cross-formational gravity flow of groundwater, a mechanism of the transport and accumulation of petroleum (generalized hydraulic theory of petroleum migration). *In:* ROBERTS, W. M., III, & CORDELL, C. J. (eds) *Problems of Petroleum Migration, A.A.P.G. Stud. Geol.* **10**, 121–67.

F. W. JONES & J. A. MAJOROWICZ, Institute of Earth and Planetary Physics and Department of Physics, University of Alberta, Edmonton, Alberta T6G 2J1, Canada.

Ancient fluid flow within foreland terrains

Harry J. Bradbury & Grant R. Woodwell

SUMMARY: Dewatering of the front Ranges of the Canadian Cordillera in the Campanian and Maastrichtian was achieved by expulsion of connate fluids from thrust sheets during active translation. Each major thrust sheet behaved as a separate hydrodynamic unit. Flow was consistently from hinterland to foreland sections of basal-sheet aquifers. The rate of fluid expulsion exceeded that of rock translation during hydrofracturing. Two flow systems, separated in time, utilized basal-sheet aquifers. Fluid flow at the margin of foreland basins in the Southern Pyrenees in the Oligocene was focused within both the crystalline infrastructure and the sedimentary suprastructure. Ascending metamorphic fluids passed from shear zone conduits in the basement into and through the basin margin, creating ghost vertical and lateral fluid fronts within the sedimentary pile. Connate fluids escaped via cleavage network channelways. Ponding of metamorphic and connate volumes of the fluid budget occurred at high levels close to the leading edge of natural strain within the basin margin, where excess fluid pressures were achieved.

Fluid migration in modern aquifers can be directly monitored. In ancient terrains, where primary migration has long since ceased, a different methodology must be employed in order to provide information on fluid sources, scales of transport, directions of flow, pathway geometries and related aspects of the fossil flow system.

Palaeoflow systems have been successfully modelled in various geotectonic settings (e.g. Lister 1972; Norton & Knapp 1977; Norton & Knight 1977; Sleep & Wolery 1978; Norton & Cathles 1979; Rosenblat 1985; Woodbury & Smith 1985). However, a lack of data from natural rocks limits the usefulness of such models at present. Accurate reconstruction of ancient flow systems will only be possible if methods are developed to identify former flow paths and to recognize and quantify the effects of fluid passage through rocks. The potential economic significance of direct analysis of ancient flow systems can be inferred (e.g. Tissot & Welte 1980; Lobato *et al.* 1983 Garven & Freeze 1984) and provides a considerable impetus to describe the plumbing system of different geological terrains better. The aim of this paper is to provide an outline description of the ancient flow systems of selected foreland terrains.

Stable isotopes

Information on ancient fluid flow can be gathered from the rocks through which fluids passed. There are three direct approaches to the study of fluid–rock interactions: petrology, fluid inclusions and stable isotope geochemistry. All these approaches have been adopted in the present study. In this paper emphasis is placed on the application of stable isotopes.

Isotopic data are presented in standard δ notation. The δ value is the difference in isotope ratio between a sample and a standard, expressed in parts per thousand or per mil. For example, a sample with a $\delta^{18}O$ value of $+10.0$ per mil is enriched in ^{18}O by 10 per mil, or 1% relative to the standard. The δ value is defined as follows:

$$\delta^{18}O = 10^3 \left\{ \frac{(^{18}O/^{16}O)_{sample}}{(^{18}O/^{16}O)_{std}} - 1 \right\}$$

$$\delta^{13}C = 10^3 \left\{ \frac{(^{13}C/^{12}C)_{sample}}{(^{13}C/^{12}C)_{std}} - 1 \right\}$$

The standard for oxygen is standard mean ocean water (SMOW) and the standard for carbon is the Cretaceous Pee Dee Belemnite (PDB), where (Friedman & O'Neal 1977)

$$\delta^{18}O(SMOW) = 1.03086\,\delta^{18}O(PDB) + 30.86$$

In the present study a high-speed diamond-tipped drill was used to prepare samples for CO_2 extraction for mass spectrometry. The CO_2 was released by reaction with H_3PO_4 as described by McCrea (1950), and was then analysed for both $\delta^{13}C$ and $\delta^{18}O$ using a Nuclide 6-60 ratio mass spectrometer. Isotopic results were reproducible to ± 0.1 per mil or better for carbon and ± 0.2 per mil or better for oxygen.

It will be necessary to consider certain aspects of the interpretation and application of isotopic data before examination of different types of flow systems on various scales. The basic method employed is defined here as J-curve analysis.

J-curve analysis: use of isotope pairs

J-curve analysis is defined as the interpretation of the characteristic patterns of isotopic data derived from analysis of rock matrix and sites of

From GOFF, J. C. & WILLIAMS, B. P. J. (eds), 1987, *Fluid Flow in Sedimentary Basins and Aquifers*, Geological Society Special Publication No. 34, pp. 87–102.

secondary precipitation which are displayed in δx–δy space for isotopes from two systems (e.g. $\delta^{13}C$–$\delta^{18}O$ space or D/H–$\delta^{18}O$ space). A full discussion of the interpretation of $\delta^{13}C$–$\delta^{18}O$ diagrams is provided by Bradbury & Rye (1986a), and so only the essential points will be outlined here. Throughout, the discussion will be restricted to a simple system in which calcite from a carbonate rock exchanges carbon and oxygen isotopes with an isotopically lighter fluid. This is the system which has been studied within all the carbonate sequences discussed later in the paper. For this system, Chai (1975) has shown that significant isotopic exchange will take place only if calcite enters and leaves the fluid by some process of material dissolution and recrystallization.

Secondarily precipitated calcite within any given rock volume may show a range of carbon and oxygen isotopic values since the exchange process and the degree of shift in the rock isotopic composition during exchange may be governed by variations in temperature, in the concentration of carbon-bearing species in the fluid and/or in volumetric fluid-to-rock ratios in the system. The influence of each of these controls can be investigated independently in order to assess which factor has dominated in the exchange process which resulted in the J-curve data sets presented later.

Variable-temperature system

Fractionation data provided by Bottinga (1968) and O'Neal (1963) allow the construction of a $\delta^{13}C$–$\delta^{18}O$ diagram which describes the systematics of calcite precipitated over a wide temperature range from a mineralizing fluid which is predominantly H_2O with H_2CO_3. The results of calcite exchange over a temperature range of 350–100°C with a fluid which maintains a constant isotopic composition of zero per mil for both carbon and oxygen is shown as a calibrated curve in Fig. 1. A different fluid composition would merely displace the curve; however, the convex shape with respect to the oxygen axis would be preserved as would the relatively high angle of the slope, both of which are characteristic of temperature-controlled isotopic exchange.

Fluid composition and fluid-to-rock ratios

The dominant carbon-bearing species in solution in circulating groundwaters is commonly bicarbonate. Speciation under the conditions considered in most of this paper is indeterminate and is simply taken to be CO_2. Species selection does not affect the modelling undertaken. The effects of variable XCO_2 and of volumetric fluid-to-rock ratio can be considered using Fig. 1. Continuous exchange between a carbonate rock of initially fixed isotopic composition and an isotopically lighter fluid will cause a progressive shift in the isotopic signature of the rock towards lighter values. The extent of this shift will be a function of the degree of exchange which in turn reflects relative isotope abundances within the fluid and rock reservoirs and consequently volumetric fluid-to-rock ratios. Rock exchange paths in low XCO_2 systems, under isothermal conditions, are

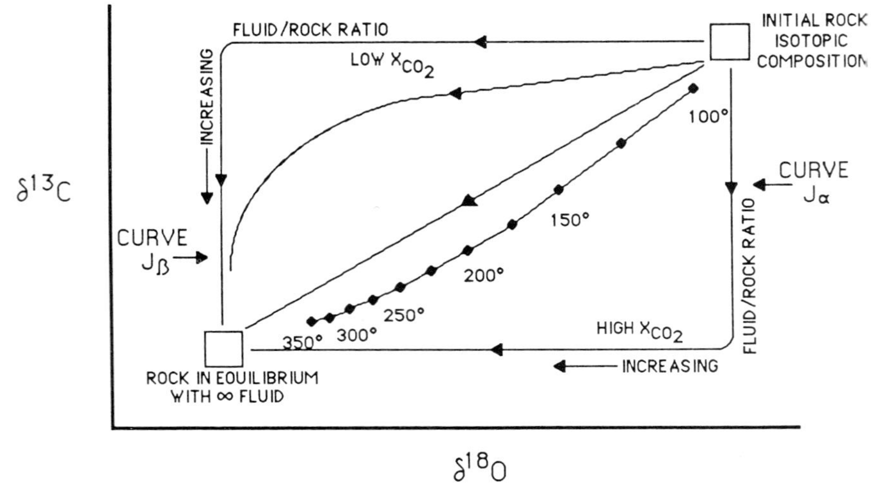

FIG. 1. Non-dimensionalized $\delta^{13}C$–$\delta^{18}O$ diagram. The temperature-calibrated curve represents isotopic values of calcite precipitated from a mineralizing fluid in the temperature range 100–350 °C. Curves J_α and J_β represent calcite isotopic exchange pathways for systems with high XCO_2 and low XCO_2 respectively. The curved and straight exchange pathways represent exchange pathways for possible intermediate fluid compositions.

characterized by pure oxygen shift where the rock isotopic reservoir dominates the low fluid-to-rock ratio part of the system for the least-exchanged materials and by pure carbon shift where the fluid isotopic reservoir dominates the high fluid-to-rock ratio part of the system for the most-exchanged materials (Bradbury & Rye 1986a). In this system, variations in fluid-to-rock ratios lead to the type of curved exchange pathway labelled as J_β in Fig. 1. In contrast, high XCO_2 systems show carbon isotopic exchange at low fluid-to-rock ratios succeeded at higher fluid-to-rock ratios by a marked shift towards lighter oxygen signatures. In this case rock exchange pathways follow the type of curve labelled J_α in Fig. 1. A suite of exchange curves for differing mole fractions exists between these end-member cases.

To summarize, the extent to which temperature variability controls an isotopic exchange process for calcite can be determined by comparing the isotopic data set with the calculated temperature-dependent exchange pathway as plotted in $\delta^{13}C$–$\delta^{18}O$ space. In addition, the plotted isotopic data set yields information about the fluid composition with respect to the concentration of carbon-bearing species in solution. Since the amount of isotopic depletion of the exchange rock material is dependent on the volume of fluid available for exchange, a quantitative fluid-to-rock ratio can also be obtained by using isotopic mass-balance equations.

Quantitative analysis of fluid-to-rock ratios and calculation of the initial isotopic signature of the fluid phase

Bradbury & Rye (1986a) employed the following closed-system mass-balance equation for calcite-water exchange:

fluid-to-rock ratio

$$= n_o \frac{\delta^{18}O^f_{cc} - \delta^{18}O^i_{cc}}{\Delta^{18}O^f_{cc-H_2O} + \delta^{18}O^i_{H_2O} - \delta^{18}O^f_{cc}}$$

$$= [n_c/XCO_2] \frac{\delta^{13}C^f_{cc} - \delta^{13}C^i_{cc}}{\Delta^{13}C^f_{cc-CO_2} + \delta^{13}C^i_{CO_2} - \delta^{13}C^f_{cc}}$$

where the oxygen terms are as follows: n_o is the number of moles of oxygen per mole of calcite; $\delta^{18}O^f_{cc}$ and $\delta^{18}O^i_{cc}$ are the final and initial oxygen isotopic compositions of the rock calcite; $\Delta^{18}O^f_{cc-H_2O}$ is the oxygen isotopic fractionation between water and calcite; $\delta^{18}O^i_{H_2O}$ is the initial oxygen isotopic composition of the fluid.

A full discussion of the use of this expression is unnecessary in the present paper. What is relevant is the combined application of J-curve analysis and this expression. As argued more completely elsewhere (Bradbury & Rye 1986a), exchange profiles of the J_β type reach equilibrium for oxygen exchange beyond the inflection point where the carbon signature is shifted. This end-point for exchange is theoretically the final oxygen value for calcite in equilibrium with an infinite amount of fluid. This point thus represents the equilibrium balance between the initial fluid and initial rock isotopic compositions for the maximum degree of exchange at fixed temperature. In terms of the mass-balance equation, the end-point for oxygen exchange therefore equates with the sum of the fractionation factor and the initial fluid isotopic composition. For J_β-type profiles it is possible to solve the mass-balance equation (and thereby quantify fluid-to-rock ratios) irrespective of whether or not the temperature of exchange is known. If this temperature is determined, it is possible to back-calculate the initial isotopic composition of the fluid phase which exchanged with the rock. The latter application is of considerable importance for studying incremental shifts in fluid isotopic compositions from place to place in rock volumes which no longer contain these fluids.

Aquifer flow within thrust sheets

The geometric and kinematic development of the McConnell Thrust Sheet is typical of allocthons in active or formerly active foreland terrains. The palaeofluid circulation history of the McConnell provides an example of dynamic aquifer flow.

Physical system

The McConnell Thrust, which has a strike length of 410 km and a maximum displacement of 40 km, is a major thrust at the easternmost margin of the Front Ranges of the southern Canadian Cordillera. Movement on the McConnell, which is constrained to an 8 m.y. period commencing approximately 72 m.y. BP (Elliot 1976), transported thick-bedded Palaeozoic carbonate units progressively NE during net crustal convergence, A-subduction of the evolving foreland and subsidence of the Alberta Basin (Beaumont 1981).

Like other thrusts in the belt, the McConnell is listric in form, is asymptotic to a master décollement at depth and has a stair-step topology. A high-level ramp is exposed at Mount Yamnuska (Figs 2 and 3). Translation has juxtaposed middle Cambrian micritic limestones in the hanging wall against upper Cretaceous sandstones and shales in the foot wall.

Physical evidence of fluid–rock interactions is seen only within a limited volume of the hanging-

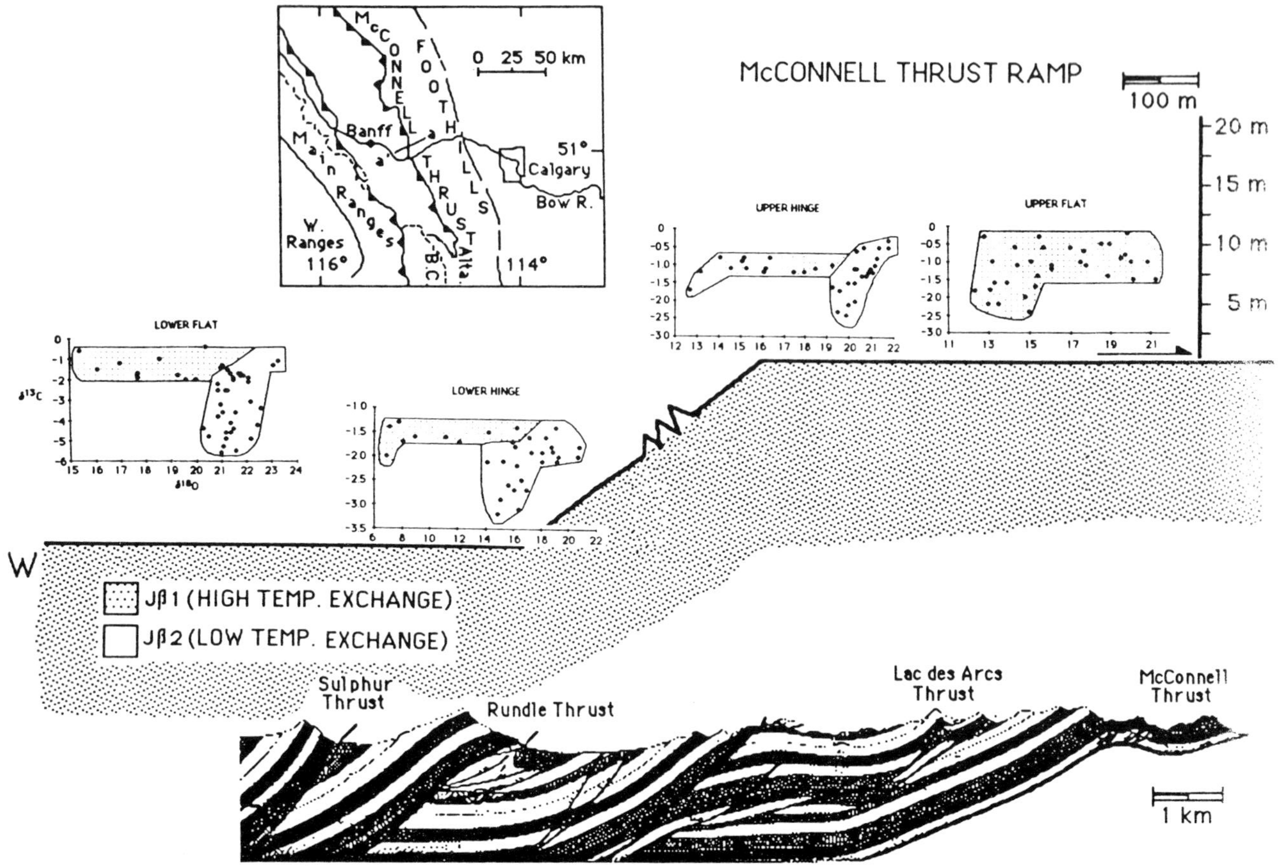

FIG. 2. Diagrammatic cross-section of the thrust ramp exposed at the eastern margin of the McConnell Thrust. Isotopic data sets for rock samples representing different parts of the ramp structure have been placed in their relative positions on the section. The cross-section represents transect aa′ on the sketch map above and displays the major thrust faults of the eastern main ranges.

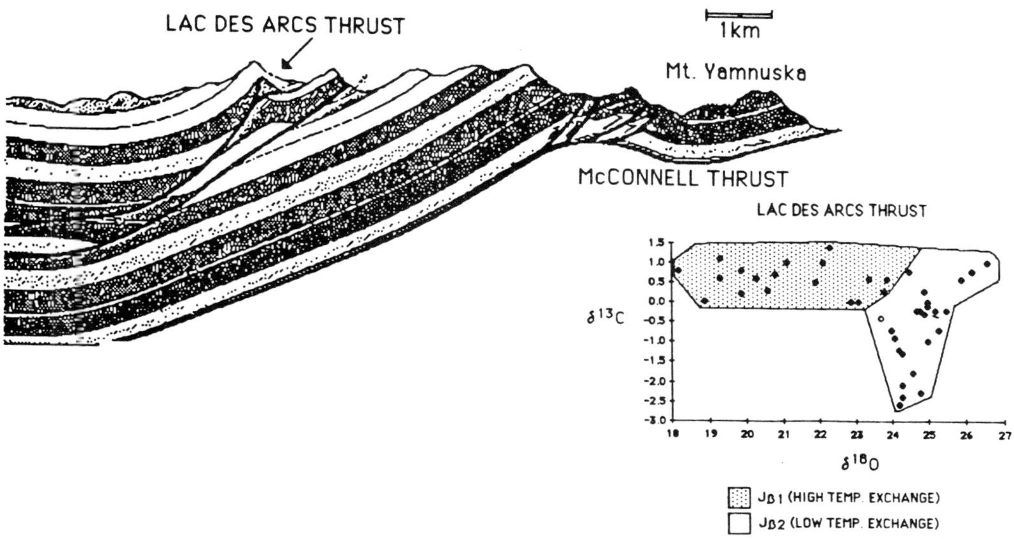

FIG. 3. Isotopic data set for the Lac des Arcs thrust. The approximate sample location relative to the ramp at Mount Yamnuska is shown by the arrow on the cross-section. A high-temperature–low-temperature exchange history has been preserved which is similar to results observed for the McConnell Thrust.

wall carbonates up to 15 m from the fault plane. Solution cleavage (sites of net calcite loss) occurs only sporadically at the erosional tip line but increases in density along the upper flat towards the upper hinge of the ramp. Cleavage is best developed between the upper and lower hinge of the ramp where it is associated with asymmetric minor folds and secondary thrusts. Solution cleavage is poorly developed within the lower flat region. Everywhere within the lowermost 10–15 m thickness of the hanging-wall carbonate sequence there is evidence of secondary precipitation within vein materials (sites of net calcite gain). The regions of highest concentration of veins are the ramp hinges. Veins occur in conjugate arrays at a high angle to bedding and are commonly offset in an *en échelon* manner. Cross-cutting relationships indicate two distinct periods of vein development. The first phase generated veins throughout the 15 m thick carbonate section. The second phase generated veins only within the ramp and lower hinge–flat regions.

Price (1967) and Jamison & Spang (1973) have provided evidence to show that vein formation was contemporaneous with translation of the McConnell Sheet. Study of calcite twin lamellae during palaeostress analysis (Jamison & Spang 1973) has demonstrated that the McConnell sole propagated in stair-step trajectory and the topology remains essentially unmodified by subsequent regenerative thrust imbrication during foot-wall collapse.

Aquifer system

There is a direct correspondence between the width of the zone in which veins occur and that in which fluid–rock interactions can be shown to have occurred on the basis of stable isotope geochemistry; a significant shift in the isotopic composition of the rock matrix occurred only within this volume. This physical system is interpreted as a bounded aquifer. The lower boundary was probably a mechanical and permeability boundary. The upper boundary is not defined by variations in any lithofacies characteristic. It is a structural and palaeothermal boundary (Woodwell & Bradbury, in preparation). Hydrofracturing, isotopic exchange and advective fluid flow are considered to have been coeval and probably mutually self-catalysing processes within the aquifer.

Following similar lines of isotopic argument to Bradbury & Rye (1986), the aquifer can be shown to have been an open system for fluid flow throughout its history. Thermal considerations (Woodwell 1985; Woodwell & Bradbury, in preparation) provide evidence for a minimum lateral distance of 2–6 km for the source region of fluids within the same section of the lithostratigraphy deeper in the thrust stack.

Multiple-pass flow system

Isotopic data for the McConnell Sheet is presented in Fig. 2. $\delta^{13}C$–$\delta^{18}O$ plots are positioned relative to sample locations. Data points are carbon–oxygen values for exchanged matrix calcite, calcite cement or vein calcite. Data populations define two separate J_β curves in the case of the lower flat, lower hinge and upper hinge, and only one J_β curve for the upper flat. J_{β_1} curves incorporate data from early-formed veins with consistent mean pressure-corrected fluid inclusion homogenization temperatures of 340°C. J_{β_2} curves incorporate data from late-formed veins with homogenization temperatures of 215°C.

Data from each J_β curve comprise an isothermal population. Structural and fluid inclusion studies allow the correlation of the J_{β_1} data sets. J_{β_2} data sets represent exchange during a separate, later and lower-temperature event. Within different parts of the aquifer, fluid–rock interactions responsible for the generation of J_{β_1} data sets can be considered to have been approximately isochronous exchange reactions with a low XCO_2 fluid under isothermal conditions. J_{β_2} data sets represent a separate and later low-temperature stage of flow and exchange with a water-rich fluid within a more limited volume of the same aquifer.

Complete separation of J_{β_1} and J_{β_2} data sets in the $\delta^{13}C$–$\delta^{18}O$ space representative of any part of the ramp flat suggests the likelihood of two distinct periods not only of isotopic exchange but also of fluid flow occurring at discrete stages during the thermal history of the sheet. Within the same aquifer, early and later batch-flow systems are considered to have been separated in both space and time.

Directionality of palaeofluid flow

The initial isotopic composition of the fluid phase which interacted to develop each of the data populations which make up J_β curves can be calculated separately for different positions within the thrust-sheet aquifer and for both J_{β_1} and J_{β_2} data sets. Figure 2 shows that, for either J_{β_1} or J_{β_2} sets, the range of oxygen signatures preserved is progressively shifted towards heavier values between the lower and upper hinges of the ramp. For J_{β_1}, this pattern continues into the upper flat. From the type of isotopic mass-balance analysis presented earlier, it is possible to back-calculate the initial isotopic composition of the fluid phase for any part of the aquifer. For the early high-temperature exchange the initial $\delta^{18}O$ composition of fluids which interacted at the lower ramp hinge, upper ramp hinge and upper flat were respectively +2.4 per mil, +8.9 per mil and +9.4 per mil. For the later lower-temperature exchange the initial $\delta^{18}O$ composition of fluids which interacted at the lower-ramp-hinge and upper-ramp-hinge volumes were respectively +5.1 per mil and +9.7 per mil. For both the fluid flux and the exchange stages the $\delta^{18}O$ composition of the fluid phase became progressively heavier towards the erosional tip line of the thrust. If, during translation, rock volumes had simply carried rock and fracture porosity fluid volumes and continued to exchange progressively, such a pattern of incrementally heavier fluid isotopic signatures would not have been generated. If, however, fluid had pulsed out of any given rock volume in the direction of net tectonic transport but at a rate faster than the transport of rock volumes within the allochthonous mass, fluids with an already relatively heavy $\delta^{18}O$ signature from exchange at the lower hinge of the ramp would be capable of exchange with pristine carbonate at the upper hinge to produce the observed shift in J_{β_1} curves towards heavier oxygen. Similarly, pulsing fluids from the upper hinge into the present-day upper flat would incrementally shift the position of the lightest oxygen signature in the J_{β_1} curves towards a heavier value compared with values further back in the sheet.

The above conclusions suggest that palaeofluid flow within the aquifer was directed from the hinterland towards the foreland and parallel to the direction of thrust-sheet translation. In this model the timing of high-temperature fluid–rock interactions from place to place would not have been strictly isochronous but systematically younger towards the thrust tip as a given fluid batch passed from hinterland to foreland. The same arguments would apply independently to J_{β_2} curves and low-temperature exchange.

Hydrodynamics

The $\delta^{18}O$ signature of the high-temperature fluid which entered the lower hinge region was calculated to have been +2.4 per mil. This value is consistent with its origin as an unmixed connate fluid. However, as noted, thermal modelling indicates that the lateral transport distance for this fluid within the aquifer system may have been 2–6 km. Formation water from the Alberta Basin (Hitchon & Friedman 1969) has connate $\delta^{18}O$ values as light as −10 per mil. Furthermore, the variation in connate fluid isotopic signatures from place to place within the Basin is similar in magnitude to that calculated for the palaeofluid regime of the McConnell Sheet over the same lateral distance. Within the McConnell, it seems probable that the pattern of incrementally lighter fluid signatures towards the hinterland is repeated down-section towards the fluid source region.

Therefore the value of +2.4 per mil calculated for the fluid isotopic composition may itself have been part of a series of incrementally heavier signatures along the entire length of the aquifer towards the foreland. If this is the case, it would indicate that a flow system had been in existence throughout the propagation–translation history of the McConnell Thrust, and that the evolution of the mechanical and the fluid flow systems had been intricately linked. In the 2 km lateral section examined, the changes in the initial fluid composition during high-temperature flow are entirely consistent with such a model. Flow was unidirectional within an aquifer structurally above the thrust plane. Comparable flow did not occur in the foot-wall sequence; the flow regime was asymmetric with respect to the décollement.

The mechanical and physicochemical roles of fluids in both rigid–plastic and viscous models of thrust-sheet translation are well known (e.g. Hubbert & Rubey 1959; Elliot 1976, 1980; Gretener 1977; Chapple 1978; Fyfe *et al.* 1978; Wiltschko 1979; Suppe 1981; Suppe *et al.* 1981). All models require localization of fluids within a high-strain zone (most commonly envisaged to be at the thrust surface itself). Methods of retention of fluids during translation, fluid access routes and the pumping mechanisms required for fluids to feed surfaces of motion are all interrelated problems which require further research.

From the analysis above, it is possible to provide a hydrodynamic model for the McConnel Thrust which has a bearing on these problems. Figure 4 is a two-stage model where case I represents the high-temperature flow field and case II represents the low-temperature flow field. The present-day upper flat (A) can be considered as the toe of the thrust sheet before propagation of the sole. By analogy, hydrofracturing at A can be considered to have occurred within the equivalent of a crack-tip microfracture network whereas that within the hinterland portion of the aquifer occurred within a volume akin to a crack process zone (Peck 1983) which incorporated earlier crack-tip volumes. It is known from the results of the stable isotope study that synkinematic fluid-rock interactions were not simply restricted to the McConnell sole thrust itself. Some model of progressive tip-line migration related to fluid flow would adequately explain the initial geometry of the aquifer and attendant flow system. Additionally, such a model would provide a mechanism for continued, although discontinuous, fluid–rock interactions within the aquifer throughout translation. This would be a dilatency pumping mechanism with repeated cycles of periodic or spasmodic stress build-up, hydrofracturing, fluid pressure drop and localized mass transfer coupled with pulsing of fluids towards the foreland. This would conform to models of stick–slip motion. The effective permeability increase provided by fracture porosity would allow continued migration of the hinterland portion of the aquifer volume. This would be a self-driving mechanism provided that fluids continued to gain access to the thrust tip.

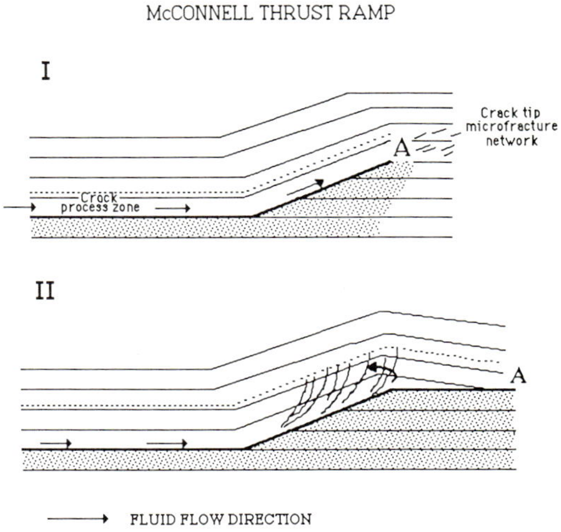

FIG. 4. Two-stage model of the development of the McConnell Thrust ramp. The dotted lines represent the limits of the palaeo-aquifer and the arrows indicate fluid flow direction. Case I represents the early high-temperature fluid flow at a time when the region which is now the upper flat of the ramp (A) was located at the thrust tip. Case II represents the later low-temperature fluid flow where A has migrated further E and has been isolated from further fluid exchange by the channellizing effect of the ramp and its cleavage network.

The high-temperature fluid flow aquifer system was laterally extensive. Its development would best fit the type of model given above. In contrast, the lower-temperature flow field was spatially restricted. The most plausible explanation for the occurrence of a single high-temperature J_β set within the upper flat would be that this rock volume had been physically separated from the region of more restricted low-temperature fluid flow at the time when this flow occurred. This would require localization of secondary fluid flux to the steep ramp wall and suggests that the present-day tip-line rock volume had passed beyond this region because of translation before low-temperature circulation had been established.

The restricted volume in which evidence for low-temperature fluid circulation is found corresponds directly to the localization of strain associated with solution cleavage development. A possible model to explain this correspondence would link low-temperature flow to the terminal stages of translation, drainage and eventual strain hardening of the ramp wall, locking of the McConnell sole and wholesale uplift with cooling of the McConnell Sheet at the start of Foothills imbrication. The cleavage imprint would have provided a channellized flow regime through which fluids could have been dissipated into the sheet, locally accommodating secondary thrusting in the upper hinge of the ramp.

The possibility of leakage from the primary aquifer, and especially back-flow from the ramp during the final stages of fluid flux, is suggested by isotopic data from the lower flat (Fig. 2). As implied in Fig. 2, carbonate was collected from some distance into the hanging wall away from the McConnell sole at the upper boundary of the aquifer system. The initial $\delta^{18}O$ fluid composition for low-temperature fluid flow in this region falls at the end of a sequence of progressively heavier values, implying a flow path line and direction from the lower hinge to the upper hinge followed by back-circulation into the sheet as shown in Fig. 4.

Flow within thrust belts and foreland basin margins

Deep circulation of fluids associated with continuous recharge is considered to be characteristic of thermally driven geodynamic systems such as batholithic complexes (e.g. Taylor 1977) and zones of thinning of continental (Wickham & Oxburgh 1985) and oceanic (e.g. Sleep & Wolery 1978) lithosphere. Such models may be of economic importance (e.g. Sawkins 1984). However, as yet we know little about reaction-induced incremental changes in isotopic compositions within the above settings and, whereas the first-order pattern of path lines can be successfully modelled (e.g. Norton & Cathless 1979; Garven & Freeze 1984), no natural data on pathway mapping has been published to support such models. Furthermore, it is uncertain whether or not the concept of large-scale continuous flow cells penetrating significant thicknesses of geological section is appropriate or generally applicable for other geodynamic settings, such as foreland margins, where various driving mechanisms are likely to cooperate during advective flow at different scales. Attention will be given to fluid flow within the Front Ranges of the Canadian Rockies and within the Southern Pyrenean Foreland Margin.

Intra-sheet and inter-sheet flow

The McConnell Thrust Sheet provides an example of bounded aquifer flow coeval with tectonic translation. Elsewhere within the hinterland portion of the same section of the Front Ranges thrust belt, identical carbonate lithofacies of the Cambrian Eldon and Pika formations are repeated in sheets of comparable size and with similar kinematic histories. The Lac des Arcs thrust sheet is structurally above the McConnell Thrust Sheet in the imbricate stack. The relative timing of movement on Front Ranges and Foothills thrust sheets is well established on the basis of stratigraphic considerations of hanging-wall and foot-wall sequences, structural superposition, patterns of migrating molassic depocentres and related data (e.g. Elliot 1976; Beaumont 1981; Price 1981). Movement on the Lac des Arcs commenced and ended before that on the McConnell.

A study comparable with that of the McConnell Sheet has shown that the Lac des Arcs similarly had a basal-sheet aquifer in which fluid–rock interactions were localized. Moreover, the Lac des Arcs preserves evidence on both high- and low-temperature exchange histories corresponding to early and late hydrofracturing respectively (Fig. 3). Similar studies in the hanging-wall–foot-wall sequences of other dominant thrusts such as the Rundle and Sulphur Mountain (Fig. 2) further demonstrate that basal-sheet open-system aquifer flow was common, although in these cases variations in temperature and/or fluid type in the system H_2O–CO_2–CH_4 have led to more complex isotopic data sets (Woodwell 1985).

A number of important facts emerge from the isotopic studies within both the Lac des Arcs and McConnell sheets. Firstly, the palaeoplumbing system of each tectonic sheet apparently evolved in a similar manner with long-range high-temperature fluid flux followed by a separate pulse of low-temperature fluid circulation more restricted in its extent. Each thrust sheet records essentially the same type of fluid–rock interaction history, but the flow systems for the two sheets were active at different times. The fact that both the Rundle and the Sulphur Mountain sheets similarly record separate fluid–rock interaction histories during separate and earlier periods of fluid flow confirms that intra-sheet flow predominated and that different fluid volumes migrated through different rock volumes in the imbricating sequence through time. There is a clear link between propagation of a thrust and active dewatering of the same thrust sheet. Each thrust sheet appears to have acted as an entirely separate hydrodynamic unit. No evidence indicates the former existence of deep fluid circulation cells across composite tectonic units. Fluid flow appears to have been unidirectional throughout. Fluid source regions were probably dewatered without replenishment by secondary recharge through overlying sheets. The consistently close association between dewatering and thrust activity suggests that rapid fluid expulsion due to hydrofracturing caused by overpressuring was probably related to the advancement of the toe of each individual thrust sheet. The fluid budget may have been a limiting mechanical factor for continued advance of the toe. The isotopic data profiles for the Lac des Arcs circulation history have been preserved; there is no evidence for masking of the effects of earlier flow systems by later flow systems during imbrication. This implies that little or no fluid communication existed between adjacent thrust sheets during movement on a lower décollement. This would be consistent with restriction of flow to newly activated aquifers tectonically deeper than the passively transported upper sheets at each stage without significant leakage of the fluid fraction back towards the hinterland. Locked formerly active décollements were apparently sealed and isolated from later flow systems.

Fluid flow within a foreland basin margin

Modern geophysical and geological studies in the Pyrenees (e.g. Cámara, personal communication) (Fig. 5) have demonstrated the importance of coupling between basement and cover during Alpine uplift of the reactivated Hercynide metamorphic core and substance of its southern foreland basins. Formerly deep-buried crystalline complexes were exhumed as basement-cored duplexes prograded generally southwards carrying separate basins which had initiated at the duplex leading edge (Fig. 5). The Jaca Basin (Fig. 5) was delimited in the S by the External Sierras Belt, which was an active emergent thrust front in the Upper Eocene (Priabonian) with only local activity continuing after its burial by the Lower Miocene (Aquitanian) molasse. The northern margin of the Jaca Basin was finally exhumed in the late Oligocene (Chattian) (e.g. Stroock & Bradbury 1984) during rapid dewatering of the basin-wall carbonates. Consideration will be given to the fluid generation and migration history of the northern margin of the Jaca Basin.

Flow system of crystalline basement

Beneath the northern margin of the Jaca Basin the basement is segmented with crystalline lenses bounded by shear zones which are discretely spaced (Fig. 5). The entire infrastructure to the basin carbonates is a crystalline duplex with a floor thrust at the level of the Permo–Triassic and a roof close to the base of the Cretaceous carbonates in the section under consideration. As in the case of cover thrusts in the Rockies, basement shear zones first initiated furthest from the foreland and were sequentially generated closer to the foreland through time. Total displacement diminished from approximately 40 km on early higher thrusts to 10 km on lower later thrusts within the duplex. Shear zones are curviplanar between roof and floor (Fig. 5). This form is more exaggerated in the case of earlier zones because of continued foot-wall modification.

Duplex formation spanned much of the Upper Cretaceous to Palaeogene history of sedimentation in the Jaca Basin (e.g. Labaume & Seguret 1984). Because of this, any contribution made by fluid–rock interactions within or beneath the crystalline complex is of importance to the fluid budget of the evolving basin margin.

Research on the shear-zone products derived from a granodiorite precursor (e.g. Losh & Bradbury 1984; Losh 1985) has allowed characterization of the fluid flow systems of the crystalline duplex. In brief, the following conclusions can be drawn. Uplift and excavation of shear zones was accompanied by reworking and heterogeneous incorporation of different deformation products as rocks crossed the brittle–ductile transition on their way to the surface. Products of both ductile and brittle regimes are nowadays closely intercalated but represent tectonic juxtaposition of rocks from widely varying crustal levels by a continued process of self-cannibalization of the original granodiorite (*cf.* Sibson 1977). Material balance in retrograde assemblages demonstrates a net silica loss from ductile products. Silica-flooded zones are exposed in shear zones above the quartz brittle–ductile transition. Minimum volumetric fluid-to-rock ratios of 30:1 can be calculated from silica production and solubility. A more detailed consideration of the fluid budget of shear zones comes from study of stable isotopic data and fluid inclusions within mylonite and cataclasite series rocks (*sensu* Sibson 1977) (Losh 1985). It can be established from preserved $\delta^{18}O$ compositions of mineral phases and from microthermometry that all ductile products exchanged isotopically with a metamorphic fluid with an initial fluid $\delta^{18}O$ of approximately +10 per mil. Despite its apparently low permeability, the anisotropic ductile fabric allowed throughgoing fluid flow from some external reservoir. Sequential restacking of the duplex by forward balancing coupled with

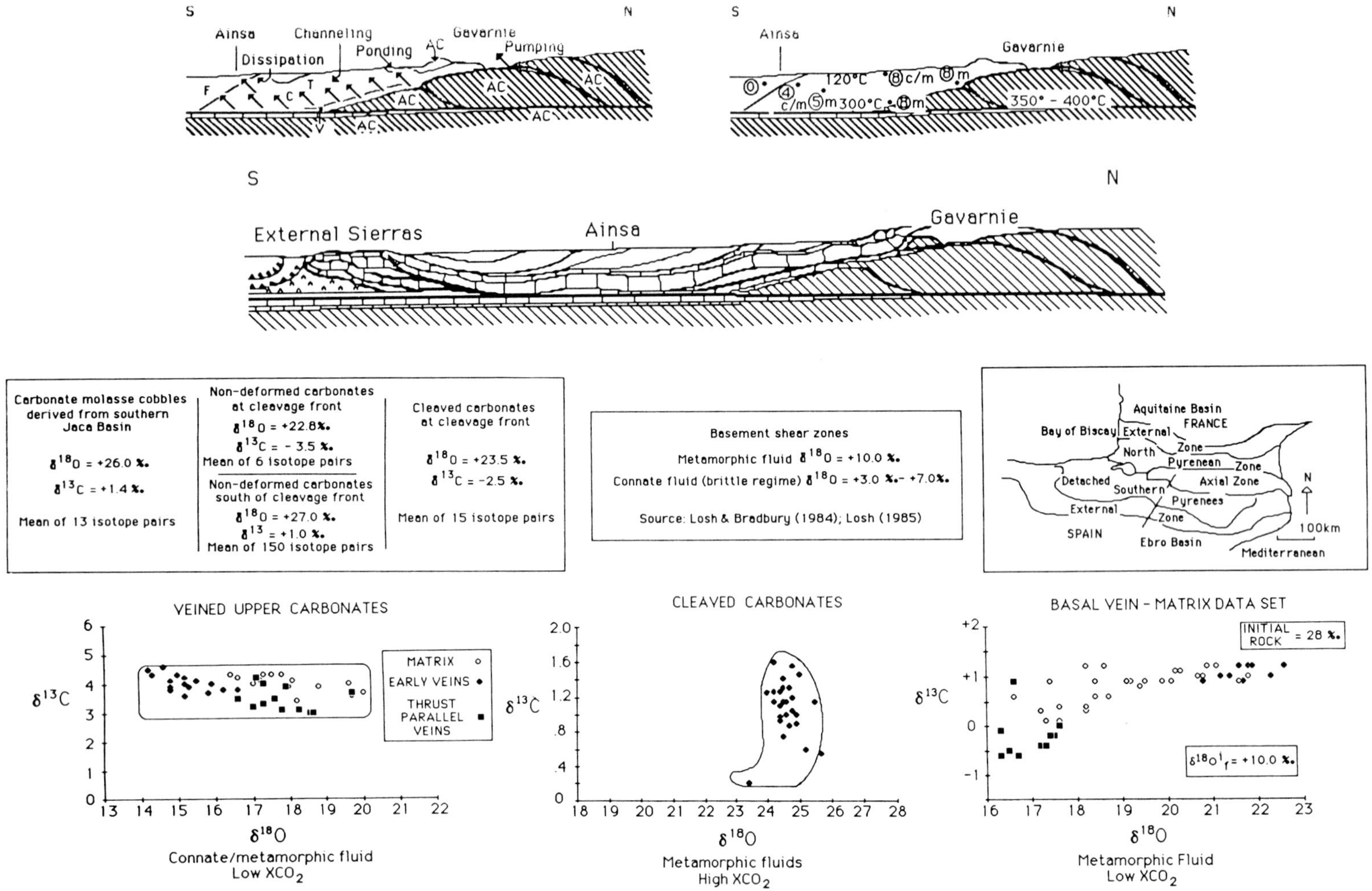

FIG. 5. Sketch map of the Pyrenees: the NE–SW transect is represented by the cross-section above. Temperatures in the upper right cross-section indicate temperatures of exchange determined by mineral compositional and fluid inclusion data. Numbers in circles represent the degree of oxygen isotopic depletion (per mil) from the respective pristine rock value. c indicates connate fluid, m indicates metamorphic fluid and c/m indicates a mixed connate and metamorphic fluid (the fluids were characterized using J-curve analysis). The upper left cross-section displays large-scale fluid circulation. AC is the aquiclude; the arrows represent fluid migration pathways within the cleavage domain. C is the domain of spaced cleavage which exists everywhere between the cleavage front (F) and the basement (shaded). T is the volume of carbonate in which thrust surface parallel veins exist (shown as a lens at the top of the section). The region marked as 'Ponding' similarly contains thrust surface parallel veins. V is the carapace of net-veined carbonate above the basement. Isotope plots: 'veined upper carbonates' from T; 'cleaved carbonates' from C; 'basal vein–matrix data set' from V. Palaeozoic basement exposed at Gavarnie: Cretaceous and Tertiary (chiefly carbonate) fill shown beneath Ainsa; Eocene–Miocene sediments exposed at or close to External Sierras.

modelled reaction stoichiometries and the transient generation–migration of reaction surfaces have allowed the fluid source region and potential volume of the external fluid reservoir to be identified (Losh 1985). The most likely source for these fluids is the overplated metasedimentary section immediately beneath the duplex.

It has been established from detailed stable isotopic analysis of all appropriate mineral phases inside and outside shear zones (see Losh 1985) that significant fluid–rock interactions were restricted to the shear zones themselves. The shear zones were effectively screened from the surrounding granodiorite during fluid flow. This implies that they acted as conduits for release of metamorphic volatiles upwards in the crystalline mass. From this data base it can be concluded that there is no evidence for fluid flow within crystalline volumes between shear zones. This appears to be consistent with the results of the independent programme of work in the Canadian Rockies reported earlier.

In contrast with the ductile regime, isotopic and geological data indicate that conduit flow within the brittle regime of the same shear zones accessed both ascending metamorphic fluids from depth and descending connate fluids from the overlying cover carbonates. Losh (1985) has related this to seismic pumping. The zone of mixing of these fluid batches broadly corresponds to the quartz brittle–ductile transition.

From these studies, it seems clear that changes in the flow pathway within the crystalline duplex would be governed by geometrical and rheological changes in each shear zone and by the establishment of new movement zones in both space and time. During imbrication the combination of a decrease in displacement resulting in less overplating and an increase in the maximum thickness of the crystalline mass resulting in early devolatilization probably led to a reduction in the total fluid volume generated over the history of duplex development, but this has not yet been established.

Dewatering of the sedimentary pile

Isotopic data from both the upper part of the crystalline duplex and the lower part of the carbonate pile indicates that metamorphic fluids and subsidiary volumes of connate fluids were expelled from the shear-zone conduits to gain access to the cover carbonates. From the known imbrication history of the basement, it seems probable that older shear zones acted as feeders to the sedimentary pile at a time when sedimentation persisted relatively close to the basement–cover interface. In contrast, fluid expulsion from younger basement shear zones occurred within carbonates which must already have partially dewatered and reached greater depths of burial.

Bradbury & Rye (1981, 1982, 1986a, b) have described in detail the flow system of the chiefly carbonate sedimentary sequence. Only certain general aspects will be presented in this paper. Ahead of master shear zones in the crystalline complex, where they emerge to become thrust faults in the cover, the oxygen isotopic composition of marine carbonates is markedly shifted by as much as 8 per mil from known values for rocks of the same stratigraphic interval further S or elsewhere in the northern Jaca Basin. Bradbury & Rye (1986a, b) have shown that isotopic exchange occurred during hydrofracturing and the preferential development of vein networks in these regions. Furthermore, it can be established from isotopic mass-balance calculations that this system was open to metamorphic fluids with an initial fluid $\delta^{18}O$ composition of $+10$ per mil (calculated from J-curve analysis). This is consistent with the values obtained independently from studies of shear-zone products.

The localization of sites of significant isotopic shift, the close correspondence between the values calculated for fluid composition and the fact that there is a direct physical link between the anisotropy of the shear-zone fabrics in the basement and the channellized flow network provided by a stylolite–cleavage system within the lower part of the carbonate pile suggest that fluids were preferentially focused within the shear zones and expelled upwards to produce exchange haloes within the carbonate immediately ahead of the shear-zone tip lines in the cover sequence.

Towards the preserved leading edge of the crystalline duplex (Fig. 5) the temperature of exchange was approximately 300°C, on the basis of mineral compositional and fluid inclusion data. Within this section and in the equivalent structural position along the length of the Jaca Basin for approximately 30 km, net-veined carbonates form a carapace which completely envelops the crystalline duplex at the lowest section in the cover sequence. A periodicity of fluid infiltration, hydrofracturing and dewatering can be recognized in some sections and is probably related to fluid pulsing from the basement (Bradbury & Rye 1986b). The entire vein network at this low level was generated during advective fluid flow out of the basement coeval with uplift. It developed at the interface between two channellized flow regimes of differing scales. Shear zones in the infrastructure are widely spaced. In the suprastructure the channellized flow regime formed by penetrative cleavage imprint is dramatically more closely spaced. Bradbury & Rye (1986b) have shown that regionally significant cleavage in the middle and upper sections of the exposed carbonate pile predominates over other features; the hydrofractured and net-veined volume is commonly restricted to the lowest 50 m of the cover sequence.

Isotopic data from different parts of the spaced cleavages and from place to place in the carbonates (e.g. Bradbury & Rye 1982) demonstrates that the anastomosing network of cleavage surfaces acted as channelways for fluids and as sites of isotopic exchange. In contrast, the lithons bounded by cleavage surfaces did not allow access of or exchange with fluids during or following cleavage development. Most isotopic data for cleavage domains fall on J_α curves, demonstrating exchange with a high-XCO_2 fluid. This would be consistent with escape of fluids upwards within the cleavage system from the underlying vein network, with modification of the fluid composition as the originally water-rich volume exchanged with more carbonate rock and mixed with more connate fluid. Mixing fronts would be expected to exist within the carbonate pile.

Vertical and lateral fluid fronts

Two types of front exist both vertically and laterally within the northern Jaca Basin margin. These are physical and geochemical fronts. Two types of physical front exist. The first are vertical and lateral boundaries limiting a volume of rock in which pervasive cleavage exists from those in which it does not (Fig. 5). Cleavage is extensively developed in the northern part of the basin wall. Vertically, this is terminated against early thrust surfaces. These are considered to be natural strain fronts. Laterally, the same cleavage domain abruptly terminates at a mappable three-dimensional surface variously inclined from approximately 5° to 20° to the horizontal in a southerly direction and inclined at a high angle to the cleavage, which commonly dips N (Fig. 5). This surface, or lateral cleavage front, is considered to have developed either as a strain field above a persistently buried blind thrust tip line or, more probably, as a larger strain field associated with uplift of the entire crystalline duplex at a relatively late stage (Bradbury & Rye 1986b).

Both the vertical and the lateral physical fronts associated with the transition from cleaved to non-cleaved rock represent changes from a regime in which fluid flow was demonstrably channellized to ones in which it was probably by porous flow. The most important change in rock composition occurs close to the lateral cleavage front. Rocks immediately adjacent to the front on both sides have average matrix oxygen isotopic compositions which are similar. However, within 100 m of the cleavage front, non-cleaved rocks to the S are unexchanged and have markedly different oxygen compositions (Fig. 5). This suggests that this physical front from channellized to porous flow regimes had some influence on exchange patterns. It seems probable that dissipation of an already depleted fluid budget across the cleavage front allowed exchange within a limited volume outside the cleaved domain.

The second type of physical front is related to the presence of thrust parallel veins. These principally exist towards the upper and frontal end of the thrust complex N of the lateral cleavage front (Fig. 5). Hoffman *et al.* (1983) and Bradbury *et al.* (in preparation) have argued that these vein types form under conditions of high fluid pressures. Fluid inclusion microthermometry suggests temperatures of vein formation of approximately 120°C compared with 300°C for the vein system close to the basement. The stable isotopic characteristics of vein materials are consistent with a fluid origin as a mixed connate–metamorphic fluid (Hoffman 1983). A zone of excess fluid pressures at the frontal end of the thrust stack may have been generated by flushing and mixing of connate and metamorphic fluid batches within the carbonate sequence. Ghost geochemical fronts exist within the carbonate pile. Specifically, calculated initial oxygen isotopic compositions of fluids at a low level in the sequence are considered to be metamorphic in origin, whereas at a high level, and laterally, transitions exist to values more consistent with connate fluids. Unlike the incremental shifts in initial fluid compositions determined for the thrust sheets of the Canadian Rockies, fluid volumes at a low level within the Jaca Basin margin were isotopically heavier for oxygen during exchange than those at a high level within the same sequence and in the direction of fluid flux. The differences between heavier and lighter pools is as great as 8 per mil. The differences in fluid isotopic compositions are the inverse of what would be expected had fluids from a relatively deep source in the sequence progressively exchanged with pristine carbonate during its escape upwards and laterally within the basin margin (*cf.* McConnell Thrust System). This is taken to indicate that fluids within lower and upper volumes had different primary origins—the former of metamorphic origin, and the latter a connate fluid. Connate fluids may have already escaped from the lower carbonate section before large volume infiltration by metamorphic fluids. Thrust-surface parallel veins appear to be concentrated within the transition from one fluid system to the other in locations where there were obvious physical barriers to escape of fluids from the tectonosedimentary pile, such as immediately beneath the non-cleaved upper-level thrust sheets. Isotopic data confirm that carbonates above the vertical cleavage front had not exchanged with fluid in communication with those within the cleavage domain. If metamorphic fluid had pulsed out of the leading edge of the crystalline duplex relatively late in the dewatering history, it could have gained access to higher levels in the pile without significant mixing with connate fluid, since at least partial dewatering would probably have occurred. Ponding of connate fluids and the augmentation of the total fluid budget by ascending metamorphic fluids would be likely to lead to excess fluid pressures in volumes in which fluids had no easy access routes for escape (e.g. where capped by an impermeable upper thrust sheet) or at the transition from one fluid type to the next. It is suggested, therefore, that the regime of thrust parallel veins represents both a geochemical boundary between fluids of differing primary origins and a mechanical boundary of excess fluid pressures.

Discussion and conclusions

Fluid flow and tectonics

Clear relationships exist between the displacement history of supracrustal thrusts and dewatering of their thrust sheets. However, pre-existing thrust sheets which were passively carried were isolated from this dewatering. Equally clear relationships exist between both ductile and brittle deformation and the release and passage of fluids through shear zones which were active at that time. Fluid flow appears to have been continued throughout the entire history of thrust translation and throughout the entire history of formation, reworking and excavation of ductile and brittle shear-zone products. Fluid migration and active tectonics are probably mutually self-catalysing processes in both suprastructural and infrastructural regimes.

Fluid origins

There is a close correspondence between the oxygen isotopic compositions calculated for fluids which escaped preferentially from shear zones within the crystalline basement of the Southern Pyrenees and that calculated for fluids which exchanged with carbonates immediately ahead of these zones. Ascending fluids ultimately migrated upwards within the sedimentary pile, but are considered to have originated as a result of metamorphic devolatilization reactions at depth. The probable source region for these fluids is the over-plated metasedimentary section beneath the crystalline duplex. Whereas the oxygen isotopic composition of fluids at their point of access to the cover is a metamorphic signature, initial fluid $\delta^{18}O$ values at a high level in the same sedimentary sequence are characteristically connate.

Thermal modelling (Woodwell 1985; Woodwell & Bradbury, in preparation) has shown that it would be possible for fluids which accessed cover thrusts within the Rockies to have been generated and to have migrated entirely within the supracrustal pile. In this case patterns of changes in fluid isotopic compositions from place to place within individual thrust sheets are entirely consistent with progressive modification of a primary connate fluid volume.

Apparently, where crystalline basement was actively being deformed (e.g. Southern Pyrenees) during dewatering, a contribution was made to the fluid budget by this basement. In contrast, where crystalline basement was not being actively deformed (e.g. Canadian Overthrust Belt) during dewatering, no such contribution was made.

Fluid histories

In the Canadian Overthrust Belt both the McConnell and the Lac des Arcs thrust sheets underwent large-scale high-temperature basal aquifer flow during active thrust-sheet migration and small-scale low-temperature basal aquifer flow during uplift in the terminal stages of sheet migration. In the Pyrenean Foreland, where aquifer flow was confined to the crystalline basement (with wholesale channellized flow within the cover), separate high- and low-temperature flow systems cannot be identified and basement involvement has led to a more complicated semicontinuous history of displacement and uplift.

Flow paths

Fluid passage through the McConnell Thrust Sheet was unidirectional in the tectonic transport direction at the time of displacement. Changes in the initial oxygen isotopic composition of fluids which interacted with rocks within different parts of the basal-sheet aquifer indicate that a given batch of connate fluid incrementally modified its oxygen signature as fluids pulsed along the length of the aquifer over considerable distances. In contrast, no such sequential change in the oxygen isotopic composition of fluids appears to have occurred as fluids passed upwards within the carbonate pile of the northern margin of the Jaca Basin. The fact that connate signatures are recognized in carbonate volumes above sequences through which metamorphic fluids have passed suggests that these connate signatures are primary and that the geochemical boundary between upper and lower regimes is a mixing front and not one across which fluids have passed and undergone exchange in their transport path.

Fluid fronts

The only type of fluid front which is recognizable in the thrust sheets of the Rockies is defined by the boundaries of the basal-sheet aquifers. However, in the Pyrenean Foreland basin margin, both lateral and vertical fronts exist between fluids of different origin. The lower and more northerly section of the carbonate pile has evidence of metamorphic fluids. The upper and more southerly section has evidence of connate fluids. The topology of the front between these fluid types may generally mimic that of the basement–cover interface.

Fluid communication

In crystalline basement beneath the Southern Pyrenees shear zones preferentially acted as conduits for fluid flow. Intervening segments of the basement were aquicludes (Fig. 6). At the basement–cover interface the scale at which channellized flow occurred varied significantly from that at separated shear zones to that at separated cleavage surfaces (see Bradbury & Rye 1986a, b).

Bradbury & Rye (1986a) have shown that flow at the scale of cleavage surfaces occurred by preferential migration of fluids within individual stylolite or spaced cleavage surfaces with screening of this fluid from that within the immediately surrounding matrix and from separated fluid volumes in adjacent cleavage–stylolite surfaces. On this scale, communication only occurred as fluid entered or left such surfaces.

On a larger scale, fluids within each basal-sheet aquifer within the Canadian Overthrust Belt were isolated from any resident connate fluids within the pore spaces of carbonates above each aquifer. At least from an isotopic viewpoint, fluids within different parts of the same aquifer at the same time were not in communication.

Scales of fluid transport

In the case of the Southern Pyrenees, the minimum transport distance for metamorphic fluids was 5 km. The transport distance for connate fluids in the Canadian Overthrust Belt is considered to have been between 2 and 6 km.

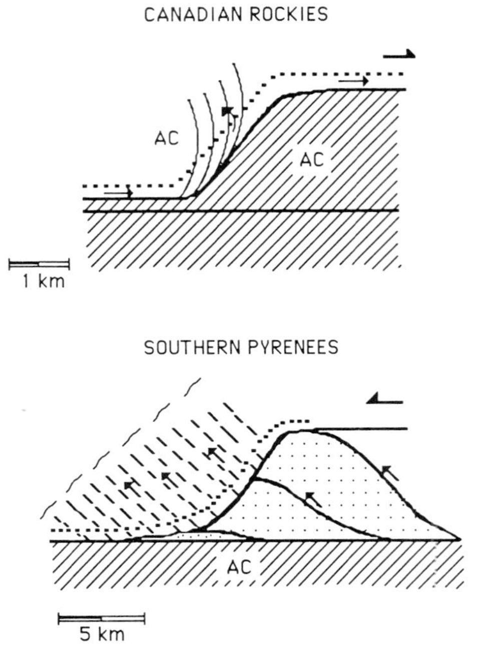

FIG. 6. Cartoon comparison of the flow histories of the southern Canadian Rockies and the southern Pyrenees. The flow regime in the Rockies is restricted to a basal aquifer which was active during thrust-sheet migration and involved two separate fluid pulses. In the Pyrenean Foreland the crystalline basement served as an aquifer (via shear zones) while channellized flow occurred throughout the strained section of the cover. Note the similar gross geometry, opposed translation directions, opposed attitudes of cleavage (full lines, Rockies; broken lines, Southern Pyrenees) and difference in scale.

Fluid volumes and flow rates

These subjects cannot be treated adequately here since proper consideration cannot be given to thermal models for the geological situations addressed. However, the results of some of these studies have been presented elsewhere (e.g. Losh 1985; Woodwell 1985; Bradbury & Rye 1986b). Minimum fluid-to-rock ratios for isotopic exchange of the rock matrix at the base of the Jaca Basin carbonate sequence were of the order of 1.5:1 (Bradbury & Rye 1986a). Minimum values of 30:1 were calculated for basement shear zones. These differences probably reflect the change in channellized systems from fluid focusing within basement shear zones to flow within a greater volume in the cover cleavage network. Losh (1985) has calculated that the time-integrated metamorphic fluid volume for the entire length of an individual basement shear zone is of the order of 2×10^9–4×10^9 m^3. This is approximately 15–25% of the fluid volume available within the foot-wall sequence from devolatilization reactions. For the case of the McConnell Thrust Sheet, planar crack models by Woodwell & Bradbury (in preparation) suggest that the possible range of mean fluid flow rates within the high-temperature aquifer falls within the range for phreatic land springs (e.g. Lowell 1975), caldera hydrothermal systems (e.g. Sorey *et al.* 1978) and oceanic-vent discharge zones (e.g. Sleep & Wolery 1978) over a wide variety of residence times which compare well with those known from tritium and ^{14}C studies of hot-spring flow systems (e.g. Bedinger *et al.* 1979).

ACKNOWLEDGMENTS: Research was carried out at Yale University from 1981 to 1985. We gratefully acknowledge financial support from the National Science Foundation (NSF-EAR-8200524), Arco/Anaconda and Yale. We thank Drew Hoffman, Steven Losh and Dan Rye for discussion.

References

BEAUMONT, C. 1981. Foreland basins. *Geophys. J.R. astr Soc.* **65**, 291–329.

BEDINGER, M. S., PEARSON, F. J., REED, J. E., SNIEGOCKI, R. T. & STONE, C. G. 1979. The waters of Hot Springs National Park, Arkansas—their nature and origin. *U.S. geol. Surv. prof. Pap. 1044-C*, pp. C1–C33.

BOTTINGA, Y. 1968. Calculation of fractionation factors for carbon and oxygen exchange in the system calcite–carbon dioxide–water. *J. Phys. Chem.* **72**, 800–8.

BRADBURY, H. J. & RYE, D. M. 1981. Fluid-rock interactions in shear zone vein arrays from the Spanish Pyrenees. *British Tectonic Studies Group Abstracts, Oxford.*

—— & —— 1982. Cleavage formation within Pyrenean thrust sheets: evidence from structural and stable isotopic studies. *Geol. Soc. Am. Abstracts with Program*, Vol. 14, p. 450.

—— & —— 1986a. Dewatering of Pyrenean thrust sheets I. Stable isotope geochemistry of vein arrays. *Am. J. Sci.*, in press.

—— & —— 1986b. Dewatering of Pyrenean thrust sheets II. Large scale plumbing system. *Am. J. Sci.*, in press.

——, —— & HOFFMAN, D. Dynamic dewatering of thrust surfaces, in preparation.

CHAI, B. H. T. 1975. The kinetics and mass transfer of calcite during hydrothermal recrystallization process. Ph.D. Thesis, Yale University.

CHAPPLE, W. M. 1978. Mechanics of thin-skinned fold and thrust belts. *Geol. Soc. Am. Bull.* **89**, 1189–98.

ELLIOTT, D. 1976. The energy balance and deformation mechanisms of thrust sheets. *Phil. Trans. R. Soc. Lond. Ser. A*, **283**, 289–312.

—— 1980. Mechanics of thin-skinned fold and thrust belts Discussion. *Geol. Soc. Am. Bull.* **91**, 185–7.

FRIEDMAN, I & O'NEIL, J. R. 1977. Compilation of stable isotope fractionation factors of geochemical interest. *In:* FLEISCHER, M. (ed.) *Data of Geochemistry*, 6th edn, *U.S. geol. Surv. prof. Pap. 440-kk.*

FYFE, W. S., PRICE, N. J. & THOMPSON, A. B. 1978. *Fluids in the Earth's Crust*, Elsevier, Amsterdam.

GARVEN, G. & FREEZE, R. 1984. Theoretical analysis of the role of groundwater flow in the genesis of stratabound ore deposits. *Am. J. Sci.* **284**, 1085–174.

GRETENER, P. E. 1977. Pore pressure: fundamentals, general ramifications, and implications for structural geology (revised). *Education Course Note Series No. 4*, American Association of Petroleum Geology.

HITCHON, B. & FRIEDMAN, I. 1969. Geochemistry and origin of formation waters in the western Canada sedimentary basin—I. Stable isotopes of hydrogen and oxygen. *Geochim. cosmochim. Acta*, **33**, 1321–49.

HOFFMAN, D. 1983. Fluid movement through a small scale thrust zone. M.S. Thesis, Yale University.

——, & BRADBURY, H. J. & RYE, D. M. 1983. Dynamic dewatering of Pyrenean Thrust Sheets. *Geol. Soc. Am. Abstracts with Program*, Vol. 15, p. 596.

HUBBERT, M. K. & RUBEY, W. W. 1959. Role of fluid pressure in mechanics of overthrust faulting. I. Mechanics of fluid-filled porous solids and its application to overthrust faulting. *Geol. Soc. Am. Bull.* **70**, 115–66.

JAMISON, W. R. & SPANG, J. H. 1973. Dynamic analysis of the McConnell Thrust Plate, *EOS*, **54**, 146 (abstract).

LABAUME & SEGURET 1984. The southern onlap of the Eocene 'MT3' megaturbidite and the thrust–fold system in the Eocene turbiditic series and its shelfal Maestrichtian–Paleocene 'basement'. *Thrusting and Deformation; Field Guide*, Paul Sabatier Univ., Toulouse.

LISTER, C. R. B. 1972. On the thermal balance of a mid-ocean ridge. *Geophys. J. R. astr. Soc.* **26**, 515–35.

LOBATO, L. M., FORMAN, J. M. A., FYFE, W. S., KERRICH, R. & BARNETT, R. L. 1983. Uranium enrichment in Archaean crustal basement associated with overthrusting. *Nature, Lond.* **303**, 235–7.

LOSH, S. 1985. Fluid migration and interaction in brittle–ductile shear zones, Central Pyrenees, France. Ph.D. Thesis, Yale University.

—— & BRADBURY, H. J. 1984. Monitoring fluid-rock interactions across a mid-crustal brittle–ductile transition: possible implications for fluid circulation within a cover thrust belt. *Deformation and Thrusting Abstracts*, Paul Sabatier Univ., Toulouse.

LOWELL, R. P. 1975. Circulation in fractures, hot springs and convective heat transport on mid-ocean ridge crests. *Geophys. J. R. astr. Soc.* **40**, 351.

McCREA, J. 1950. The isotopic chemistry of carbonates and a paleotemperature scale. *J. chem. Phys.* **18**, 849.

NORTON, D. & CATHLESS, L. M. 1979. Thermal aspects of ore deposition. *In:* BARNES, H. L. (ed.) *Geochemistry of Hydrothermal Ore Deposits*, pp. 611–632, Wiley, New York.

—— & KNAPP, R. 1977. Transport phenomena in hydrothermal systems: the nature of porosity. *Am. Sci.* **277**, 913–36.

—— & KNIGHT, J. 1977. Transport phenomena in hydrothermal systems: cooling plutons. *Am. J. Sci.* **277**, 937–81.

O'NEAL, J. R. 1963. Oxygen isotope fractionation studies in mineral systems. Ph.D. Thesis, University of Chicago.

PECK, L. M. 1983. Stress corrosion cracking in Sioux Quartzite. Ph.D. Thesis, Yale University.

PRICE, R. A. 1967. The tectonic significance of mesoscopic subfabrics in the Southern Rocky Mountains of Alberta and British Columbia. *Can. J. earth Sci.* **4**, 39–70.

—— 1981. The Cordilleran foreland thrust and fold belt in the southern Canadian Rocky Mountains. *In: Thrust and Nappe Tectonics*, The Geological Society of London.

ROSENBLAT, S. 1985. A mathematical model of compaction induced migration. *J. geophys. Res.*, **90**, (B1), 779–89.

SAWKINS, F. J. 1984. Ore genesis by episodic dewatering of sedimentary basins; applications to giant Proterozoic lead–zinc deposits. *Geology*, **12**, 451–4.

SIBSON, R. 1977. Fault rocks and fault mechanisms. *J. geol. Soc. Lond.* **133**, 191–214.

SLEEP, N. & WOLERY, T. J. 1978. Egress of hot water from mid-ocean ridge hydrothermal systems: some thermal constraints. *J. geophys. Res.* **83**, 5913–22.

SOREY, M. L., LEWIS, R. E. & OLMSTED, F. H. 1978. The hydrothermal system of Long Valley Caldera, California. *U.S. geol. Surv. prof. Pap. 1044–A*, pp. A1–A60.

STROOCK, E. & BRADBURY, H. J. 1984. Transverse geological structures in the External Sierras, Southern Pyrenees. *Eos California AGU.*

SUPPE, J. 1981. Mechanics of mountain building and metamorphism in Taiwan. *Geol. Soc. China Mem.* **4**, 67–89.

——, NAMSON, J. & PYTTE, A. M. 1981. Role of fluid pressure in an active overthrust belt: Taiwan. *Geol. Soc. Am. Abstracts. with Program*, Vol. 13, p. 562.

TAYLOR, H. P., Jr. 1977. Water/rock interactions and the origin of water in granitic plutons. *J. geol. Soc. Lond.* **133**, 509–58.

TISSOT, B. P. & WELTE, D. H. 1980. *Petroleum Formation and Occurrence*, 2nd edn, Springer-Verlag, New York.

WICKHAM, S. M. & OXBURGH, E. R. 1985. Continental rifts as a setting for regional metamorphism. *Nature, Lond.* **318**, 330–3.

WILTSCHKO, D. V. 1979. Partitioning of energy in a thrust sheet and implications concerning driving forces. *J. Geophys. Res.* **84**, 6050–8.

WOODBURY, A. D. & SMITH, L. 1985. On the thermal effects of three dimensional groundwater flow. *J. Geophys. Res.* **90**, 759–67.

WOODWELL, G. R. 1985. Fluid migration in an overthrust sequence of the Canadian Cordillera. Ph.D. Thesis, Yale University.

—— & BRADBURY, H. J. Fluid flow within thrust sheets, in preparation.

HARRY J. BRADBURY & GRANT R. WOODWELL,* Department of Geology, Birkbeck College, University of London, 7–15 Gresse Street, London W1P 1PA, U.K.
*Present address: Mary Washington College, Fredricksburg, Virginia, 22401, U.S.A.

Section 3
Fluid Flow in United Kingdom Groundwater Basins

Regional groundwater flow in sedimentary basins in the U.K.

R. A. Downing, W. M. Edmunds & I. N. Gale

SUMMARY: The U.K. contains seven groundwater provinces. Regional groundwater flow systems occur in these provinces above an essentially impermeable basement of varying age. The pattern of flow in the U.K. is strongly influenced by the distribution of the Permo–Triassic sandstones. Intermediate flow systems are encouraged by the scarp and vale topography that has developed as a consequence of the alternating sequence of aquifers and aquicludes in the Mesozoic.

Regional changes in groundwater chemistry reveal the direction of flow paths, and isotopic ratios and inert gas contents give an indication of the residence time of water in the flow systems. The composition of the groundwater in deep systems has been modified by shale-membrane filtration. Density settling may be a feature in thick relatively homogeneous aquifers.

Groundwater flows from areas of high fluid potential (i.e. high mechanical energy per unit of mass) to areas of low fluid potential (Hubbert 1940, p. 797). The general direction of flow is reflected by the form of the water table and the piezometric surface which are portrayed by lines of equal potential. However, the groundwater potential field, and hence the flow field, is three-dimensional and maps of the water table and piezometric surface are merely two-dimensional surfaces through the field (Hubbert 1940, p. 910), in practice commonly based on data derived from different depths in an aquifer.

High hydraulic heads are associated with upland areas that are recharge areas, and low hydraulic heads with valleys or discharge areas. With increasing depth the permeability of rocks steadily decreases and at a depth varying from region to region they are effectively impermeable. Between this depth and the water table, groundwater is circulating, under natural conditions, in a state approximating to dynamic equilibrium. The three-dimensional form of the flow paths in the saturated zone is controlled to a significant extent by the topography and the geology.

This picture has been recognized for many years but in the last 20 years it has become possible to analyse groundwater flow patterns in increasing detail by mathematical models. Hubbert (1940, p. 926) described the form of flow paths between upland sources and valley sinks in an isotropic homogeneous permeable medium and pointed out that in such a medium stagnant water does not exist, unless it has different physical properties (such as high mineralization), and that the region of most intense flow is where flow lines converge towards the outlet. Tóth (1962) developed a mathematical model to account for the general features of flow from upland recharge areas to valley sinks, and Freeze & Witherspoon (1966, 1967) numerically analysed more complex systems incorporating continuous horizons of different permeabilities. These studies all confirmed that groundwater flow is controlled by both the form of the topography and the geology including the depth to the essentially impermeable base of the flow system. Basically they revealed that there are local, intermediate and regional flow systems that encompass entire drainage basins (Tóth 1963). The various systems are arbitrarily defined by the depth of the flow lines: local systems have recharge and discharge zones in adjacent areas; intermediate systems are more extensive and encompass one or more local systems; regional systems extend across entire groundwater basins from the principal watershed to the lowest groundwater sink. In the U.K. groundwater flow in aquifers at outcrop is usually in local systems while shallow flow in confined aquifers may be considered to be in intermediate systems.

A knowledge of regional groundwater flow is important in many fields, including hydrocarbon, mineral and geothermal exploration, and for planning the safe disposal of wastes, in particular toxic wastes. The purpose of this paper is to consider regional groundwater flow patterns in sedimentary basins in the U.K. The U.K. has not been extensively or systematically explored for oil, except possibly in the Carboniferous of the E Midlands of England, and as a consequence groundwater potentials in deep formations are not generally known. In these circumstances greater reliance has to be placed on other evidence, particularly on regional changes in groundwater quality, including that of stable and radioactive isotopes, lithological variations and geological structure, and also on thermal data, such as heat flow, sub-surface temperatures and geothermal gradients. Of these, geochemical data are currently the most useful in the U.K. Penetration of fresh groundwater into deep

From GOFF, J. C. & WILLIAMS, B. P. J. (eds), 1987, *Fluid Flow in Sedimentary Basins and Aquifers*, Geological Society Special Publication No. 34, pp. 105–25.

aquifer systems usually marks important flow paths and the time-scales along the flow lines can sometimes be recognized using isotopic ratios.

Geological factors influencing hydrogeology in the U.K.

Although at present the U.K. forms part of the stable foreland of NW Europe, in the past it has lain at the cross-roads of major orogenic events. In particular, the Precambrian, Caledonian and Variscan orogenies have affected to varying degrees the nature and properties of pre-Mesozoic sediments and the Alpine orogeny has influenced the structure, particularly of southern Britain.

Precambrian rocks form the deep basement below a cover of Palaeozoic sediments, but the oldest known rocks beneath a large part of the British Isles comprise sediments and volcanic rocks ranging in age from Precambrian to Silurian which form part of the complex Caledonian orogenic belt. Groundwater flow in these rocks is primarily restricted to fractures and fissures within 100 m of the ground surface. At depth, flow is probably limited and confined to fault zones.

Erosion of the continent formed by the Caledonian orogeny gave rise to the thick detrital deposits of the Old Red Sandstone which formed in internal molasse basins in northern Britain and on coastal alluvial plains in southern Britain. The thick sandstones and siltstones cropping out in S Wales and the Welsh Borderlands and occurring beneath cover under southern Britain may appear to be potential aquifers, but diagenetic changes including extensive cementation have reduced their porosities and they have low permeabilities. In northern Britain, however (e.g. in the Midland Valley of Scotland), they are more permeable and even at depth may have water-bearing potential.

Major marine invasions transgressed the Old Red Sandstone continent in early Carboniferous times. On the foreland to the N of the Variscan Front, Lower Carboniferous sediments were deposited in fault-controlled basins, and these were succeeded by a widespread cover of fluvio-deltaic Silesian sediments including many sandstones and limestones. S of the Variscan Front they have been subject to intense structural deformation, but to the N they now occur in structural basins that developed as a consequence of late Carboniferous movements.

Following the Hercynian orogeny, fault-controlled depositional basins formed in Permian and Triassic times in response to tensional stresses. Sediments eroded from the uplands were deposited in the various Permo–Triassic basins which tend to be marginal areas of the major offshore sedimentary basins in the North Sea, Irish Sea and Western Approaches. The Permo–Triassic sandstones deposited in these basins are major aquifers.

During the remainder of the Mesozoic and early part of the Tertiary, detrital sediments and limestones accumulated in fault-controlled basins over much of Britain. These sediments contain many aquifers including the Jurassic limestones, the Lower Cretaceous sandstones, the Chalk and the Pleistocene Crag. All except the Crag were affected by the Alpine orogeny and have been folded into broad shallow basins as exemplified by the Hampshire and London basins.

An indication of the depth to the essentially impermeable basement, and hence the lower boundary of effective groundwater flow, can be deduced from seismic surveys. Seismic reflection profiles beneath southern Britain can be divided into three zones based on reflection character (Whittaker & Chadwick 1984). These zones are structural or tectonic units, and each can include rocks of widely differing ages. Zone 1, the upper zone, comprises cover rocks that have not undergone intense structural deformation. It includes the Mesozoic sequence of southern England, but N of the Variscan Front, over the E Midlands microcraton, it includes all the Phanerozoic sequence and locally even the upper part of the Precambrian. N of the microcraton, zone 1 consists of Mesozoic and Upper Palaeozoic rocks only.

Zone 2 comprises rocks that have been deformed in orogenic fold-belts. S of the Variscan Front this is the Variscan fold-belt and N of the Midlands microcraton it is the Caledonides of the Lower Palaeozoic. Over the microcraton itself this zone may be absent. Zone 3 is tentatively identified as metamorphic and crystalline basement.

On the basis of this interpretation of the deep structure of southern Britain, the base of the Mesozoic can be regarded as the limit of significant groundwater flow S of the Variscan Front. The measured permeability of less than 0.1 mD for Devonian siltstones immediately below the Triassic at Southampton may not be atypical of the Palaeozoic, although the Carboniferous Limestone extending from Sussex to the Severn Estuary (Whittaker 1985) may have some fracture permeability. N of the Variscan Front the Palaeozoic basement of the Midlands microcraton, and the Wales–Brabant massif, is also essentially impermeable, despite the fact that it has not suffered intense deformation. An exception to this could be the Coal Measures in the syncline between the Oxfordshire and the Warwickshire coalfields (Whittaker 1985). Further N, however, in central and northern England, the Lower Palaeozoic and the Devonian can probably be regarded as the impermeable basement in most areas, but in Scotland the base of the Upper Palaeozoic is probably the boundary as parts of the Old Red Sandstone have significant permeability (Browne *et al.* 1985).

This definition of the impermeable basement is necessarily arbitrary. It is meant to indicate the lowest level at which groundwater flow exists for all practical purposes. It does not mean that flow does not exist below this level for clearly it does, but mainly in fractures, possibly concentrated in zones of limited width.

Regional groundwater flow in the U.K.

If the general definition of the impermeable basement in Britain is accepted, the Upper Palaeozoic (i.e. the Carboniferous in northern England, and the Carboniferous and Devonian in Scotland) and the Mesozoic and Tertiary sequences fall naturally into a number of groundwater provinces that are defined by probable regional flow paths arising from the distribution of fluid potential that is controlled by upland divides and lowland, coastal or submarine groundwater sinks. Such a groundwater province can be regarded as a region in which sub-surface drainage, possibly from a series of aquifers, flows to a common outlet.

In England and Wales four such groundwater provinces (Fig. 1) can be recognized (Rodda *et al.* 1976): (1) Eastern Province draining to the North Sea; (2) Hampshire Province draining to the English Channel; (3) Severn Province draining to the Bristol Channel; (4) NW Province draining to the Irish Sea. Additional provinces are represented by the Midland Valley of Scotland, the Orcadian Devonian Basin of NE Scotland and the basins of Northern Ireland (Fig. 1). Recognition of these provinces assumes regional hydraulic continuity, i.e. water moves through deposits of widely-differing permeabilities including those normally regarded as aquicludes. Most of the water infiltrating at outcrops is discharged to the rivers draining the outcrops through local flow systems. The remainder flows through the aquifers below overlying beds of low permeability and ultimately finds a route to surface outlets by cross-formational flow. This water is moving in intermediate and deep regional flow systems.

The form of intermediate flow systems is influenced by the characteristic scarp and vale

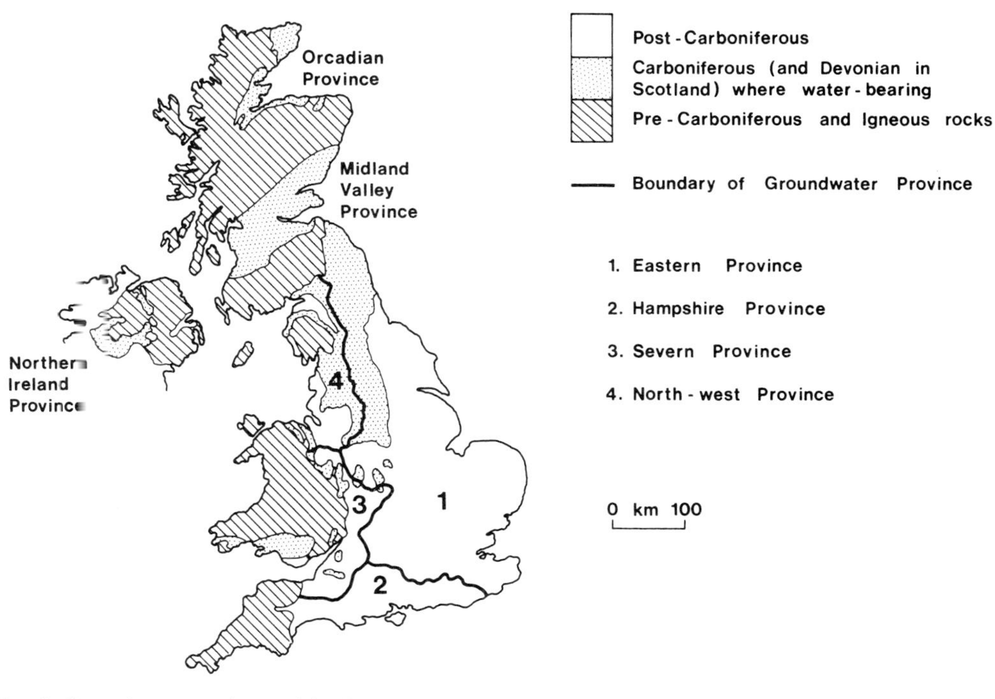

FIG. 1. Groundwater provinces of the U.K.

topography of eastern and southern England produced by the alternating sequence of gently dipping aquifers and aquicludes in the Mesozoic. Groundwater flowing down-gradient in the aquifers tends to discharge in the low-lying clay belts by upward leakage through the overlying confining clays. Deeper regional flow lines are not influenced by these areas of low hydraulic head and continue to major sinks in the lower reaches of major valleys or around the coast.

It is essential to keep in perspective the relative rates of flow in local systems on the one hand and in intermediate and regional systems on the other. Flows in the two groups are of different orders of magnitude, reflecting the different aquifer properties. The permeability of a good aquifer at outcrop can be as much as 1000 D, while at a depth of 1–2 km the permeability of the same formation may be reduced to about 100 mD or even less. At depths where lateral movement of groundwater is slow, differences in salinity and temperature, and hence density and viscosity, become significant factors controlling the direction and rate of movement of water. Significant volumes of water discharge from extensive confined aquifers by cross-formational flow through overlying confining beds.

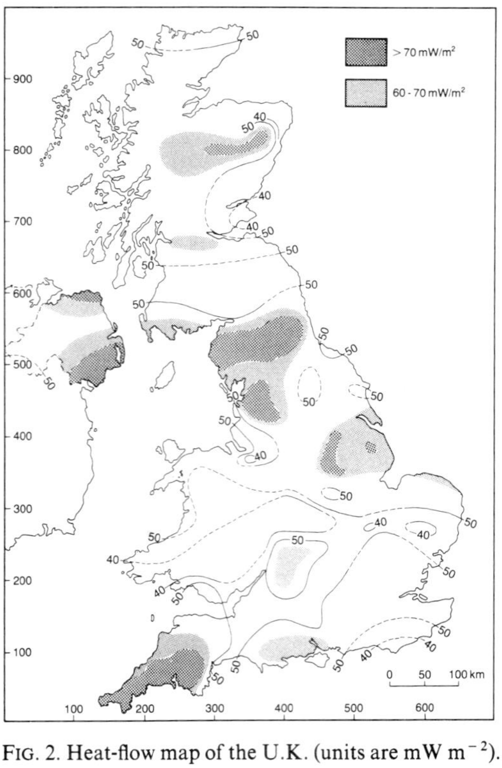

FIG. 2. Heat-flow map of the U.K. (units are mW m^{-2}).

Heat-flow field in the U.K.

Some general implications regarding regional groundwater flow can be drawn from the heat-flow pattern (Fig. 2). The main thermal anomalies are associated with heat produced by the radioactive granites of Cornwall, the Lake District, Weardale and the Eastern Highlands of Scotland which are superimposed on a background averaging 52 mW m^{-2} and ranging from 40 to 60 mW m^{-2} (Wheildon & Rollin 1986). Values are also high where thick sedimentary sequences occur, as in the E Midlands, the Wessex and Worcester basins and part of the Midland Valley of Scotland. It has been postulated (Richardson & Oxburgh 1978) that these anomalies could be caused by enhanced heat production in the basement, but slow upward seepage of groundwater on a regional scale is probably a more significant factor. Certainly, the local anomaly in the E Midlands associated with the Eakring Anticline is caused by convective transfer of heat by groundwater rising in the Carboniferous Limestone (Bullard & Niblett 1951) and discharging into the Upper Carboniferous and ultimately the Mesozoic by cross-formational flow (Gale *et al.* 1984b). The anomaly in the Carboniferous of Lancashire (Fig. 2) may have a similar explanation, and other cases have been recognized during deep exploration drilling. These examples indicate that flow does occur in the Carboniferous at considerable depths.

Geochemical evidence for regional groundwater flow

Over geological time-spans, it is probable that formation waters have been replaced several times and only in very exceptional cases has water remained static for any length of time. Penecontemporaneous connate waters may have been seawater, as in most marine sediments, freshwater in some continental or lacustrine sediments and brackish/saline water in sediments formed in closed basins. The present-day groundwater chemistry thus only represents a transient condition and all the other numerous transient conditions which existed in geological times must not be ignored in reconstructing flow history from the geochemistry. These transient conditions would have included changes in pressure and temperature (geothermal gradient) as well as chemistry in response to tectonic events and erosion.

Sequential down-gradient changes in groundwater chemistry have been observed in most of the U.K. aquifer systems and have been interpreted as a response to the evolution of water along flow lines. The principal geochemical changes include the following: (1) mineral dissolution and the attainment of saturation with various carbonate and silicate minerals; (2) various redox reactions, especially the complete reaction of oxygen derived from outcrop (Edmunds *et al.* 1984); (3) cation exchange reactions and water softening (Ineson & Downing 1963); (4) mixing with saline water.

The distribution of fresh water well down-gradient of recharge zones is perhaps the best clue that groundwater flow systems are active at the present time. If total mineralization (especially chloride) provides the best conservative index of the degree of removal and uptake of solutes, stable isotopes ($\delta^{18}O$, $\delta^{2}H$) provide the best means of studying the behaviour of the water molecule. Stable isotope ratios preserve palaeoclimatic information and provide evidence of marine water components as well as fractionation due to temperature and water–mineral interactions. The important evidence for each province is reviewed below for both solutes and water. A summary of the isotopic compositions of formation waters in the U.K., updated from Edmunds (1986), is given in Fig. 3. All data are referred to the world meteoric line and to SMOW. It is important to notice that (1) most formation waters lie along the meteoric line, (2) waters recharged during colder (Pleistocene) periods can be distinguished, (3) only one example of a residual connate (*sensu stricto*) water has been found to date, in the Chalk of E Anglia, and (4) waters from the Wessex Basin have an internally consistent isotopic relationship ($\delta^{2}H = 1.87\delta^{18}O - 25.5$) and differ from other areas.

Eastern Province

The province comprises a Carboniferous to Pleistocene sequence with a general regional flow to the E to outlets in the North Sea (Fig. 4). It includes the Carboniferous basins of the E Midlands and NE England, the Permo–Triassic

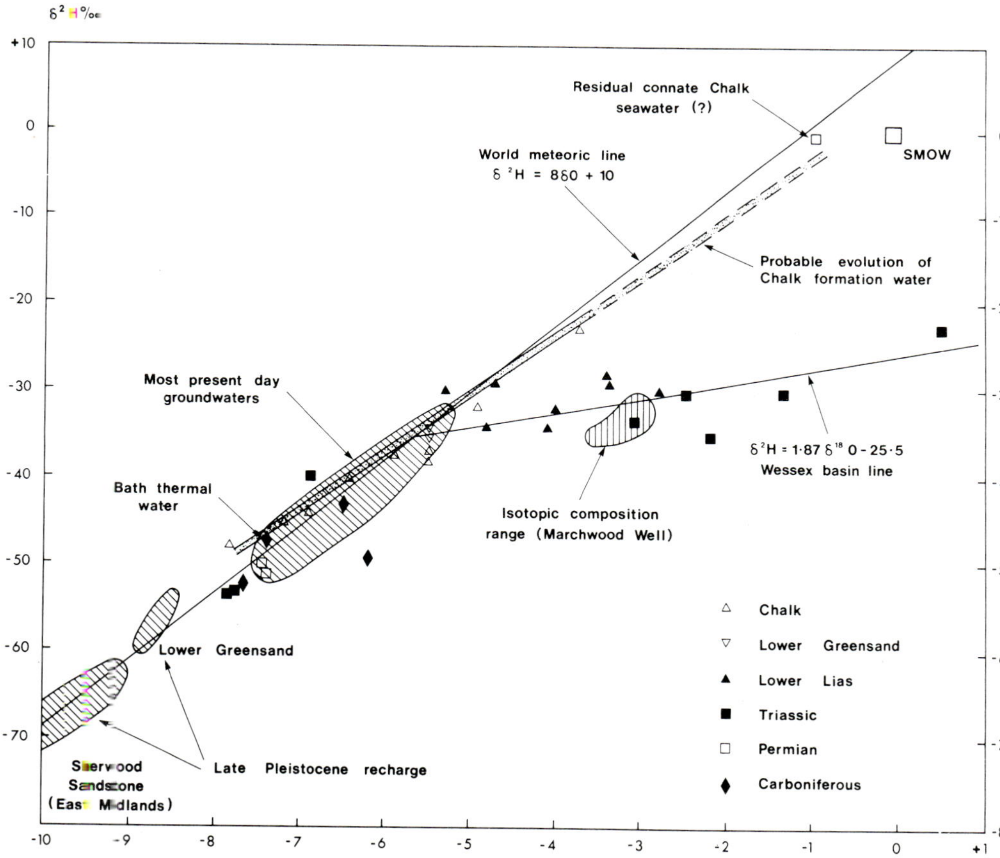

FIG. 3. Isotope compositions ($\delta^{18}O$ and $\delta^{2}H$) for formation waters in the U.K.

basin of E Yorkshire and Lincolnshire and the Cretaceous and Tertiary London Basin (Fig. 5). The regional groundwater divide coincides with the Pennine watershed in the N and then continues through the W and S Midlands before turning E along the southern boundary of the London Basin.

E Midlands

Upper Palaeozoic

This region has been extensively explored for hydrocarbons and as a consequence the hydrogeology is relatively well known. The Carboniferous rocks which crop out in the Pennines are overlain by Mesozoic sediments forming the E Yorkshire and Lincolnshire Basin. They dip to the E beneath the North Sea and attain thicknesses of more than 2000 m along the coast. Quaternary glacial deposits, predominantly tills, cover extensive areas.

Deposition of the Carboniferous rocks was controlled by growth faults resulting in a basin-and-shelf system generally thickening to the N away from the London–Brabant Massif and towards the Central Pennine Basin (K. Smith & N. J. P. Smith, personal communication, 1984). In the E Midlands the Carboniferous thickens to

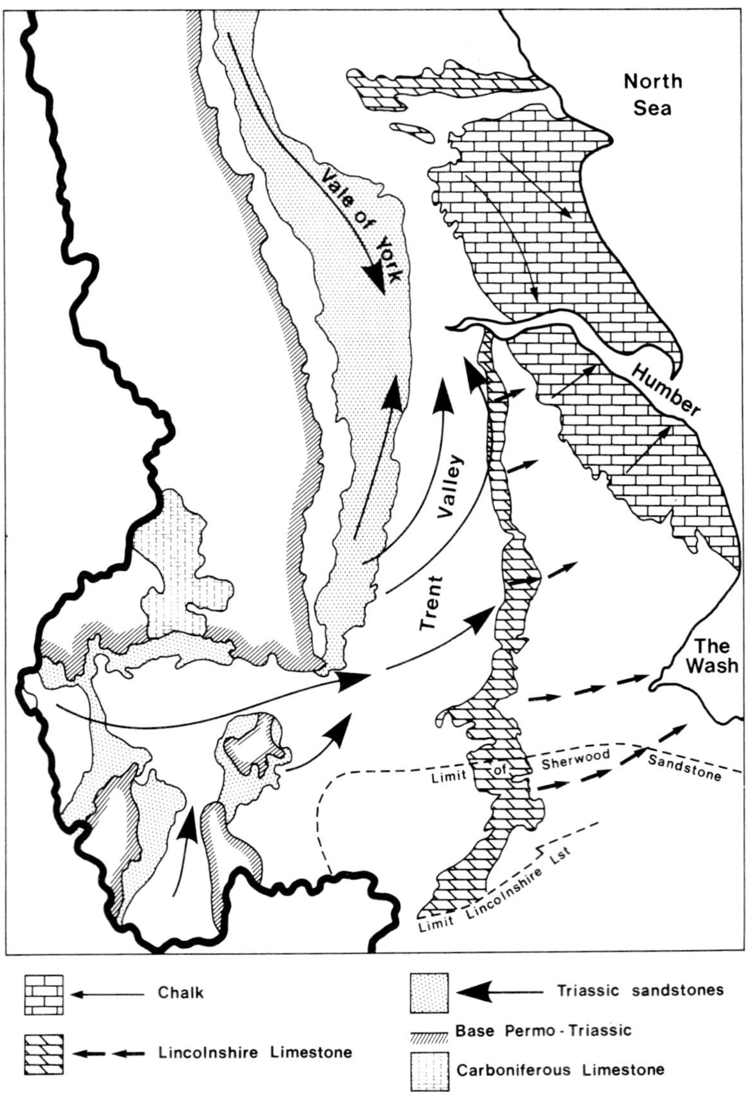

FIG. 4. Principal aquifers and directions of groundwater flow in eastern England.

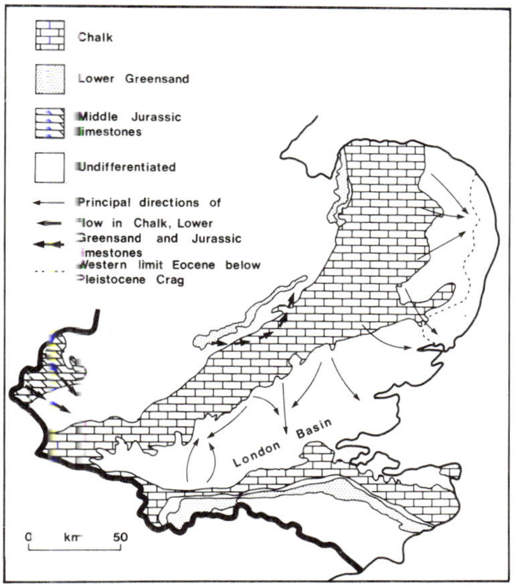

FIG. 5. Principal aquifers and directions of groundwater flow in East Anglia and the London Basin (prior to groundwater development).

more than 4 km in the Gainsborough Trough and to over 3 km in the Widmerpool Gulf (S of Nottingham), and is possibly in excess of 2 km around Holderness (in the Humber Trough) beneath 2 km of Mesozoic rocks.

The Carboniferous sequence is divided into three major lithological groups: the Carboniferous Limestone, the Millstone Grit and the Coal Measures. The lithology of the Carboniferous Limestone ranges from massive shelf limestones passing laterally into marginal reef facies and thin inter-bedded limestone and shale basinal facies. The basinal facies tend to be conformable with the overlying Millstone Grit but a break in sedimentation is associated with the shelf facies. The basinal areas continued to thicken more rapidly than the shelf areas during deposition of the Millstone Grit which comprises shales and associated sandstones. The sedimentary break increases southwards, and the Millstone Grit thins and is absent to the S and E of Lincoln but is up to 1 km thick in the basins. In late Namurian (Millstone Grit) and early Westphalian (Coal Measures) times, fluvio-deltaic deposits showed only marginal thickening over the basinal areas, but eventually all but obliterated the shelf-and-basin sedimentary pattern. The Coal Measures are usually conformable with the Millstone Grit and thicken to the W and N to over 1400 and 700 m respectively. In the SE of the region they directly overlie the Carboniferous Limestone and are absent around the Boston–Spalding area. The sandstones of the Millstone Grit tend to be thicker and more extensive than those of the Coal Measures which take the form of channel, levee and bar deposits.

The surface of the Carboniferous is an eroded peneplain dipping to the E exposing mainly Coal Measures (at all levels in the sequence) to the

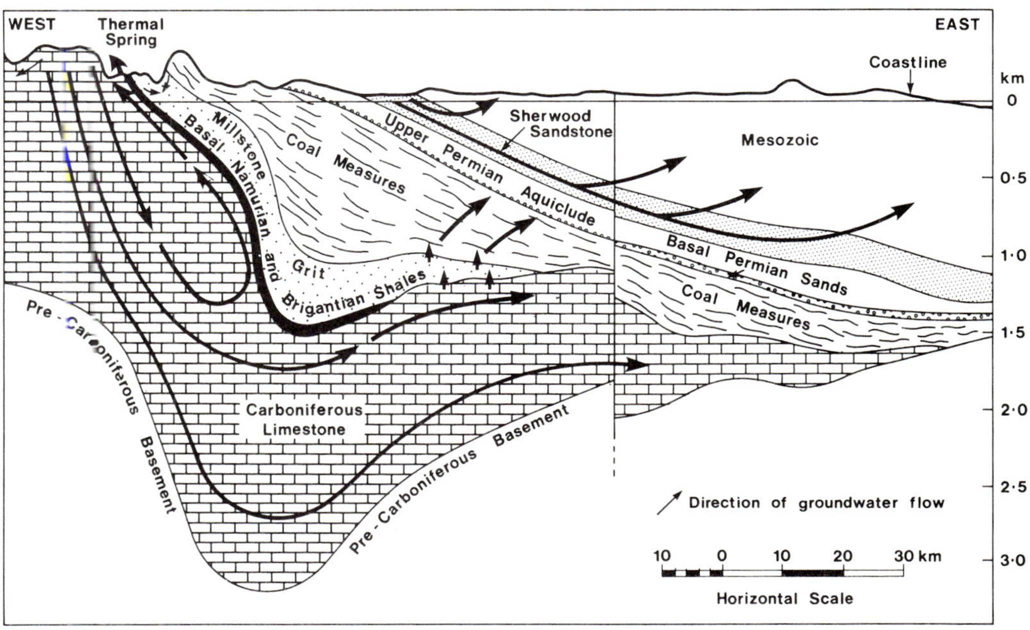

FIG. 6. Schematic cross-section through the E Midlands showing the principal directions of groundwater flow.

overlying rocks (Fig. 6). In the S of the area the Coal Measures are fluvial continental sediments and probably have moderate aquifer properties permitting movement of water into the overlying Mesozoic aquifers. To the N of Grantham, an up-faulted block exposes Carboniferous Limestone and Millstone Grit to the pre-Permian surface. This is a potential area for enhanced groundwater discharge into the overlying Permian. Because of the faulting, the Carboniferous Limestone inlier in the Pennines is not connected to the limestone at depth and water is probably flowing in preferred paths determined by faults. In general, the Carboniferous rocks have low porosities and permeabilities which are enhanced locally by zones of fracturing, particularly in the Carboniferous Limestone (Gale *et al.* 1984b); more permeable sandstones occur in the Upper Carboniferous. Early emplacement of oil in these sandstones may have reduced the effects of the diagenetic processes that reduce permeability (Hawkins 1978).

The regional groundwater flow pattern has largely been deduced from the chemistry of numerous groundwater samples collected during exploration for hydrocarbons (Downing & Howitt 1969). In general, the salinity increases upwards from the Carboniferous Limestone through the Millstone Grit to the Coal Measures and the salinity in all three lithological units increases to the N (with total mineralization of more than 99 $g\,l^{-1}$ in the Carboniferous Limestone, more than 180 $g\,l^{-1}$ in the Millstone Grit and more than 200 $g\,l^{-1}$ in the Coal Measures). The origin of these brines is considered to be the result of the diagenesis of formation waters which were concentrated by filtration through argillaceous beds acting as semipermeable membranes (Downing & Howitt 1969). Where groundwaters are less saline than present-day seawater, dilution by meteoric waters has occurred indicating a more active dynamic flow system. The flow is controlled by the hydraulic head in the Derbyshire Dome which drives water from the Carboniferous Limestone upwards into the Millstone Grit, where it is not restricted by the Brigantian Shales, and then into the Coal Measures. A schematic cross-section of the flow paths is shown in Fig. 6. The flow pattern is supported by pressure data from drill-stem tests in the Millstone Grit, which indicate a pressure 'ridge' coinciding with a lobe of less saline water that implies upward water movement between the edge of the Brigantian Shales and the easterly limit of the Millstone Grit.

Deep groundwater movement within the Carboniferous Limestone is also revealed by the occurrence of thermal springs around the contact with the Millstone Grit Series. At the four major discharge points, the temperatures range from 15.5°C (at Bakewell) to 27.7°C (at Buxton). Geochemical evidence (Edmunds 1971) indicates that the waters are meteoric and have been diluted in both concentration and temperature to a limited extent by shallow groundwaters. The heat-flow anomaly along the Eakring Anticline has been referred to already.

Regional groundwater movement, as indicated by pressure and hydrochemical data, shows that the Millstone Grit is hydraulically more continuous, both laterally and on an areal basis, than the Coal Measures in which sandstone bodies are smaller and more discrete. Water quality in adjacent Coal Measures sandstones can vary greatly, implying restricted water movement. Upward movement of groundwater through the Coal Measures is likely to be more restricted in the N of the region compared with the S where the formation is thinner and arenaceous facies are more pervasive. Near-stagnant conditions are indicated in the N of the E Midlands by the very saline brines found throughout the Carboniferous sequence.

It must be concluded, therefore, that saline water displaced from the Carboniferous must be recharging overlying formations S of the Humber and eventually displacing fresher water in younger formations, most probably the Basal Permian Sands and/or the Sherwood Sandstone which have high permeabilities. The area where this is believed to be occurring is indicated by the regional heat-flow anomaly in Lincolnshire (Fig. 2).

Triassic sandstones

Detailed geochemical studies of the Triassic Sherwood Sandstone have been carried out in the E Midlands (Bath *et al.* 1979; Edmunds *et al.* 1982; Andrews *et al.* 1984) which show that remarkably fresh groundwater (10 mg Cl^- l^- or less) persists to the E of the Trent Valley. This water has been clearly shown to be of Pleistocene age and the low chloride relates to periglacial recharge as shown by light isotopic composition and palaeo-temperatures derived from inert-gas analyses. Since the waters in the deepest boreholes are so fresh, it is likely that they are fresh some distance to the E; therefore if upward leakage of saline waters is occurring it must be to the E of the longitude of Lincoln.

Piezometric levels in the Sherwood Sandstone fall from 100 m above sea level (ASL) at outcrop to 10 m ASL in the Trent Valley and to below sea level in an area near the Humber. Although this depression in the water table near the Humber is accentuated by recent pumping, under natural conditions it is unlikely to have been much higher and this area is probably a natural discharge area for the Sherwood Sandstone (Fig. 4). The head in the Lincolnshire Limestone and Chalk of the Lincolnshire Wolds may reduce the easterly flow in the Sherwood Sandstone and thus the principal flow lines are believed to be towards the Humber. Nevertheless, there is evidence of flow in the Sherwood Sandstone in E Lincolnshire. At Cleethorpes, the calculated fresh-water head (Downing *et al.* 1985) is 20 m ASL. This could reflect a relict gradient towards the discharge area at the head of the Humber that pre-dates the current artificially low potential in NE Lincolnshire caused by groundwater abstraction. Alternatively, and more likely, it probably reflects regional down-gradient flow into the North Sea basin under the influence of the piezometric heads to the W. Evidence for this is provided from the geochemical data from the Cleethorpes Well where a salinity gradient occurs from 35 g l^{-1} at the top to 80 g l^{-1} at the middle and base of the Sherwood Sandstone. Isotopic evidence indicates that this saline water is meteoric and contains little if any marine component. This suggests that the salinity must be derived in part from diffusion/exchange from evaporites, and this is further supported by the low Br^-/Cl^- in this water. Evaporites are, of course, present in the underlying Permian (Downing *et al.* 1985).

The salinity gradient in the relatively homogeneous Sherwood Sandstone may indicate that density settling is occurring because of the slow movement of the water. Most of the flow in the regional easterly flow system is believed to be near the top of the formation. Small flows are probably derived by upward leakage from the Carboniferous, as discussed earlier, and this would account for some of the salinity.

N of the Humber, in the Vale of York, the flow pattern is very similar to that in the lower part of the Trent River Basin. The Triassic sandstones are the main outlet and groundwater flows S down the Vale of York along the line of the Yorkshire Ouse, with easterly flow possibly being restricted by the head in the Chalk of the Yorkshire Wolds and in the Jurassic of the Cleveland Hills. There may be some deep regional flow to the E in the Triassic sandstones, particularly along the line of the Vale of Pickering. To the N of the Vale of York, water is draining to the Tees estuary from the groundwater divide to the S of Darlington.

The flow systems in the E Midlands can be summarized as follows. There is clearly an intermediate flow system controlled by hydraulic heads in the Trent Valley and the Vale of York with a major outlet where the Trent and the Yorkshire Ouse meet the Humber Estuary. However, there is probably also a deeper flow system in the Sherwood Sandstone moving directly to the North Sea which also drains groundwater from deep Upper Palaeozoic rocks, particularly S of the Humber. Groundwater discharging from this flow system is probably responsible for the high heat flow observed over Lincolnshire and possibly partly responsible for the high values in the North Sea off the Lincolnshire coast (Andrews-Speed *et al.* 1984). However, the cause of the high heat flow recorded at shallow depths in the western North Sea, including the area off the Lincolnshire coast, may be compaction of sediments in the central part of the North Sea or hot or warm groundwaters rising along faults in the central graben, as proposed by Carstens & Finstad (1981) in the northern part of the North Sea Basin; the two opposing flow systems (the on-shore and off-shore) possibly meet near the coast.

The accumulation of oil in the Carboniferous of the E Midlands may be mainly related to palaeo-groundwater flow patterns which are likely to have had regional flow directions similar to those at the present day at different times during the Permian and Mesozoic.

The Jurassic limestones

The Lincolnshire Limestone and the overlying argillaceous deposits of the Upper Jurassic produce typical scarp-and-vale topography. Groundwater flows down-gradient in the limestone below overlying confining beds to outlets in the Fens by upward seepage and, in S Lincolnshire, possibly to buried glacial channels centred on the Wash. The very high permeability of the limestone where it is confined in S Lincolnshire could have been caused by enhanced flows through the aquifer during the Pleistocene when sea levels and the base level of drainage were lower (Downing & Williams 1969). The apparent age of the water in the limestone increases down-gradient with the chemistry reflecting different conditions at the time of recharge in Recent and Pleistocene times. However, the relationship of the chemistry of the water to the flow regime is complicated by the continuous exchange of the dissolved constituents in the water between the matrix of the rock and the fissure system (Downing *et al.* 1977).

A saline interface exists in the limestone E of the principal outlets from the confined aquifer, marking the eastern limit of rapid down-gradient flow. A pronounced redox boundary occurs some

17 km from the outcrop (Edmunds 1973) and coincides with complete reduction of oxygen in the water as well as in the rock. The saline waters in the most easterly section of the aquifer have been shown, by measurement of excess ^4He contents, to have pore fluids of similar age to that of the formation (Andrews & Lee 1980).

E Anglia

In E Anglia the Chalk is the dominent hydrogeological unit with maximum water levels falling from about 130 m ASL in the S to about 50 m ASL in N Norfolk. The Lower Greensand outcrop parallels that of the Chalk, but groundwater levels are lower (Monkhouse 1974) and flow in the area where the Lower Greensand is confined is to the N, virtually along the direction of the strike, to an outlet in the Fens N of Cambridge and ultimately to the Wash. To the E the salinity increases to over 500 mg l^{-1} and ^{14}C data imply that there are two distinct waters with the fresher younger water skirting the more saline water to the E (Evans *et al.* 1979).

Along the E Anglian coastal area, groundwater in the Chalk below the Eocene is saline and the piezometric surface is generally below sea level. Saline water is also found in synclinal areas of the Chalk, e.g. NE of Stowmarket (Woodland 1946), and it is clear that the preferred flow occurs at shallow depths.

A complete geochemical profile of interstitial waters occurring in 472 m of Chalk was obtained at Trunch, Norfolk (Fig. 7) beneath 41 m of Pleistocene deposits (Bath & Edmunds 1981). This profile shows that there is a progressive dilution of salinity towards the top of the Chalk from a connate water with marine composition (197 000 mg $Cl^- l^{-1}$). Departures from the expected dilution trend up the profile are interpreted as evidence for diagenetic reactions during burial or as a result of fresh-water circulation following sub-aerial uplift. Using a diffusion model, it has been calculated that the observed profile could have developed by diffusion alone within 3 m.y. This can be compared with a time scale of 40 m.y. from the start of uplift and marine regression in the area. It is more likely, though, that the initial post-Cretaceous fresh-water influx may not have commenced until late in the Tertiary cycle of denudation and before the Pleistocene. The stable isotope profile of interstitial water (Fig. 7) quite clearly shows progressive dilution of a water with a marine origin, in contrast with the meteoric signature of many other saline waters.

It is likely that, in E Anglia, increased groundwater circulation occurred during the period of low Pleistocene sea levels. The drainage then, and at present, was mainly to river systems W of the Eocene sub-crop, with a less important flow route to the sea in the Chalk below the Eocene and through the Pleistocene cover (Fig. 5). E Anglia illustrates particularly well the nature of groundwater flow in the Chalk which is concentrated in linear routes of high permeability that follow river valleys, especially in the upper 50 m of the aquifer. Flow away from the valley lines is restricted where Pleistocene cover is extensive (Lloyd *et al.* 1981; Bath *et al.* 1985).

Thames Basin and London Basin

The Thames Basin, incorporating the London Basin, can be regarded as a sub-province of the Eastern Province. It includes the major aquifers of the Chalk, Lower Greensand and Middle Jurassic limestones as well as the Corallian in the Upper Jurassic (Fig. 5). The Chalk of the London Basin is recharged in the Chiltern Hills, Berkshire Downs and N Downs and drained under natural conditions to outlets in the Thames Estuary and also to outlets in the Thames Valley near Windsor and Bray (Water Resources Board 1972). Where the London Clay overlies the Chalk and Lower London Tertiaries, overflowing artesian conditions existed over appreciable areas of low ground in the nineteenth century prior to extensive groundwater development.

The flow in the Chalk below the London Clay is mainly restricted to the upper more permeable 50 m of the aquifer (Water Resources Board 1972; Owen & Robinson 1978) and it is also influenced by regional variations in transmissivity which are biased towards high-permeability zones below river valleys. Morel (1979) suggested that the distribution of transmissivity in the Chalk where it is overlain by the Eocene has largely existed in its present form since the Lower Eocene.

In the central part of the London Basin the transmissivity of the Chalk is much lower than at outcrop and it is only about 10 m^2 day^{-1} along the southern limb of the syncline from the Woking area as far W as Kingsclere (Water Resources Board 1972; Morel 1979). In the confined Chalk, groundwater abstracted from boreholes usually contains some radiocarbon and therefore is generally less than 25 000 years old. It may be considerably younger than this, since groundwater in the Chalk has undergone considerable exchange with carbonate minerals (Edmunds 1976) and the groundwater has $\delta^{13}C_{HCO_3}$ values approaching those of the rock (Downing *et al.* 1979).

The stable oxygen and hydrogen isotope ratios from the London Basin are not as negative as in

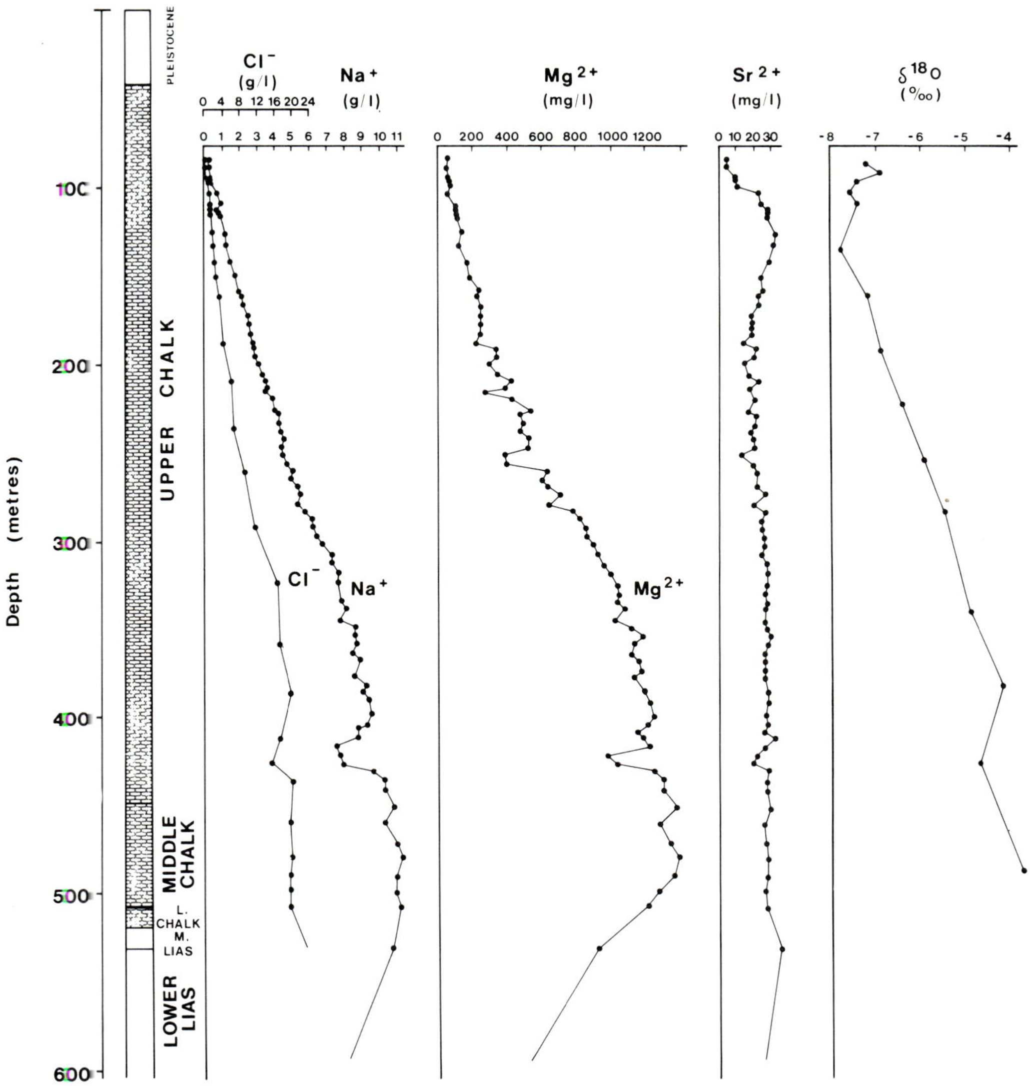

FIG. 7. Depth profiles of the chemistry of pore-water solutions in the Chalk in a borehole at Trunch in Norfolk.

the Sherwood Sandstone of the E Midlands, for example and it is probable that most of the groundwater flowing in fissures in the London Basin (as distinct from that in the matrix) is post-glacial, indicating significant flow even through the central less-permeable parts of the basin to the present day. However, it seems likely that most of the water flowing in the Chalk below the London Clay under natural conditions flowed around the central area of low permeability, preferring to take the high-permeability zones such as the Lee Valley (e.g. Water Resources Board 1972, Fig. 3.4). In the central part of the confined aquifer groundwater flow tends to be predominantly in the area N of the synclinal axis (Morel 1979).

Saline interstitial waters (up to 7000 mg Cl^- l^{-1}) are found at Faircross in the W of the basin (Edmunds 1976) from a profile which is analogous to Trunch but represents a more advanced stage in fresh water–saline water exchange. Residual interstitial slightly saline water has also been found in unconfined Chalk in the W of the basin indicating that exchange with fresh water is not complete even at outcrop (Edmunds *et al.* 1973).

Under natural conditions the Thames was the natural outlet for groundwater in the entire Thames basin. There is still probably slow upward leakage from the Lower Greensand through the overlying Gault into the Chalk and in places into the river system. Artesian conditions exist in the Lower Greensand where waters of Pleistocene age are developed for supply near Slough (Mather *et al.* 1973), and there is isotopic evidence for Pleistocene water in the Lower Greensand of Kent (Evans *et al.* 1979). The potential in the Middle Jurassic limestones of the Cotswold Hills is generally greater than 150 m ASL and attains maxima of over 200 m, generally 50 m greater than similar values for the Chalk. As a consequence, regional flow probably occurs in these limestones below the Upper Jurassic, again to outlets in the Lower Thames valley, by cross-formational flow through deposits of low permeability.

Groundwater flow in the Corallian of the Upper Thames Basin has been examined in some detail by Alexander & Brightman (1985). They have shown that water flows not only in a down-gradient direction from the outcrop but also in an up-dip direction, where the aquifer is confined, because of the high potential in the Chalk outcrop. The two flow systems meet in an area of low fluid potential along a line near the base of the Kimmeridge Clay. Support for this view is provided by stable and radioactive isotope ratios. This is a further example of flow from confined aquifers by upward leakage to discharge at the surface in the low-lying clay outcrops. However, such flow systems are not universal, for the hydraulic head in the Chalk does not appear to prevent flow in the Lower Greensand below the Gault in the London Basin as, for example, in N Kent. Whether or not high ground prevents deep groundwater flow below it depends upon the relative values for the permeability of the aquifers and the intervening confining layer.

Hampshire Province

Groundwater in this province drains to the English Channel. The Hampshire Basin and the Weald are the dominant topographical features, but at greater depths it is the disposition of sediments in the structural Wessex Basin that is important. The surface geology is formed principally by the Chalk and Tertiary rocks, but Permo–Triassic and Jurassic rocks crop out in the W and occur at depth (Fig. 8).

The Permo–Triassic sediments were deposited on a basement, ranging in age from at least Ordovician to possibly Upper Carboniferous,

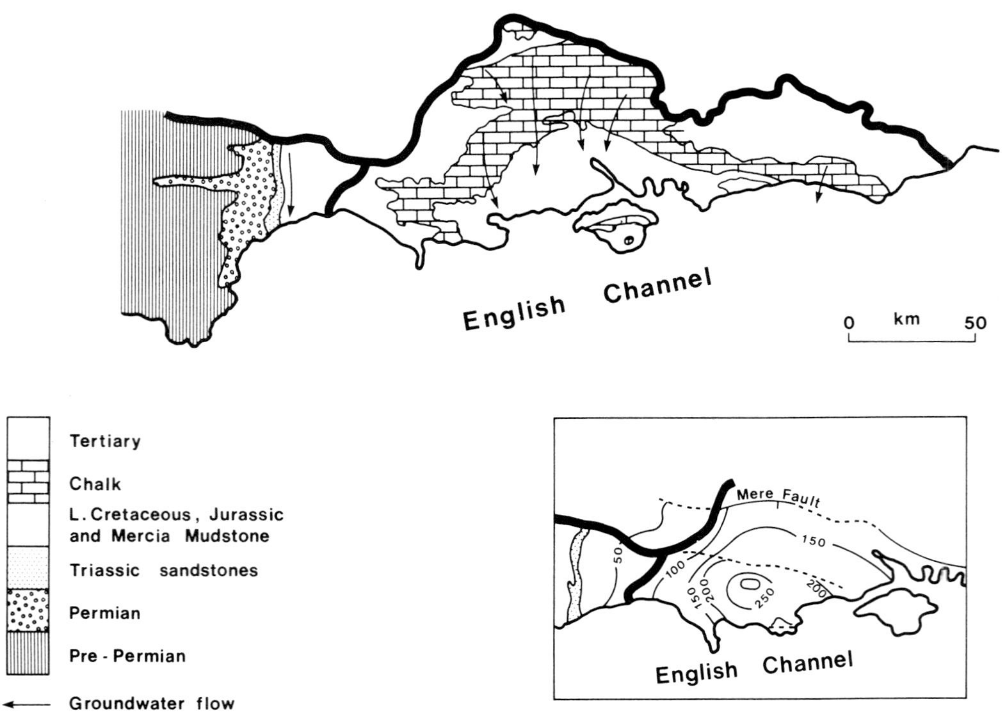

FIG. 8. Principal directions of groundwater flow in the Hampshire Province. The inset shows the salinity of groundwater in the Sherwood Sandstone in grams per litre. (After Allen & Holloway 1984.)

which was deformed by the Variscan orogeny. Rare assessments of the aquifer properties of these basement rocks in the Weald and near Southampton suggest that they have negligible permeability and this is probably the situation in the region as a whole (Allen & Holloway 1984).

The Chalk is clearly the main aquifer. Groundwater drains from the recharge areas along the higher ground to discharge points in the river systems crossing the outcrop and probably only relatively small amounts flow below the Tertiary cover (Fig. 8). In the western part of the Hampshire Basin many of the rivers flow from W to E, and although shallow groundwater flow is generally in the same direction it seems not improbable that deeper flow lines occur in the Chalk in a general N–S direction below the rivers and E–W catchment divides.

The Triassic Sherwood Sandstone is an important aquifer in the western part of the Wessex Basin. It increases in thickness in a westerly direction from about 50 m near Southampton, attaining maximum thicknesses of over 250 m NW of Dorchester and maximum depths NE of Dorchester (Allen & Holloway 1984). The maximum salinity of the groundwater coincides with the area of maximum depth and also with the distribution of salt in the Mercia Mudstone (Lott *et al.* 1982). The salinity of the water steadily increases from the outcrop in the W and also from the Mere–Portsdown–Middleton fault line which marks the northern limits of the main development of the Triassic sandstones in the Wessex Basin (Fig. 8).

Although the Triassic sandstones crop out in the western part of the Hampshire Province, the groundwater head in the Upper Cretaceous must exert an influence on the regional flow pattern. Groundwater in the Triassic sandstones outcrop is essentially drained to the S along the Otter Valley. Any water in the sandstones flowing down-dip to the E below the Mercia Mudstone is likely to be diverted to the S by the head in the Upper Cretaceous. If this is so, there is no direct recharge of the Triassic sandstones in the Wessex Basin from the outcrop. Water in the sandstones is possibly only replenished by lateral flow across the Mere–Portsdown–Middleton Fault and by vertical cross-formational flow from younger formations. This hypothesis is supported by the unusual chemical characteristics of the water.

Geochemical investigations of groundwaters, including interstitial fluids, have been carried out as part of the geothermal exploration programme of the Wessex Basin (Edmunds 1986). The salinity generally increases with depth in any one locality, reaching a maximum (around 300 g l^{-1} total mineralization) in the Permo–Triassic sandstones for which salinity contours have been drawn (Fig. 8). Groundwaters from the Triassic and overlying formations of the Wessex Basin have distinctive isotopic compositions compared with brines from other U.K. sedimentary basins (Fig. 3) with progressive relative enrichment in ^{18}O with increasing salinity. The increasing salinity is also accompanied by a decrease in the Br-to-Cl ratio indicating that halite dissolution has been an important factor in the evolution. In addition a marine starting composition for these brines is unlikely, the more so as one moves westwards in the basin. The evidence of ^{4}He values indicates that the residence time is in excess of 15 m.y., i.e. Miocene or older (J. N. Andrews, personal communication). By analogy with data from the Trunch Borehole, there would have been ample time since the Miocene for the salinity enhancement to have occurred by diffusion with evaporites in the Mercia Mudstone. The fact that waters with salinities in excess of 250 g l^{-1} occur in the Sherwood Sandstone, where the aquifer is at its greatest depth in the Wessex Basin (Fig. 8), may suggest that density settling in water that is essentially stationary is a factor enhancing the salinity. At the present time less saline waters are possibly flowing around this region in more permeable preferred flow routes.

The most likely explanation for the isotopic composition of the brine is that a highly saline brine has been diluted periodically with waters with isotopic compositions lying on the meteoric line. During the course of this dilution water in the basin underwent significant isotopic modification as a result of an ultrafiltration process, resulting in retention of the heavier isotope (Coplen & Hanshaw 1973), although the solute chemistry indicates that little if any membrane modification of ionic ratios has occurred. The driving force for this isotopic shift may have been cross-formational flow generated by the high potentials in the upland parts of the basin or tectonic pressures caused by the Alpine orogeny during which increased overpressures within the formation are likely to have occurred, combined with a rise in the geothermal gradient; the latter proposal is a factor which could have affected the Wessex Basin but not other U.K. sedimentary basins. Oxygen exchange between water and rock is unlikely to account for the isotopic shift since temperatures in the basin have never been higher than present values.

At present, water is probably entering the Triassic sandstones by vertical cross-formational flow and by lateral flow across the Mere–Portsdown–Middleton fault line and discharging by slow upward seepage in the lower parts of the Hampshire Basin. The area of this discharge appears to be reflected by the high heat flow in the Wessex Basin (Fig. 2). The Triassic sandstones are faulted against Upper Jurassic clays along the Purbeck–Isle of Wight disturbance (Colter & Havard 1981), thereby restricting lateral groundwater flow in the sandstones. The high heat flow occurs N of the disturbance and would seem to be due to the upward flow of groundwater, which was probably a factor in the migration of oil into the Triassic sandstones of the Wessex Basin.

Severn Province

This groundwater drainage basin encompasses S Wales and a large part of western England (Fig. 9), and includes a sequence from the Devonian to the Jurassic draining to outlets in the Bristol Channel. There are four major structural basins in the province—the S Wales Coalfield, the Bath–Bristol Carboniferous basin and the Permo–Triassic Worcester and Somerset basins. Because of their geographical distribution and geological structure, groundwater flow in each is an independent system.

Bath–Bristol Basin

The basin contains the best evidence in the U.K. for active groundwater circulation to a depth exceeding 2 km, resulting in the emergence of thermal springs at Bath (46.5 °C) and Bristol (24 °C). Geochemical and hydrogeological evidence shows that the thermal water has originated by recharge from the Mendip Hills (Andrews *et al.* 1982). A maximum water temperature of 80 °C (derived from geothermometric calculations) is likely to be reached during circulation to depths of between 2.7 and 4.3 km before the water rises up southerly-dipping E–W thrusts to the surface. Carbon isotope ratios suggest that water–rock equilibration has occurred within the Carboniferous Limestone, although some flow is thought to be derived from the Old Red Sandstone. The residence time of the bulk of the water is less than 10 000 years. Some of the chemistry of local shallow groundwaters is almost identical with that of the thermal waters, and thus suggests that leakage of thermal water into strata overlying the Palaeozoic may be occurring and that deep discharge is not only occurring at discrete thermal springs. If this is the case, it has important implications for groundwater flow in other areas

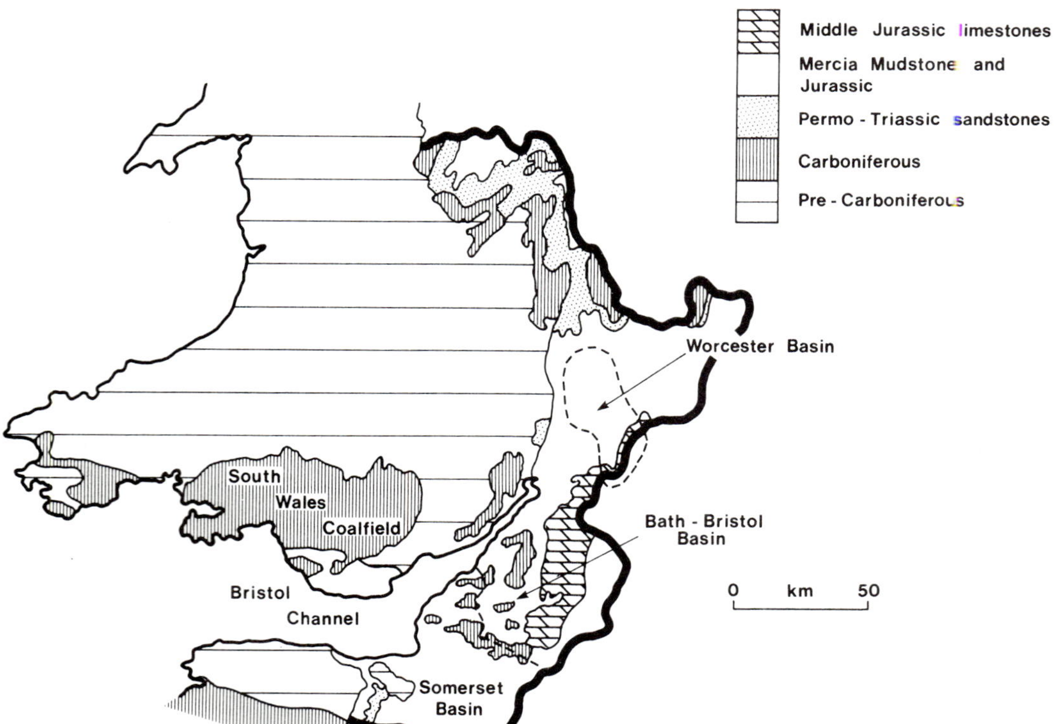

FIG. 9. Sedimentary basins in the Severn Groundwater Province.

with Carboniferous basins where undetected flow of relatively fresh water may be recharging across unconformities.

S Wales

The northern rim of the S Wales syncline consists of Carboniferous rocks and is at a higher altitude than the southern rim. This leads to the possibility that groundwater flows from N to S through aquifers below the coalfield. The aquifers in the Carboniferous are sandstones in the Coal Measures, particularly in the Upper Coal Measures, sandstones in the Millstone Grit, and the Carboniferous Limestone. In the context of regional flow attention is focused on the Carboniferous Limestone and to a lesser extent the Millstone Grit.

Thermal water, with a maximum temperature of 22°C, is found on the S crop of the coalfield, N of Cardiff, and is considered to have an analogous origin to that of Bath and Bristol. The temperature implies that the source depth is 600 m and, if some dilution is assumed, the Carboniferous Limestone appears to be the probable source rock. Dissolved inert-gas ratios suggest a recharge temperature that implies that infiltration has occurred some 500 m above the level of the spring outlet (Burgess *et al.* 1980). This supports the idea of N–S flow from Breconshire beneath the coalfield over a confined distance of some 25 km (Thomas *et al.* 1983).

In the eastern part of the S Wales Coalfield inrushes of water into mine workings occur in the lower seams some distance from outcrop. Initial flows under pressure can be of the order of 50–80 $l\,s^{-1}$ and exceptionally 200 $l\,s^{-1}$ (Ineson 1967), but they decline with time, sometimes ceasing but often continuing at a much reduced flow for many years. However, in a few cases very significant flows continue for a long time, for example 29 $l\,s^{-1}$ after 24 years, 20 $l\,s^{-1}$ after 15 years and 7 $l\,s^{-1}$ after 54 years, implying appreciable fissure permeability and storage in the source rocks for the water. In the eastern part of the coalfield the sequence between the lowest working seams and underlying sandstones in the Lower Coal Measures and Millstone Grit is very thin, amounting to about 10 m. In these circumstances water under pressure in the underlying sandstones bursts into the workings. The source of water in some cases is the Carboniferous Limestone with the water probably gaining access to the workings via faults, but this is unlikely to provide the general explanation for what is quite a widespread phenomenon and regional flow in the Millstone Grit and Lower Coal Measures sandstones is believed to be the main source of the water (Ineson 1967; M. J. Allen, personal communication, 1985).

The extent to which water flows below the coalfield from limestone and sandstone outcrops at the northern margin of the coalfield is unknown. It probably occurs to a limited extent, although the Taffs Well spring and the mine-water inflows provide the only evidence. However, one piece of indirect evidence for the presence of flow channels is the economic occurrence of haematite in the Carboniferous Limestone along the southern outcrop. One view (Williams, quoted by Squirrell & Downing 1969) is that the source of the iron was hydrothermal solutions circulating through the limestone, possibly in Tertiary times, a circulation that was associated with folding and faulting at that time. These earth movements could have fractured the limestone over extensive areas below the coalfield as recently as the Tertiary.

Worcester Basin

The Worcester Basin is a Permo–Triassic basin lying to the E of the Malvern Hills. The form of the basin is not evident from the surface geology because of the overlap of the Lower Jurassic. However, below the Jurassic rocks a thick Permo-Triassic sequence is preserved in two sub-basins arranged *en échelon* below Worcester and Winchcombe respectively (Smith & Burgess 1984). Groundwater flow in the basin has recently been considered by Black & Barker (1983) and Smith & Burgess (1984). The regional flow pattern is controlled by piezometric levels in the high ground surrounding the basin to the W, N and E. Along the outcrops of the Permo–Triassic sandstones to the W and N, where direct recharge is possible, groundwater levels are 60–100 m above sea level. Below the high ground of the Cotswold Hills the water level in the Jurassic limestones is 160–200 m above sea level. These high potentials around the rim of the basin induce groundwater flow through the Permo–Triassic sandstones at depth before rising vertically through the Mercia Mudstone to outlets in the Severn Valley. An analytical model has indicated how the flow in the Permo–Triassic sandstones is controlled by the ratio of the horizontal permeability of the Permo–Triassic sandstones to the vertical permeability of the Mercia Mudstone (Black & Barker 1983). At present the only measurements of piezometric pressure in the centre of the basin are from drill-stem tests in the Kempsey Borehole where the pressure surface in the Sherwood Sandstone was about 20 m above ground level (40 m above sea level). An interpretation of the regional hydrogeology appears to support the view that artesian conditions exist in the argillaceous Mercia Mudstone and Lower Lias in low-lying areas (Black & Barker 1983).

As well as the groundwater flow pattern from the high ground surrounding the Severn Valley to outlets in the valley, there is evidence for a deep N–S regional flow path in the Triassic sandstones across the Worcester Basin. A number of high heat-flow values have been calculated over the southern limb of the Worcester Basin, below the high ground of the Cotswold Hills, and this implies that groundwater is rising up this limb and flowing S into the Wessex Basin across the E–W faults that occur in the N of the latter basin.

An interesting feature of the Permo–Triassic in the Worcester Basin is the anomalously low salinity of the groundwater for the depth at which it occurs, although this can only be an observation at present as data are restricted to the Kempsey Borehole. Samples collected from drill-stem tests or by extracting formation water from cores indicated a total mineralization of about 6000 mg l^{-1} for the Wildmoor Sandstone and a range of 22 000–29 000 mg l^{-1} for the deeper Bridgnorth Sandstone (Smith & Burgess 1984). At Stratford-upon-Avon, some 22 km from the nearest outcrop, the water is potable. These facts imply that there is a significant groundwater flow down the Severn Valley that is possibly associated with a relatively narrow valley surrounded by high ground.

Somerset Basin

The Somerset Basin drains to the Bristol Channel and lies between the Mendip Hills and the Quantock Hills; the south-eastern boundary is part of the main surface-water divide of the S–W peninsula which in this area is formed by Jurassic rocks. The hydrogeology of the basin is not known in any detail, although a reconnaissance study has been made by Alexander & Noy (1981).

The recharge areas comprise Devonian, Carboniferous and Jurassic rocks forming the high ground at the periphery of the basin. The low-lying central region is composed of the Lias and Mercia Mudstone and these probably overlie Permo–Triassic sandstones at depth. It seems probable that water circulates from the marginal areas through these sandstones to outlets in the Bristol Channel or by upward flow through the overlying mudstones to the local river system.

Alexander & Noy (1981) used a mathematical model to evaluate the various possible forms of the regional flow paths and showed that they are closely related to the ratio of the permeability of the deep permeable sandstones to that of the overlying mudstones. High ratios, which might be expected to prevail, result in long flow paths in the sandstones from the margins of the basin. The presence of relatively thick halites in the Mercia Mudstone, with evidence of only small areas where they have been dissolved, suggests limited groundwater circulation in the mudstones.

NW Province

This province, lying to the W of the Pennines, extends from the Southern Uplands to the Weaver–Severn watershed in the S. The principal features are the Cheshire Basin in the S, the central region on the flanks of the Pennines (forming the eastern margin of the Irish Sea Basin), the Lower Palaeozoic massif of the Lake District and, in the N, the Carlisle Basin and its extension into the Vale of Eden (Fig. 10). In the main this province mirrors the northern part of the Eastern Province, the major difference being the absence of the thick Jurassic and Cretaceous sequences.

The Permo–Triassic sandstones in the Clwyd, Dee and Mersey valleys, in W Lancashire and around the Solway Firth form the main outlets for groundwater. Direct evidence for flow from Upper Palaeozoic rocks into the Permo–Triassic sandstones is limited. During the course of detailed groundwater resource assessments of the Triassic sandstones in the Fylde of Lancashire in the early 1970s it was concluded that an acceptable water balance requires a flow of some 30 000 m^3 day^{-1} into the sandstones from the Carboniferous to the E along a front of perhaps 15 km (Oakes & Skinner 1975). This view was supported by Sage & Lloyd (1978) who interpreted hydrochemical data to draw the same conclusion. Groundwater flow in the Fylde follows three routes—in the S to the Ribble, in the centre to the River Wyre and in the N to rivers draining to Morecambe Bay; the available evidence appears to preclude significant flow to the W below the Mercia Mudstone, but a slow flow probably does occur. The occurrence of oil at Formby implies that near the coast there is or has been a rising groundwater discharge zone originating in the Carboniferous. Apart from the possibly anomalous heat-flow value at Kirkham, heat flow in the Fylde is not above average but the data are sparse and more detailed information could change the picture near the coast.

Further S in the Cheshire Basin the flow direction is in a general north-westerly direction across the basin from the high ground in the S and SE of the basin. The Permo–Triassic sandstones lie at depths of more than 3 km in the

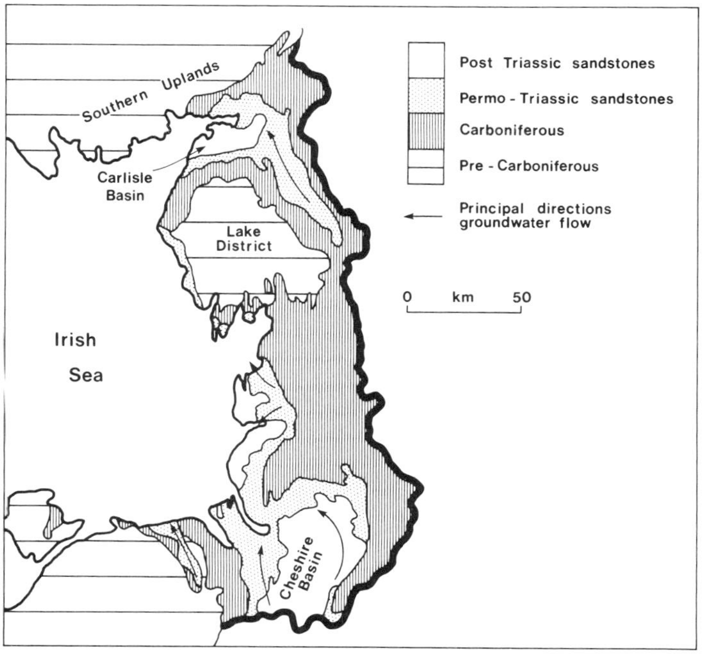

FIG. 10. Principal directions of groundwater flow in the NW Province.

centre of the basin and, although the total transmissivity is considered to exceed 10 D m (Gale *et al.* 1984a), flow probably tends to follow peripheral routes, around the deeper area, where the permeability is better and is enhanced by fractures.

The Upper Palaeozoic rocks below the deeper parts of the Cheshire Basin are likely to have very low permeabilities, and water flowing W in the Carboniferous sequences of the Pennines probably flows laterally into the Triassic sandstones. The heat flow in the Cheshire Basin tends to be low, implying that rising groundwater flow is not a feature. This supports the regional flow directions just described, with the water moving mainly around the deeper parts of the basin in essentially horizontal directions to outlets in the Mersey estuary.

Midland Valley of Scotland

The Midland Valley graben between the Highland Boundary and Southern Upland faults forms a distinct groundwater province predominantly in Upper Palaeozoic rocks (Fig. 1). The detailed geology of the region is complex, but in general terms sedimentary and volcanic rocks of the Old Red Sandstone crop out adjacent to the boundary faults and are overlain by Carboniferous rocks in the centre (Bennett & Harrison 1980). The hydrogeology has been examined in a number of areas in some detail, but the regional form of the water table and piezometric surface is incompletely known. It must be complex with many local flow systems related to individual outcrops and catchment areas. In general, natural regional flow paths would be expected to be related to the major surface divides with drainage to the main rivers, the Forth, Clyde and Tay. In the central part of the Midland Valley heat flows for several sites are above average at 60 mW m^{-2}, but interpretation of the data is not possible at this stage.

The Old Red Sandstone, particularly the Knox Pulpit Formation, could have significant permeability at depth below the central region (Browne *et al.* 1985). Because the Carboniferous is an interbedded sequence of sandstones, limestones and mudstones, there must be a preference for lateral flow in the sandstones and limestones, particularly through fractures. Dewatering to allow deep mining of coal has modified natural groundwater flow paths but, with the reduction of mining, groundwater levels are recovering in many areas.

Orcadian Province

Extremely thick Old Red Sandstone molasse deposits, forming the on-shore part of the

Orcadian Basin, are found in NE Scotland, the Orkneys and Shetlands. They consist predominantly of sandstones, generally with low permeabilities, and they are important as aquifers only locally, for example on the southern shore of the Moray Firth (N. Robins, personal communication, 1985). Nevertheless, they represent a distinct groundwater province (Fig. 1).

Some 130 km E of Aberdeen, in the Buchan Field of the North Sea, Devonian sandstones are oil bearing. The reservoir is a fluviatile sandstone about 600 m thick at a depth of over 2700 m. The sandstones are well cemented with low matrix permeabilities, but they are extremely fractured. Fractures make a major contribution to the productivity, although there may also be a considerable contribution from the matrix (Butler *et al.* 1976). In the present context, the principal interest in the development of oil in the Buchan Field is that it demonstrates that significant fluid flows are possible in the Old Red Sandstone at appreciable depths where fractures, probably associated with major faults, are important controls of permeability.

Northern Ireland

In Northern Ireland groundwater is derived from Carboniferous, Permian, Triassic and Cretaceous rocks. The Devonian rocks are very indurated with little primary permeability. They are of little importance for water supply and could probably be regarded as part of the impermeable basement which is represented mainly by the Lower Palaeozoic and the Precambrian crystalline basement.

A prominent feature of the eastern part of Northern Ireland is a deep sedimentary basin in which deposition has been virtually continuous since the Carboniferous and which can be regarded as a continuation of the Midland Valley of Scotland (Bennett & Harrison 1980). Two deep Triassic sub-basins exist containing thick Triassic and Permian sandstones. These rocks can have significant primary porosity and permeability, at depths of 1–2 km, particularly in the upper part of the Sherwood Sandstone, although in some areas and at greater depths the properties have been destroyed by secondary cementation. Nevertheless, deep regional flow paths probably exist in the Permo–Triassic sandstones which are over 2 km deep in the centre of the basins.

The waters in the sandstones in these deep areas are very saline with values of up to 200 g l^{-1}. Stable isotope ratios indicate they are of meteoric origin and low temperature gradients in the sandstones imply that deep groundwater circulation is still occurring (Bennett 1983).

Conclusions

Four groundwater provinces can be recognized in England and Wales, and three others occur in Scotland and Northern Ireland. Conceptual models for regional groundwater flow systems in these provinces are supported by data for several areas, particularly hydrochemical and heat-flow data. It is clear that the distribution of the Permo–Triassic sandstones exerts a dominant influence. These aquifers are major drains for sub-surface flow in pre-Jurassic rocks, leading to outlets in the low-lying areas formed by the Mercia Mudstone in, for example, the lower parts of the Trent, Yorkshire Ouse, Severn, Mersey, Weaver and Eden river systems. The preservation of salt deposits in the Mercia Mudstone in several areas implies that vertical cross-formation flow in the mudstones is not important and that deep horizontal flow paths are a feature in the Triassic sandstones.

Intermediate flow systems have formed in response to the scarp-and-vale topography of the Mesozoic rocks of eastern and southern England. Examples can be recognized in the Corallian of the Thames Valley, the Lincolnshire Limestone and the Lower Greensand of Cambridgeshire. The influence of the Chalk escarpment, such a dominant feature of eastern and southern England, is not entirely clear. In some areas it may prevent flow beneath it but more probably flow is only restricted.

The flow systems recognized are only the onshore parts of more extensive aquifer systems that extend beneath the shallow seas surrounding the U.K. Preferred flow routes are probably important avenues of groundwater flow whether they are solution fissures, fault fractures or more permeable layers or vertical zones in both sandstones and mudstones. Outlets from confined aquifers are probably by means of preferred flow routes through the confining beds. In thick homogeneous aquifers where flow velocities are very small, density settling may be a factor that enhances the salinity, as possibly in the Sherwood Sandstone of Lincolnshire and Wessex. In other areas argillaceous rocks increase the salinity by membrane filtration.

The drainage basins in the U.K. are small. This, together with the fact that low-lying ground formed by shales between aquifers acts as sinks for groundwater flow in intermediate flow systems, means that flow paths are of limited length. A consequence is that saline groundwater occurs in aquifers at relatively small distances down-gradient of outcrop. Very extensive confined aquifers containing fresh water do not occur. The Lower Greensand and the Chalk in the London Basin are the only exceptions, but even here the scale is small.

A feature that has been remarked upon in this paper is the apparent coincidence of high heat flows in the E Midlands and Wessex with the occurrence of oil fields in these regions. The link is possibly regional groundwater flow patterns, but it could be with palaeo-flow systems that had similar patterns to those of the present day.

The flow systems discussed in this paper have in the main related to natural systems, but in a number of areas they have been modified by extensive development of groundwater since the latter part of the nineteenth century. For example, the main outlet for groundwater flow in the London Basin is now the well field in the Chalk. Coal-mine drainage has also reversed the hydraulic gradient in sediments overlying the workable Coal Measures. Drainage of the Kent Coalfield has produced a downward potential in the Lower Greensand and presumably induced leakage from the Chalk into the Lower Greensand. Similar situations exist in other coalfields where downward potentials have developed in Permian and Triassic rocks.

The data currently available with a bearing on regional groundwater flow have been reviewed in this paper. For a number of areas the interpretation is conceptual and is based on limited generalized data. A better understanding of regional flow systems in the U.K. can only come from the careful collection and analysis of pressure and temperature measurements in deep aquifers, the chemical analysis of formation waters from cores and production wells, and the measurement of aquifer properties from core samples and deep-well tests. The measurement of bottom-hole temperatures and geothermal gradients in deep exploration boreholes and the calculation of heat flow can throw considerable light on groundwater flow patterns and could have significant bearing on interpreting secondary migration paths of oil.

ACKNOWLEDGMENTS: This review has drawn on the work of many colleagues in the British Geological Survey and this is gratefully acknowledged. The authors also thank Dr J. A. Barker, Mr D. A. Gray and Mr M. Price for kindly reviewing the paper in manuscript. The paper is published by permission of the Director of the British Geological Survey.

References

ALEXANDER, J. & BRIGHTMAN, M. A. 1985. A hypothesis to account for groundwater quality variations in the Corallian of the Thames Valley. *In: Hydrogeology in the Service of Man, Mem. 18th Congr. of the International Association of Hydrogeologists*, pp. 54–63.

—— & NOY, D. J. 1981. Hydrogeological reconnaissance study: the Somerset Basin. *Rep. ENPU81–15*, Institute of Geological Sciences.

ALLEN, D. J. & HOLLOWAY, S. 1984. The Wessex Basin. *Investigation of the Geothermal Potential of the U.K.*, British Geological Survey.

ANDREWS, J. N. & LEE, D. J. 1980. Dissolved gases as indicators of groundwater mixing in a Jurassic Limestone aquifer. *Proc. 3rd Symp. on Water–Rock Interaction, Edmonton, Alberta, 1980*, Alberta Research Council.

——, BALDERER, W., BATH, A. H., CLAUSEN, H. B., EVANS, G. V., FLORKOWSKI, T., GOLDBRUNNER, J., IVANOVICH, M., LOOSLI, M. & ZOJER, H. 1984. Environmental isotope studies in two aquifer systems: a comparison of groundwater dating methods. *Proc. Symp. on Isotope Hydrology. In: Water Resources Development*, IAEA, Vienna.

——, BURGESS, W. G., EDMUNDS, W. M., KAY, R. L. F. & LEE, D. J. 1982. The thermal springs of Bath. *Nature, Lond.* **298**, 339–43.

ANDREWS-SPEED, C. P., OXBURGH, E. R. & COOPER, B. A. 1984. Temperatures and depth-dependent heat flow in western North Sea. *Bull. Am. Assoc. petrol. Geol.* **68**, 1764–81.

BATH, A. H. & EDMUNDS, W. M. 1981. Identification of connate water in interstitial solution of Chalk sediment. *Geochim. cosmochim. Acta*, **45**, 1449–61.

——, DOWNING, R. A. & BARKER, J. A. 1985. The age of groundwaters in the Chalk and Pleistocene sands of north-east Suffolk. *Rep. WD/ST/85/1*, British Geological Survey.

——, EDMUNDS, W. M. & ANDREWS, J. N. 1979. Palaeoclimatic trends deduced from hydrochemistry of a Triassic sandstone aquifer, United Kingdom. *In: Isotope Hydrology 1978*, IAEA, Vienna.

BENNETT, J. R. P. 1983. The sedimentary basins in Northern Ireland. *Investigation of the Geothermal Potential of the U.K.*, British Geological Survey.

—— & HARRISON, I. B. 1980. *Explanatory Notes for the International Hydrogeological Map of Europe, Sheet B3, Edinburgh*, UNESCO, Paris.

BLACK, J. H. & BARKER, J. A. 1983. Hydrogeological reconnaissance study of the Worcester Basin. *Rep. ENPU 81–3*, Institute of Geological Sciences.

BROWNE, M. A. E., HARGREAVES, R. L. & SMITH, I. F. 1985. The Upper Palaeozoic Basins of the Midland Valley of Scotland. *Investigation of the Geothermal Potential of the U.K.*, British Geological Survey.

BULLARD, E. C. & NIBLETT, E. R. 1951. Terrestrial heat

flow in England. *Mon. Not. R. astr. Soc. geophys. Suppl. No. 6*, pp. 222–38.

BURGESS, W. G., EDMUNDS, W. M., ANDREWS, J. N., KAY, R. L. F. & LEE, D. J. 1980. The hydrogeology and hydrochemistry of the thermal water in the Bath–Bristol Basin. *Investigation of the Geothermal Potential of the U.K.*, British Geological Survey.

BUTLER, M., PHELAN, M. J. & WRIGHT, A. W. R. 1976. Buchan Field: evaluation of a fractured sandstone reservoir. *Trans. 4th European Formation Evaluation Symp.*, Society of Professional Well Log Analysts.

CARSTENS, H. & FINSTAD, K. G. 1981. Geothermal gradients of the Northern North Sea Basin, 59–62°N. *In:* ILLING, L. V. & HOBSON, G. D. (eds) *Petroleum Geology of the Continental Shelf of North West Europe*, pp. 152–61, Institute of Petroleum, London.

COLTER, V. S. & HAVARD, D. J. 1981. The Wytch Farm Oil Field, Dorset. *In:* ILLING, L. V. & HOBSON, G. D. (eds) *Petroleum Geology of the Continental Shelf of North-West Europe*, Institute of Petroleum, London.

COPLEN, T. B. & HANSHAW, B. B. 1973. Ultrafiltration by a compacted clay membrane, I, Oxygen and hydrogen isotopic fractionation. *Geochim. cosmochim. Acta*, **37**, 2295–310.

DOWNING, R. A. & HOWITT, F. 1969. Saline groundwaters in the Carboniferous rocks of the English East Midlands in relation to the geology. *Q. J. eng. Geol.* **1**, 241–69.

—— & WILLIAMS, B. P. J. 1969. *The Groundwater Hydrology of the Lincolnshire Limestone*, Water Resources Board, Reading, Berks.

——, ALLEN, D. J., BIRD, M. J., GALE, I. N., KAY, R. L. F. & SMITH, I. F. 1985. Cleethorpes No 1 Geothermal Well—a preliminary assessment of the resource. *Investigation of the Geothermal Potential of the U.K.*, British Geological Survey.

——, PEARSON, F. J. & SMITH, D. B. 1979. The flow mechanism in the Chalk based on radio-isotope analyses of groundwater in the London Basin. *J. Hydrol.* **40**, 67–83.

——, SMITH, D. B., PEARSON, F. J., MONKHOUSE, R. A. & OTLET, R. L. 1977. The age of groundwater in the Lincolnshire Limestone, England, and its relevance to the flow mechanism. *J. Hydrol.* **33**, 201–16.

EDMUNDS, W. M. 1971. Hydrogeochemistry of groundwaters in the Derbyshire Dome with special reference to trace constituents. *Rep. 71/7*, Institute of Geological Sciences.

—— 1973. Trace element variations across an oxidation–reduction barrier in a limestone aquifer. *In: Proc. Tokyo Symp. on Hydrogeochemistry and Biogeochemistry*, Vol. 1, pp. 500–26, Clarke, Washington, DC.

—— 1976. Chemical variation of pore waters from the Cretaceous Chalk of southern England. *Proc. Symp. on Water–Rock Interaction, Prague, 1974*, pp. 266–7, Geological Survey, Prague.

—— 1986. Geochemistry of geothermal waters in the UK. *In:* DOWNING, R. A. & GRAY, D. A. (eds) *Geothermal Energy—The Potential in the United Kingdom*, HMSO, London.

——, BATH, A. H. & MILES, D. L. 1982. Hydrochemical evolution of the East Midlands Triassic sandstone aquifer, England. *Geochim. cosmochim. Acta*, **46**, 2069–81.

——, LOVELOCK, P. E. R. & GRAY, D. A. 1973. Interstitial water chemistry and aquifer properties in the Upper and Middle Chalk of Berkshire, England. *J. Hydrol.* **19**, 21–31.

——, MILES, D. L. & COOK, J. M. 1984. A comparative study of sequential redox processes in three British aquifers. *In: Hydrochemical Balances of Freshwater Systems*, pp. 55–70, Publication 150, International Association of Hydrological Sciences.

EVANS, G. V., OTLET, R. L., DOWNING, R. A., MONKHOUSE, R. A. & RAE, G. 1979. Some problems in the interpretation of isotope measurements in United Kingdom aquifers. *In: Isotope Hydrology 1978*, IAEA, Vienna.

FREEZE, R. A. & WITHERSPOON, P. A. 1966. Theoretical analysis of regional groundwater flow, 1, Analytical and numerical solutions to the mathematical model. *Water Resour. Res.* **2**, 641–56.

—— & —— 1967. Theoretical analysis of regional groundwater flow, 2, Effect of water-table configuration and sub-surface permeability variation. *Water Resour. Res.* **3**, 623–34.

GALE, I. N., EVANS, C. J., EVANS, R. B., SMITH, I. F., HOUGHTON, M. T. & BURGESS, W. G. 1984a. The Permo–Triassic aquifers of the Cheshire and West Lancashire basins. *Investigation of the Geothermal Potential of the U.K.*, British Geological Survey.

——, HOLLIDAY, D. W., KIRBY, G. A. & ARTHUR, M. J. 1984b. The Carboniferous rocks of Lincolnshire, Nottinghamshire and southern Humberside. *Investigation of the Geothermal Potential of the U.K.*, British Geological Survey.

HAWKINS, P. J. 1978. Relationship between diagenesis, porosity reduction and oil emplacement in late Carboniferous sandstone reservoirs, Bothamsall Oilfield, East Midlands. *J. geol. Soc Lond.* **135**, 7–24.

HUBBERT, M. K. 1940. The theory of ground-water motion. *J. Geol.* **48**, 785–944.

INESON, J. 1967. Groundwater conditions in the Coal Measures of the South Wales Coalfield. *Hydrogeol. Rep. No 3, Institute of Geological Sciences.*

—— & DOWNING, R. A. 1963. Changes in the chemistry of ground waters of the Chalk passing beneath argillaceous strata. *Bull geol. Surv. GB No 20*, pp. 176–92.

LLOYD, J. W., HARKER, D. & BAXENDALE, R. A. 1981. Recharge mechanisms and groundwater flow in the Chalk and drift deposits of southern East Anglia. *Q. J. eng. Geol.* **14**, 87–96.

LOTT, G. K., SOBEY, R. A., WARRINGTON, G. & WHITTAKER, A. 1982. The Mercia Mudstone Group (Triassic) in the western Wessex Basin. *Proc. Ussher Soc.* **5**, 340–6.

MATHER, J. D., GRAY, D. A., ALLEN, R. A. & SMITH, D. B. 1973. Groundwater recharge in the Lower Greensand of the London Basin—results of tritium nc carbon-14 determinations. *Q. J. eng. Geol.* **6**, 41–52.

MONKHOUSE, R. A. 1974. *An Assessment of the Groundwater Resources of the Lower Greensand in the Cambridge–Bedford region*, Water Resources Board, Reading, Berks.

MOREL, E. H. 1979. A numerical model of the Chalk aquifer in the Upper Thames Basin. *Tech. Note* 3, Central Water Planning Unit, Reading, Berks.

OAKES, D. B. & SKINNER, A. C. 1975. The Lancashire Conjunctive Use Scheme groundwater model. *Tech. Rep. 12,* Water Research Centre, Medmenham, Berks.

OWEN, M. & ROBINSON, V. K. 1978. Characteristics and yield of the fissured Chalk. *In: Thames Groundwater Scheme*, Institution of Civil Engineers, London.

RICHARDSON, S. W. & OXBURGH, E. R. 1978. Heat flow, radiogenic heat production and crustal temperatures in England and Wales. *J. geol. Soc. Lond.* **135**, 323–7.

RODDA, J. C., DOWNING, R. A. & LAW, F. M. 1976. *Systematic Hydrology*, Newnes–Butterworths, London.

SAGE, R. C. & LLOYD, J. W. 1978. Drift deposit influences on the Triassic Sandstone aquifer of NW Lancashire as inferred by hydrochemistry. *Q. J. eng. Geol.* **11**, 209–18.

SMITH, I. F. & BURGESS, W. G. 1984. The Permo–Triassic rocks of the Worcester Basin. *Investigation of the Geothermal Potential of the U.K.*, British Geological Survey.

SQUIRRELL, H. C. & DOWNING, R. A. 1969. *Geology of the South Wales Coalfield, 1, The Country around Newport (Mon), Mem. geol. Surv. GB*, HMSO, London.

THOMAS, L. P., EVANS, R. B. & DOWNING, R. A. 1983. The geothermal potential of the Devonian and Carboniferous rocks of South Wales. *Investigation of the Geothermal Potential of the U.K.*, British Geological Survey.

TÓTH, J. 1962. A theory of groundwater motion in small basins in central Alberta, Canada. *J. geophys. Res.* **67**, 4375–87.

—— 1963. A theoretical analysis of groundwater flow in small drainage basins. *J. geophys. Res.* **68**, 4795–812.

WATER RESOURCES BOARD 1972. *The Hydrogeology of the London Basin.* Water Resources Board, Reading, Berks.

WHEILDON, J. & ROLLIN, K. E. 1986. Heat flow. *In:* DOWNING, R. A. & GRAY, D. A. (eds) *Geothermal Energy—the Potential in the United Kingdom*, HMSO, London.

WHITTAKER, A. (ed.) 1985. *Atlas of Onshore Sedimentary Basins in England and Wales*, Blackie, Glasgow.

—— & CHADWICK, R. A. 1984. The large-scale structure of the Earth's crust beneath southern Britain. *Geol. Mag.* **121**, 621–4.

WOODLAND, A. W. 1946. Water supply from underground sources of the Cambridge–Ipswich District. *Wartime Pamphlet No. 20, Pt. 10, Geol. Surv. G.B.*

R. A. DOWNING, W. M. EDMUNDS & I. N. GALE, British Geological Survey, Wallingford, Oxon OX10 8BB, U.K.

Fluid flow and diagenesis in the East Midlands Triassic sandstone aquifer

A. H. Bath, A. E. Milodowski & G. E. Strong

SUMMARY: The flow system, hydrochemistry and isotope hydrology of the freshwater aquifer in the E Midlands Triassic sandstone are reviewed. Isotopic data suggest a mean flow velocity of around 0.6 m.y. over the last 30 k.y., in contrast with the present-day velocity of around 0.2 m.y. The difference is attributed to changes in hydraulic gradients over this period; fluctuations in recharge and gradients would have influenced meteoric water influx and diagenesis over the last 10^7 years or more. The diagenetic modifications to the arkosic sandstone are described and are divided into those taking place before and during compaction, probably with saline pore water, and those occurring as a result of meteoric water influx following uplift and denudation.

The relation between the kinetics of diagenetic reactions and the groundwater flow velocity is discussed. Relevant experimental data for reaction kinetics are summarized. The hydrochemistry and flow velocity of the present regime are used to deduce mass transfers within and outside the aquifer. These are equivalent to uniform porosity increases of only about 0.1% per million years for each of the dolomite and K-feldspar dissolution reactions. Both these dissolution reactions are incongruent, and precipitate calcite and kaolinite respectively. The spatial arrangement of reactant and product minerals in incongruent reactions is expected to reflect interplay between dissolution/precipitation kinetics and water flow rate. Scanning electron micrographs show the proximity of K-feldspar and kaolinite. Constraints can be placed on the flow rates responsible by considering chemical balances for aluminium and silicon and their maximum production rates by feldspar dissolution.

This approach involves many simplifications and assumptions, but with further experimental kinetic data it offers the possibility of deducing information on past flow systems from detailed measurements of diagenetic mineralogy and geochemistry in clastic sediments.

Studies of the present-day hydrogeological conditions and hydrochemistry of aquifers provide abundant information on the directions, rates and mechanisms of flow. The geochemical reactions taking place between water and rock in the aquifer can be deduced from the evolution of the water chemistry. These reactions and the resulting mineralogical alterations are also the most recent phase of a continuous sequence of diagenetic changes which are controlled by fluid flow and mineralogy. Conversely, diagenetic mineralogy of sediments in basins is being applied increasingly to understanding and predicting the combination of depositional, burial and fluid conditions which have determined particular reservoir properties (McDonald & Surdam 1984). These interpretations rely on geochemical models for water–rock interaction and on present analogues for past diagenetic changes. Experimental geochemical data for these reactions and their rates are also an important contribution.

The purpose of this paper is to summarize our knowledge of water movement in one of the Permo–Triassic sandstone aquifers of England and also to describe the sequence of diagenetic reactions including those occurring now. The example used is the Triassic sandstone aquifer of the E Midlands which contains freshwater from outcrop to at least 20 km down gradient in the confined aquifer on the western rim of the North Sea Basin.

Comparison with other Permo–Triassic basinal sequences demonstrates how the present-day groundwater chemistries are controlled by the present flow conditions and also indirectly by the past flows and diagenetic history.

E Midlands Triassic aquifer

The aquifer comprises the Sherwood Sandstone which forms a recharge zone in its outcrop extending northwards from Nottingham. The uniform easterly dip of about 1 in 50 takes the aquifer into confined conditions below Mercia Mudstone. Permian marls and dolomitic limestones form a relatively impermeable base and apparently isolate the Triassic aquifer from the underlying brines of the Carboniferous strata. The aquifer provides a very important source of freshwater and is heavily pumped at boreholes in the unconfined and confined aquifer at depths up to 500 m about 20 km from outcrop. The mean intergranular porosity is around 30% with a mean intergranular hydraulic conductivity of 3 m day^{-1} (Lovelock 1977). Fissuring may locally

From GOFF, J. C. & WILLIAMS, B. P. J. (eds), 1987, *Fluid Flow in Sedimentary Basins and Aquifers,* Geological Society Special Publication No. 34, pp. 127–140.

enhance permeability, as proven by field pumping tests and borehole flow logs (Williams *et al.* 1972); however, it is probable that a low frequency of fissure interconnection causes regional flow to be controlled by intergranular conductivity.

The natural hydraulic gradients are low (Fig. 1), although owing to the heavy abstraction from the aquifer the hydraulic pressure distribution has been disturbed, resulting in severe drawdowns at several of the confined aquifer sites and a possible readjustment of regional flows as shown in Fig. 2. The present topography of the land surface comprises a gentle slope eastwards towards the North Sea coast; however, topographic highs occur locally at Retford and Lincoln (as illustrated in the section in Fig. 3), corresponding to Mercia Mudstone and Lincolnshire Limestone scarps. These highs in relatively low-permeability formations might be sufficient to give groundwater heads which propagate into the underlying aquifer and influence the flow as suggested in Fig. 3. This possibility will be examined in the light of the hydrochemical and isotopic evidence.

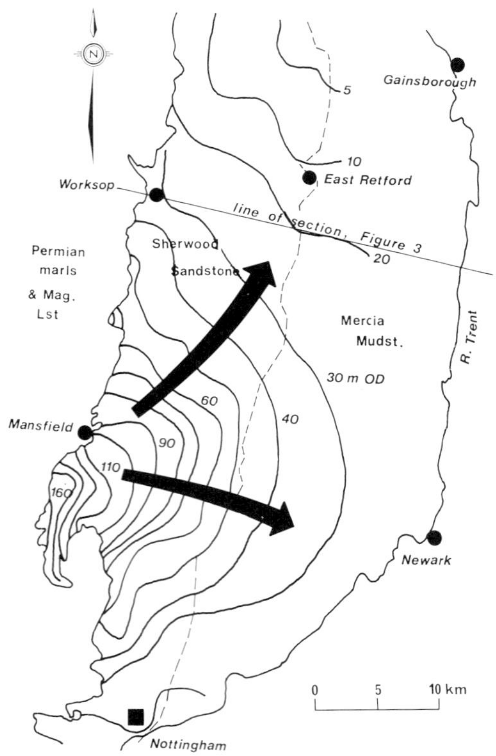

FIG. 1. Interpolated contours of the natural potentiometric surface of groundwater in the Triassic sandstone aquifer of the E Midlands. The line of the section in Fig. 3 is indicated. (After Land 1966; Institute of Geological Sciences 1981.)

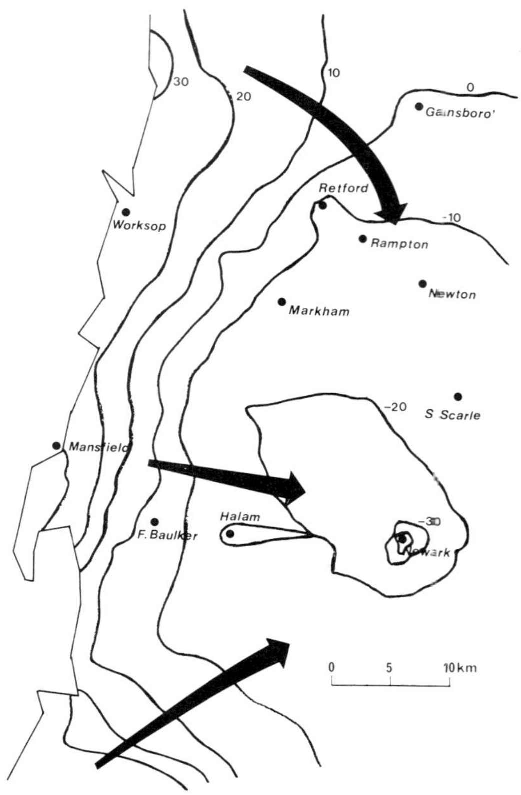

FIG. 2. Contours of the potentiometric surface computed by a finite-element simulation of steady-state abstraction under the present pumping regime (W. Balderer, personal communication). The locations of pumped groundwater samples for isotopic studies reported by Andrews *et al.* (1984) are shown.

Hydrochemistry, flow and transit times

The hydrochemistry of the Triassic aquifer has been described fully by Edmunds *et al.* (1982); the dominant chemical changes reflect anthropogenic inputs and natural reactions with the carbonate and sulphate minerals in the formation. The major chemical features define three aquifer zones:

1 the unconfined aquifer which is contaminated with introduced nitrate and chloride, contains dissolved oxygen and is already close to or at chemical equilibrium with dolomite and calcite;
2 the shallow confined aquifer which is virtually

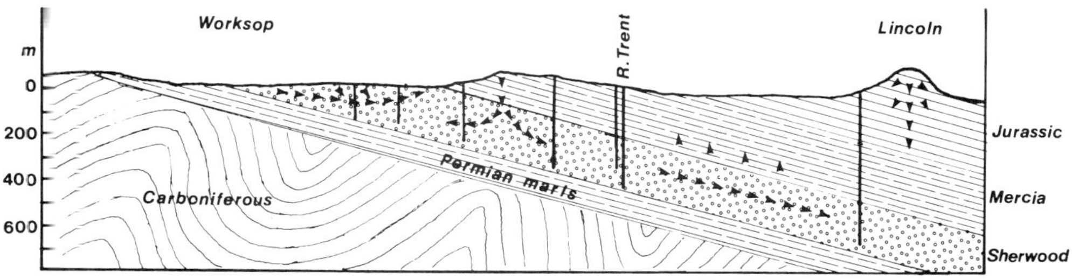

FIG. 3. Cross-section between Worksop and Lincoln showing geological strata and possible groundwater flow directions in the Sherwood Sandstone aquifer (stippled).

free from nitrate, becomes reducing and has lower sulphate and chloride concentrations;
3 the deeper confined aquifer which is reducing and may contain sulphide as well as increasing sulphate and also has low chloride levels.

However, borehole and analytical records show that a borehole at Lincoln, drilled in 1908 into the Sherwood Sandstone at a depth of 671 m, produced moderately saline water at 26 °C dominated by calcium, sulphate and chloride. The increasing salinity to 3300 mg TDS l^{-1} led to the abandonment of this groundwater source. This indicates that the freshwater penetrating some 20 km into the confined aquifer merges into progressively more mineralized water of unknown origin down-dip. An example of conditions even deeper in the basin is given by water sampled from the Sherwood Sandstone in the British Geological Survey–Department of Energy Geothermal Project borehole at Cleethorpes, which contains about 44 000 mg $Cl^- l^{-1}$ at a depth of 1300 m; water from the underlying less-transmissive Permian sands at Cleethorpes is dramatically different, containing about 135 000 mg $Cl^- l^{-1}$ (Downing *et al.* 1985).

Thermonuclear tritium and ^{14}C contents, as well as stable oxygen and hydrogen isotope ratios and dissolved inert gas and radiogenic helium contents have been applied to the problem of estimating groundwater ages (Andrews & Lee 1979; Bath *et al.* 1979). An international cooperative study also evaluated the applicability of a wider range of isotopic techniques by measurements on this aquifer (Andrews *et al.* 1984). The relevant data from this study are summarized in Fig. 4. Both studies have underlined the problem of quantitatively interpreting these data, but they provide valuable information on the approximate average residence times and on flow directions and mixing. The isotopic data—^{14}C, $^{18}O/^{16}O$, $^{2}H/^{1}H$ and ^{4}He—confirm that late-Pleistocene-age groundwater up to 30 k.y. old is being pumped from the third and deepest freshwater aquifer zone. Moreover, the remarkably low chloride concentrations found in the deeper two zones are lower than would occur as a result of present climatic and rainfall conditions. Stable isotope and dissolved inert gas data both indicate significantly lower temperatures at the time of recharge of these deeper fresh groundwaters.

The distribution of the results of various isotopic and chemical measurements indicates qualitative and semiquantitative features of flow within the aquifer and adjacent formations. The very low salinities are convincing evidence against significant upwards leakage of Carboniferous brines through the Permian—at least in the accessible part of the aquifer and at Lincoln where the hydrochemistry with high $CaSO_4$ is very unlike that of Coal Measures solutions (Downing & Howitt 1969).

The relatively sharp changes in chloride and ^{14}C contents, stable oxygen and hydrogen isotopes and inert gas contents around the second zone indicate that down-dip flow in the aquifer is not steady but may be influenced by the local development of hydraulic pressure and vertical water movement in overlying Mercia Mudstone which forms the high ground just E of its feather edge (Fig. 3). Thus natural flow in the Triassic is suggested to occur in two regimes, meeting and mixing in the second zone. The more active flow system in the outcrop sandstone contributes to the deeper confined regime. Of course this flow system has been substantially disturbed and readjusted by the pumping over the last century (Fig. 2). The combined isotopic data give an indication of the overall mixing from different sources and of different ages since the ^{85}Kr, ^{3}H and ^{39}Ar isotopes (half-lives of 10.76 years, 12.43 years and 269 years respectively) are sensitive indicators of hydrogeologically 'recent' water, whereas ^{14}C (half-life 5730 years) and ^{4}He (accumulating radiogenically from uranium and thorium decay) indicate long time-scales. Thus it is possible to construct a diagram such as Fig. 5 in which the relative contributions of waters of various age ranges to each pumped mixture are indicated.

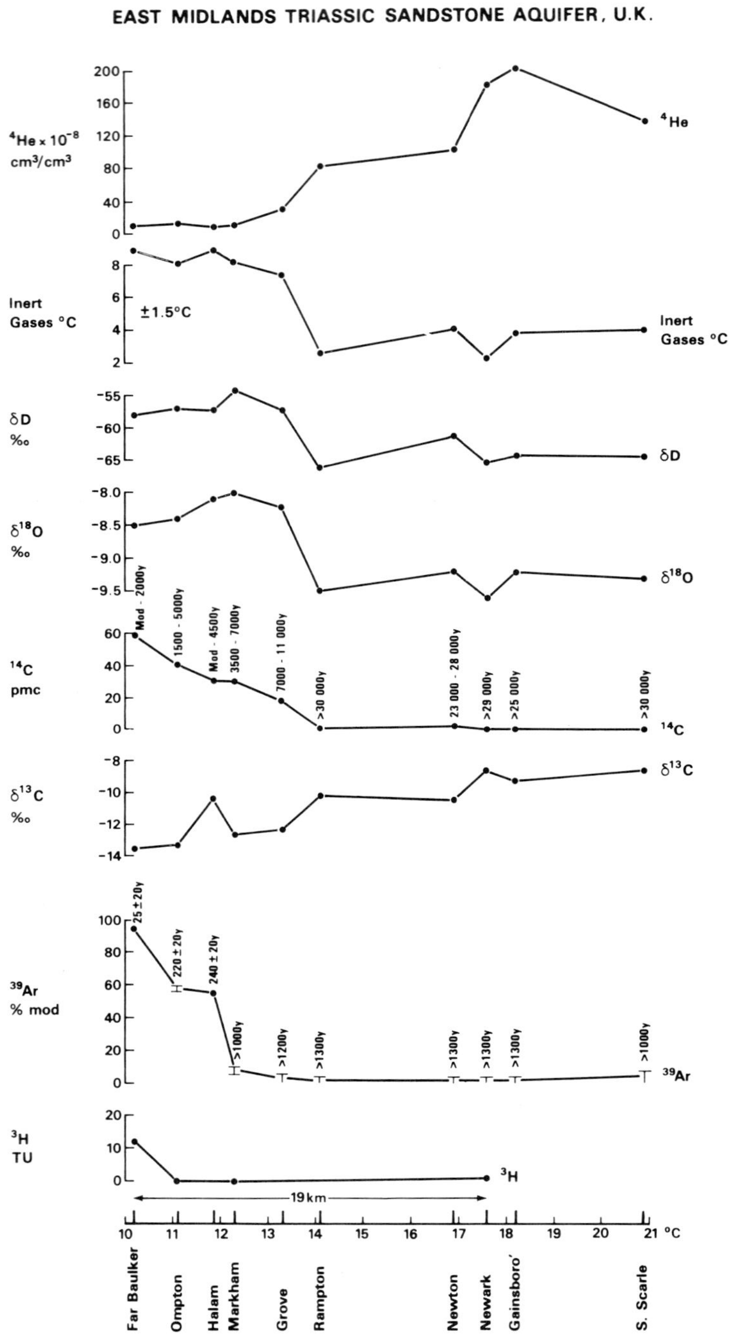

FIG. 4. Compilation of isotopic and inert gas data reported for the Triassic sandstone aquifer by Andrews *et al.* (1984). The data are plotted against the groundwater temperature on the bottom axis since this is the best indicator of average depth and position in the aquifer from which groundwater is being pumped.

It is not known whether the groundwater reported at Lincoln derives from the same late-Pleistocene recharge episode or whether it is a mixture with much older basinal formation water. The pre-pumping discharge pathways for water flowing down-gradient are uncertain but could be by upwards leakage through the overlying sediments which have low topography towards the E coast (except at the Jurassic limestone outcrop at Lincoln). If this were the case, the history of Pleistocene and post-glacial flow through the aquifer has been determined by interplay between the progressive reduction of topographic relief by glacial effects and fluctuations in sea level which may provide the constant-head boundary as the basin dips below the North Sea. Sea level reached a minimum of about −90 m O.D. at the Devensian maximum glaciation around 18 k.y. ago. Therefore the present-day regime, with relatively little surface relief, may be the culmination of a period of waning flow following the last period of high surface relief, high gradients and rapid water flow before glaciation restricted recharge, represented by the predominance of late-Pleistocene dates around 30 ka.

The isotopic data indicate a mean flow velocity of around 0.6 m.y. over the last 30 k.y., whereas the present-day natural hydraulic gradient and conductivity data give velocities of around 0.2 m.y. This supports the idea of flow waning owing to reduction in topographic contrast and hydraulic gradients following glacial scouring and sea-level changes.

Past flow systems in the Triassic sandstone

It is suggested that the peak of freshwater movement within this formation would have taken place at a period of maximum topographic relief and minimum sea level prior to erosion by glacial action. This aquifer has been exposed to meteoric water influx since uplift and progressive erosion of overlying Jurassic clays some time

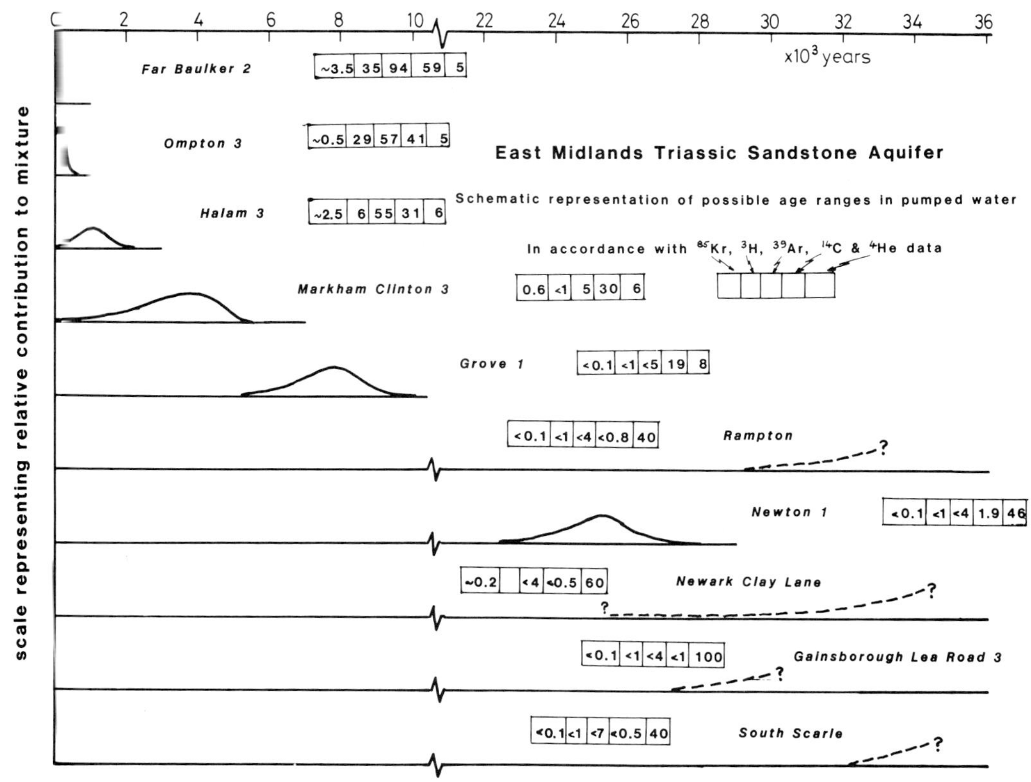

FIG. 5. Diagrammatic representation of age distributions in pumped groundwater samples, consistent with the interpretation of isotopic data shown in boxes. Units for these data are disintegrations per minute (^{85}Kr), tritium ratio (^{3}H), percentage of modern (^{39}Ar and ^{14}C) and 10^{-8} cm^3 (cm^3 H_2O)$^{-1}$ (^{4}He). The samples are arranged vertically in order of down-gradient position in the aquifer, and the distribution curves indicate relative magnitudes of components with different residence times; the vertical scale in each case is arbitrary.

during or after the Cretaceous. Therefore freshwater may have penetrated this rim of the sandstone in the North Sea Basin over the last few tens of million years (<60 m.y.). Prior to uplift, relatively static brines which evolved during burial must have occupied the closed-system basin similarly to solutions found presently at depths of several kilometres in Triassic Sandstones of the Wessex Basin (Allen & Holloway 1984). However, the E Midlands Triassic was buried to a depth of probably no more than 1 km. The brines would have evolved by diffusive mixing and advection induced by tectonic adjustments through the burial history of the sandstones. This extended throughout the Jurassic—some 70 m.y. at least of compaction and reaction. Very early post-depositional solutions entering the sandstones would have comprised saline evaporitic or playa waters of the environment in which the Mercia Mudstone was deposited. These brines probably would have displaced the fresher water remaining in the fluviatile sandstones.

Therefore the present and Pleistocene–?Tertiary meteoric water influx is probably the first freshwater flow system to be established in the sandstone since deposition. The present water–rock reactions are an example of freshwater diagenesis which might provide an analogue for the processes and rates of change in past diagenetic episodes in other clastic sediments.

Diagenesis, reaction kinetics and groundwater velocity

The evolution of water chemistry is due to diagenetic interaction with the aquifer mineral phases, and these reactions depend on reaction kinetics and groundwater velocity. The rate of dissolution of a mineral phase, not in chemical equilibrium with solution, is proportional to the ratio of reacting surface area to solution volume and to a rate constant which is characteristic of that phase (Helgeson *et al.* 1984; Lasaga 1984). However, the rate falls dramatically as solid–solution equilibrium is approached and solute transport away from the reacting interface becomes rate controlling. In some cases the dissolution rate may also depend on other solute concentrations, particularly that of H^+ (i.e. pH), when these solutes are involved in the dissolution mechanism; the classic example is that of feldspar hydrolysis (Helgeson *et al.* 1984). Thus a general expression of dissolution rates is (Lasaga 1984)

$$\text{rate} \propto \frac{A}{V} k_\theta \, a_{H^+}^{n_\theta}$$

where A is the reacting surface area of mineral phase θ, V is the relevant volume of solution per unit mass of rock, K_θ is the rate constant for mineral θ, a_{H^+} is the concentration of H^+ and n_θ is an exponent which varies between zero (for phases dissolving independently of H^+) and unity. The dependence on pH is marked for aluminosilicates (Helgeson *et al.* 1984) and carbonates for which the dissolution mechanism itself may change with a_{H^+} (Plummer *et al.* 1978). However, the pH dependence of aluminosilicate hydrolysis is compatible with a single mechanism via proton–cation exchange and activated complex formation (Aagard & Helgeson 1982). Plummer *et al.* (1978) define three regions with varying pH dependence for calcite dissolution: (i) below pH 3.5 the logarithm of the rate depends linearly on pH; (ii) between pH 3.5 and 5.5 the pH dependence decreases; (ii) above pH 5.5 the dissolution rate rapidly decreases as the pH increases and is also dependent on the partial pressure of CO_2.

Some of the available kinetic data for phases relevant to the Triassic Sandstone are summarized in Fig. 6, after recalculation to a common rate expression in moles $(\text{kg } H_2O)^{-1}$ a^{-1} and assuming particles of diameter 100 μm in a matrix of porosity 0.3 at 25 °C. Figure 6 also shows some dissolution rate data for calcite, dolomite and gypsum which were calculated from observed rates of mass transfer in two N American limestone aquifers (Floridan and Madison) (Plummer & Back 1980). The two points to be made about these data are the ranking of experimental dissolution rates, which cover 10 orders of magnitude with carbonates fastest and quartz slowest, and the even greater range of apparent difference between experimental and field data for carbonates. However, the field data represent unknown effective A/V ratios which must be very low in the case of fissure-flow systems. For example, a single fissure in a 1000 cm^3 cube of rock has a surface area 3000 times less than a similar cube comprising 100 μm spheres in a matrix of porosity 0.3.

The reaction rates can be converted into other representative information. Lasaga (1984) calculates the mean lifetimes of 1 mm crystals being dissolved; these vary from 34×10^6 years for quartz to 100 years for anorthite. Another calculated comparison is the time to reach saturation assuming zero initial dissolved concentration and that the rates do not change as equilibrium is approached. Thus with this simplified model quartz dissolution achieves saturation (6 mg SiO_2 l^{-1} at 25 °C) in static solution ($A/V \approx 165$ m^2 l^{-1}, $\phi \approx 0.3$) after 1 year, whereas calcite achieves saturation (about 500 mg

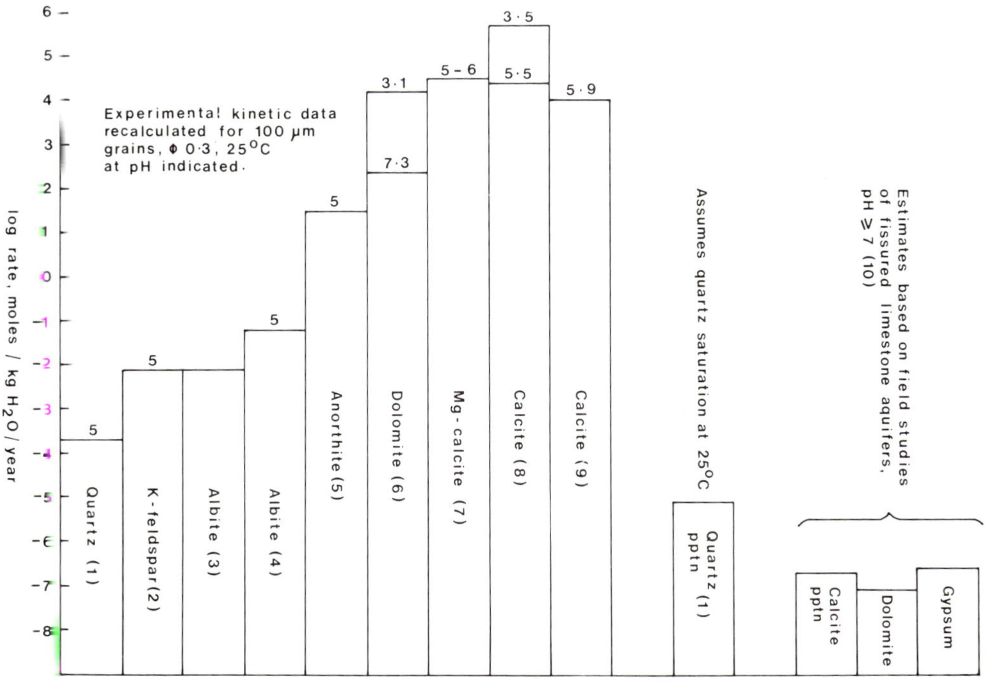

FIG. 6. Schematic compilation of experimental kinetic data for dissolution reactions from various sources recalculated to standard conditions and units. Data for quartz precipitation and data calculated from observed mass transfers in field studies in fissured limestone aquifers are also shown for comparison. Source references are as follows: (1) Rimstidt & Barnes 1980; (2) Busenberg & Clemency 1976; (3) Paces 1973; (4) Holdren & Berner 1979; (5) Fleer 1982; (6) Busenberg & Plummer 1982; (7) Rauch & White 1977; (8) Plummer *et al.* 1978; (9) Plummer & Wigley 1976; (10) Plummer & Back 1980.

$CaCO_3$ l^{-} at 25 °C, pH ≈ 7 and $p_{CO_2} \approx 10^{-1}$ bar) after about 5 s. Since dissolution rates fall as equilibrium is approached, these calculations are underestimates but nevertheless represent relative time-scales for approach to near-equilibrium solution concentrations. An alternative way of stating this is that dissolution rates are such that equilibrium will be achieved after a 1 m flow path through a matrix comprising the pure mineral only (100 μm particles, $\phi \approx 0.3$) if the velocity is less than 1 m a^{-1} for quartz or less than 20 cm s^{-1} for calcite. Although these values are of only qualitative significance for reactions occurring close to equilibrium, they show the sensitivity to flow velocity of hydrochemical evolution, mass transfer within the aquifer and diagenetic changes.

Diagenetic petrology of the E Midlands Triassic sandstone

A petrological examination of polished thin sections and stubs prepared from samples of Sherwood Sandstone taken from a borehole at Gamston [NGR SK 703766] was undertaken using optical microscopy, scanning electron microscopy (SEM) and cathodoluminescence techniques. The samples are dominantly poorly sorted fine- to medium-grained feldspathic sandstones (10%–20% K-feldspar) with a modal grain size of around 300 μm (range 50–1000 μm). They are composed of detrital quartz, K-feldspar, subordinate lithic fragments (ferruginous mudstone clasts, granitic fragments and rare carbonate pellets) and minor amounts of muscovite and biotite. These sandstones are grain supported and the abundant intergranular clay is mainly authigenic in origin. Petrographic evidence indicates that the sandstone suffered only minor effects from burial and compaction (simple grain suturing, minor stress fracturing of grains and deformation of micas and mudstone clasts).

The precipitation of iron oxide coatings on grain surfaces is the earliest diagenetic event seen. Biotite grains also show oxidation, with hematite grains being formed along cleavages. These features would have occurred near the surface under arid or semi-arid conditions. Authigenic illite or mixed-layer illite–smectite was precipitated either as grain-coating boxworks or pore-bridging filaments and plates throughout much of the rock and appears to be illite 'boxworks' may be the attachments of illite filaments to grain surfaces which became detached during drying of the core (McHardy *et al.* 1982). Minor dissolution of detrital K-feldspar preceded illite formation, as is evident from the presence of illite in secondary porosity within K-feldspar. K-feldspar overgrowths are common but are generally small and usually extensively corroded or pitted. These overgrowths are later than illite and can be seen superimposed on illite-coated grains (Fig. 7). Overgrowths are usually absent where illite precipitation fills the pores. Patchy idiotopic intergranular authigenic dolomite (Fig. 8) is a common cement and may also replace framework grains (particularly mudstone clasts) and sporadically partially fill secondary porosity in K-feldspar. Dolomite cement is an early diagenetic feature. The sometimes displacive fabric indicates formation before significant burial. Later dissolution must have occurred since the bulk of the dolomite is now corroded. Authigenic kaolinite is abundant and appears to be associated with extensive corrosion and replacement of K-feldspar. It occurs as vermicular books or irregular plates either occluding intergranular porosity (Fig. 8) or replacing corroded feldspar (Fig. 9) and mica, and is best developed adjacent to corroded K-feldspar grains. Kaolinite precipitated later than dolomite and is commonly seen on corroded dolomite substrates (Fig. 10). Rare euhedral authigenic quartz is sometimes enclosed by pore-filling kaolinite. This quartz is not corroded (Fig. 11) and occasionally fine apparently amorphous silica (chalcedony?) encrusting some corroded feldspar grains occurs. Volumetrically these late silica phases are insignificant.

Calcite is also present as minor intergranular poikilotopic cements (Fig. 12) and also occurs within the secondary porosity in K-feldspar. However, its relationship to other authigenic phases is uncertain. Cathodoluminescence shows that in some cases it may be seeded on rare detrital carbonate as overgrowths. Both cathodoluminescence and backscattering electron micrographs show that the calcite may have an overgrowth of later calcite against open porosity. Qualitative probe analysis indicates that the later overgrowth contains magnesium.

The diagenetic sequence within the Gamston borehole samples is summarized below.

1 Early precipitation of iron oxide grain coatings, oxidation of biotite and precipitation of intergranular and replacive dolomite.
2 Burial and concomitant compaction.
3 Precipitation of illite and minor feldspar dissolution.
4 Precipitation of quartz and feldspar overgrowths.
5 Extensive dissolution of K-feldspar and dolomite with abundant precipitation of kaolinite and minor coprecipitation of silica. Two minor calcite phases were identified but their textural

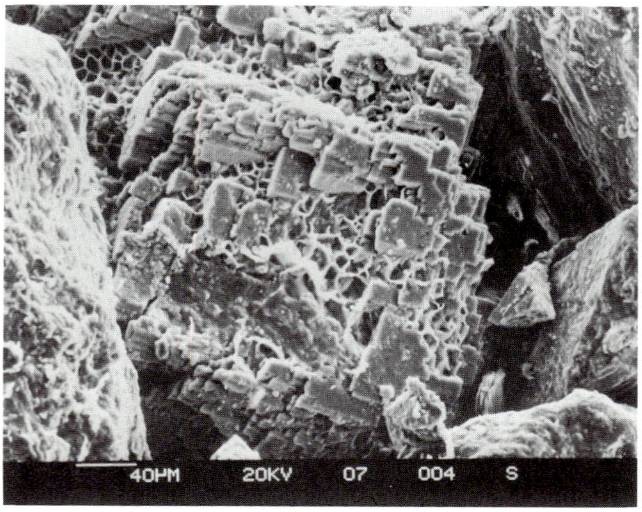

FIG. 7. Scanning electron micrograph showing corroded overgrowths of K-feldspar on an illite boxwork coating of detrital K-feldspar grain.

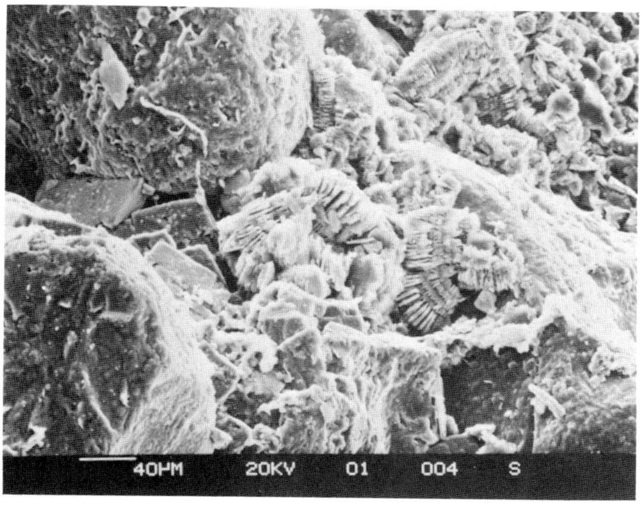

FIG. 8. Scanning electron micrograph showing kaolinite books and plates filling intergranular porosity and earlier corroded idiotopic rhombs of dolomite cement.

relationship with the above diagenetic sequence could not be resolved.

The resultant porosity in the aquifer seems mostly to be due to the preservation or rejuvenation (by dolomite removal) of primary intergranular porosity supplemented by the development of secondary porosity by K-feldspar corrosion and dissolution. The precipitation of authigenic clays, in particular abundant kaolinite, reduces the secondary porosity and may reduce the permeability of the sandstone. The last stage of alteration in the above sequence may be due to meteoric water influx which continues to the present day. No significant meteoric alteration occurred prior to burial, since K-feldspar corrosion and removal, mica alteration and kaolinite precipitation can be seen to post-date evidence for compaction.

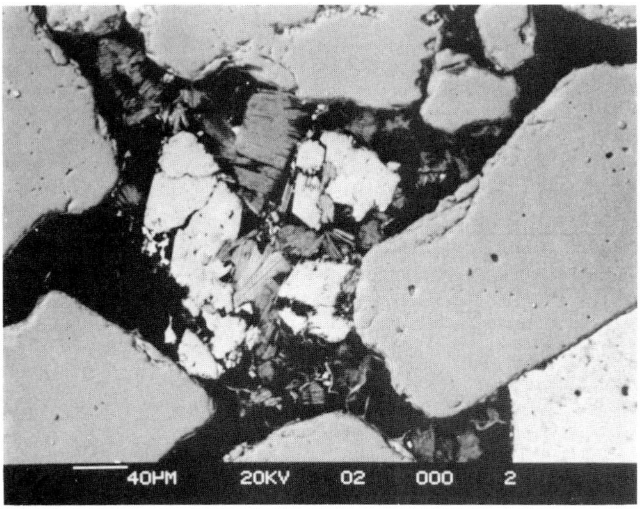

FIG. 9. Backscattering electron micrograph of a polished section showing corroded remnants of detrital K-feldspar with kaolinite infilling secondary intragranular porosity or replacing feldspar.

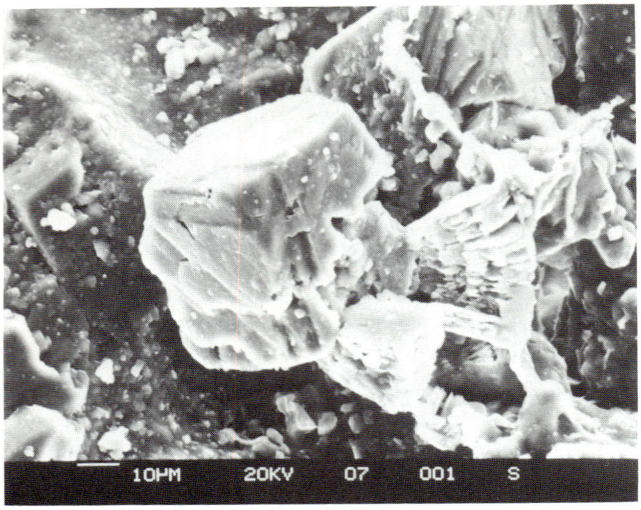

FIG. 10. Scanning electron micrograph showing kaolinite growing on the corroded surface of a dolomite crystal.

Hydrogeochemical evolution of the aquifer formation

The present flow regime provides a means of estimating the rates at which meteoric water influx can modify porosity. The simplest approach is that of calculating the net mass transfers by groundwater flowing out of the aquifer. Since the average velocity is 0.6 m a^{-1} and the average aquifer thickness is 200 m with a porosity of 0.3, the discharge from each kilometre-wide segment of aquifer is 3.6×10^7 l a^{-1}. The dominant mass transfer indicated by hydrochemistry is that of carbonate, for which dolomite is the primary source in the matrix. An average dissolved concentration of 200 mg HCO_3^- l^{-1} corresponds to the loss of about 6×10^9 g dolomite from each kilometre-wide segment per 10^3 years. The average of several analyses of the Gamston core is 2% carbonate, and therefore the net carbonate loss is equivalent to a uniform removal of around 3% of this content in 10^6 years. The carbonate is actually removed from the recharge area with a reaction front moving progressively down-gradient. However, the stoichiometry of dissolved

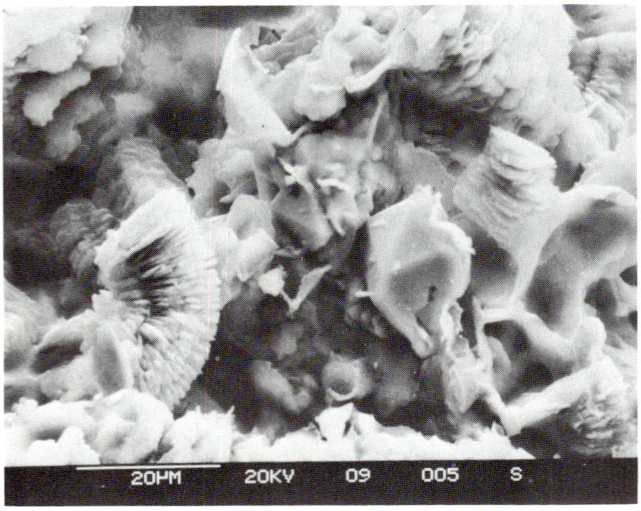

FIG. 11. Scanning electron micrograph showing rare late-stage authigenic quartz covered with authigenic kaolinite.

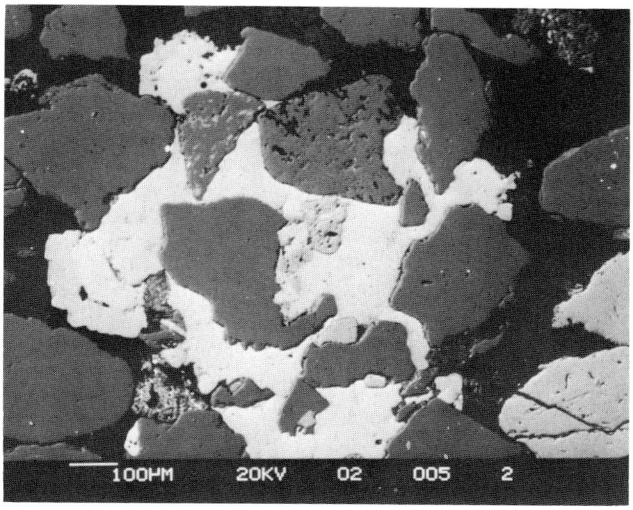

FIG. 12. Backscattering electron micrograph of a polished section showing limited poikilotopic calcite cement with a thin zone of later calcite (darker against open porosity).

Ca^{2-}, Mg^{2+} and HCO_3^- indicates that incongruent dissolution of dolomite with calcite precipitation is taking place (Edmunds *et al.* 1982). Progressively increasing SO_4^{2-} also signifies dissolution of gypsum or anhydrite, from which the parallel influx of Ca^{2+} is balanced by calcite precipitation. Active calcite precipitation is supported by measurements of the ^{13}C and ^{14}C distributions between solution and carbonates (Bath *et al.* 1979; Evans *et al.* 1984). Therefore the mass transfer includes redistribution of carbonate within the aquifer in addition to the estimated net loss. The redistribution can be calculated using a simple one-step mass-transfer model linked to a speciation model (WATEQF-ISOTOP) (Plummer *et al.* 1976; Reardon & Fritz 1978). Around 2×10^9 g of calcite is precipitated per 10^3 years in each kilometre-wide segment in addition to the 6×10^9 g of dolomite removed per 10^3 years. The overall effect is that of a 'rolling' front of carbonate removal advancing down-gradient at about 1 m 10^3y, ahead of which is a zone of secondary calcite precipitation. Unfortunately there is inadequate drill-core sampling to test the validity of this model. Very long time-scales of the order of 10^7 years are necessary for significant porosity changes by carbonate removal in a hydrological and chemical regime similar to the present one.

Similar calculations of net mass transfer show that about 3×10^8 g SiO_2 are removed every 10^3 years per kilometre-wide segment of aquifer. Therefore the removal of silica is an order of magnitude slower than that of carbonates. However, the incongruent dissolution of K-feldspar with kaolinite precipitation is identified as an important diagenetic reaction. Thus the net SiO_2 loss is representative of a much more extensive aluminosilicate reaction:

$$2KAlSi_3O_8 + 2H^+ + 9H_2O \longrightarrow Al_2Si_2O_5(OH)_4 + 2K^+ + 4Si(OH)_4$$

This alteration is written to be isochemical for aluminium, i.e. to conserve aluminium locally in solid phases. The fate of the $Si(OH)_4$ must be known if the extent of the alteration is to be estimated from the net SiO_2 lost in solution. Only minor secondary SiO_2 precipitation is detected in SEM studies; therefore the rate of the alteration and its implication for porosity development can be estimated by assuming that all $Si(OH)_4$ arises from this alteration and that most of it remains in solution. From the dissolved concentration of 9 mg SiO_2 l^{-1}, it can be shown that reaction with each litre of groundwater causes a volume decrease in the matrix of 0.0268 cm^3. For a groundwater velocity of 0.6 m a^{-1} and a 20 km length of aquifer, this is equivalent to a uniform change of 0.1% in the porosity of 0.3 per 10^6 years. Again, this emphasizes the long time-scales required for significant porosity increase to occur under the meteoric water flow regime. The rate of change will be underestimated if secondary silica has precipitated elsewhere along the flow path; the observed SiO_2 concentration of 9 mg l^{-1} is supersaturated with respect to quartz.

The relative kinetics of diagenetic reactions and their interplay with the velocity of water flow must have an important influence on the fabric of the altered matrix. For example, the faster dissolution rate of K-feldspar relative to quartz (Fig. 6) is likely to buffer $Si(OH)_4$ and inhibit quartz dissolution. SEM evidence confirms the localization of K-feldspar alteration to kaolinite over a scale of 100–500 m. Although the kinetics of kaolinite precipitation are not yet documented experimentally, the evidence suggests that the rate of kaolinite precipitation, which is a second-order reaction proportional to dissolved aluminium and silicon species, exceeds that of quartz precipitation. In the simple case whereby the altering K-feldspar and precipitating kaolinite are regarded as a 'microcosm', i.e. a system closed to influx of solutes and losing only K^+ and $Si(OH)_4$, the reaction kinetics indicate a limiting maximum water velocity. If localized (e.g. within 1 mm) precipitation of kaolinite requires, for example, dissolved $Al(OH)_3$ to be kept at a typical concentration of 14 $\mu g\, l^{-1}$ by feldspar dissolution, the rate of this defines a limiting water velocity. Thus aluminium conservation and the observed alteration fabric might occur when the flow velocity does not exceed the order of 3 m a^{-1}, in this type of matrix. The same calculation for sustaining dissolved SiO_2 at saturation, i.e. 6 mg l^{-1}, suggests a limiting flow of 1 cm a^{-1} which is obviously lower than experienced; therefore a simple model ignoring SiO_2 transport is inappropriate.

In contrast with aluminosilicate alteration mineralogy, secondary calcite does not appear to precipitate adjacent to the dissolving dolomite in this case. The high dissolution rates for carbonates suggest maintenance of equilibrium over very small distances, by analogy with what is found for aluminosilicates. However, transport of dissolved species away from dolomite is occurring, and this suggests that the precipitation of calcite may be inherently slower than dissolution, or that precipitation is inhibited. A clue is provided by the SEM evidence of later overgrowth on earlier calcite (Fig. 12). Nucleation of this calcite may be restricted to sparse pre-existing calcite or may be inhibited elsewhere by clay minerals.

Although the assumptions and simplifications of this example are vast, it has illustrated how diagenetic mineral fabric, consideration of chemical equilibrium and experimental kinetic data might be combined to deduce information on present and past water flow regimes.

Conclusion

Present groundwater flowing in the E Midlands Triassic sandstone has a residence time of around 30 k.y. at the sampling points about 20 km into the confined aquifer on the rim of the North Sea Basin. However, this freshwater flow regime has persisted for many millions of years, although the timing of uplift and denudation is uncertain and the hydrodynamic conditions have fluctuated.

The groundwater chemistry matches the diagenetic changes which are occurring in the aquifer matrix owing to this period of prolonged freshwater influx. Bulk mass-transfer calculations show the rate of these changes and illustrate the very long times (upwards of 10^7 years) necessary for significant porosity modification to occur whether by bulk removal in solution or by incongruent dissolution of aluminosilicate or carbonate phases.

Using diagenetic mineralogy in the Triassic as an example, it is suggested that alteration mineral fabrics such as those for K-feldspar transforming to kaolinite reflect the groundwater flow velocity amongst other factors. Experimental data on precipitation kinetics are necessary to explain these phenomena fully, but there is a paucity of data. However, the rapid precipitation of kaolinite adjacent to corroding K-feldspar suggests that the precipitation rate is higher than that for dissolution.

If it is possible to deduce information such as this from the active diagenetic changes, evidence of past flow regimes in clastic aquifers or reservoirs can be extracted from their diagenetic mineralogies. Fundamental data, particularly for the influence of water chemistry, e.g. pH and salinity, on reaction rates and for precipitation reactions, are needed to improve the rigour of the approach.

ACKNOWLEDGMENTS: Studies of the E Midlands Triassic sandstone aquifer were initiated at the British Geological Survey by Dr W. M. Edmunds. The isotopic studies summarized have formed part of an international cooperative study coordinated by the International Atomic Energy Agency, and the many coworkers in this project are acknowledged for their contributions. The Severn–Trent and Anglian Water Authorities provided access to pumped groundwater samples and data. The manuscript was word-processed by Mrs L. S. Towle. This paper is published by permission of the Director, British Geological Survey (Natural Environment Research Council).

References

AAGARD, P. & HELGESON, H. C. 1982. Thermodynamic and kinetic constraints on reaction rates among minerals and aqueous solutions. I. Theoretical considerations. *Am. J. Sci.* **282**, 237–85.

ALLEN, D. J. & HOLLOWAY, S. 1984. The Wessex Basin—investigation of the geothermal potential of the U.K. *Report,* British Geological Survey, Keyworth, 80 pp.

ANDREWS, J. N. & LEE, D. J. 1979. Inert gases in groundwater from the Bunter Sandstone as indicators of age and palaeoclimatic trends. *J. Hydrol.* **41**, 233–52.

——, BALDERER, W., BATH, A. H., CLAUSEN, H. B., EVANS, G. V., FLORKOWSKI, T., GOLDBRUNNER, J. E. IVANOVICH, M., LOOSLI, H. & ZOJER, H. 1984. Environmental isotope studies in two aquifer systems—a comparison of groundwater dating methods. *Isotope Hydrology 1983, Proc. Vienna Symp.*, pp. 535–76, International Atomic Energy Agency, Vienna.

BATH, A. H., EDMUNDS, W. M. & ANDREWS, J. N. 1979. Palaeoclimatic trends deduced from the hydrochemistry of a Triassic sandstone aquifer, United Kingdom. *Isotope Hydrology, Proc. Vienna Symp.*, 1978, Vol. 2, pp. 545–68, International Atomic Energy Agency, Vienna.

BUSENBERG, E. & CLEMENCY, C. V. 1976. The dissolution kinetics of feldspars at 25°C and 1 atm CO_2 partial pressure. *Geochim. cosmochim. Acta*, **40**, 41–9.

—— & PLUMMER, L. N. 1982. The kinetics of dissolution of dolomite in CO_2–H_2O systems at 1.5 to 65°C and 0 to 1 atm PCO_2. *Am. J. Sci.* **282**, 45–78.

DOWNING, R. A. & HOWITT, F. 1969. Saline groundwaters in the Carboniferous rocks of the English East Midlands in relation to the geology. *Q. J. eng. Geol* **1**, 241–69.

——, ALLEN, D. J., BIRD, M. J., GALE, I. N., KAY, R. L. F. & SMITH, I. F. 1985. Cleethorpes No. 1 geothermal well—a preliminary assessment of the resource. Investigation of the geothermal potential of the U.K. *Report,* British Geological Survey, Keyworth, 67 pp.

EDMUNDS, W. M., BATH, A. H. & MILES, D. L. 1982. Hydrochemical evolution of the East Midlands Triassic sandstone aquifer, England. *Geochim. cosmochim. Acta*, **46**, 2069–81.

EVANS, G. V., OTLET, R. L., WASSELL, L. L. & BATH, A. H. 1984. Verification of the presence of carbon-14 in secondary carbonates within a sandstone aquifer and its hydrological implications. *Isotope Hydrology 1983, Proc. Vienna Symp.*, pp. 557–89, International Atomic Energy Agency, Vienna.

FLEER, V. N. 1982. The dissolution kinetics of anorthite ($CaAl_2Si_2O_8$) and synthetic strontium feldspar ($SrAl_2Si_2O_8$) in aqueous solutions at temperatures below 100°C with applications to the geological disposal of radioactive nuclear wastes. *Ph.D. Thesis*, Pennsylvania State University, University Park, PA.

HELGESON, H. C., MURPHY, W. M. & AAGARD, P. 1984. Thermodynamic and kinetic constraints on reaction rates among minerals and aqueous solutions. II. Rate constants, effective surface area, and the hydrolysis of feldspars. *Geochim. cosmochim. Acta,* **48**, 2405–32.

HOLDREN, G. R. & BERNER, R. A. 1979. Mechanism of feldspar weathering. 1. Experimental studies. *Geochim. cosmochim. Acta*, **43**, 1161–71.

INSTITUTE OF GEOLOGICAL SCIENCES 1981. *Hydrogeological Map of the Northern East Midlands (Scale 1 : 100 000)*, Hydrogeology Unit, Institute of Geological Sciences, London.

LAND, D. H. 1966. Hydrogeology of the Bunter sandstone in Nottinghamshire. Water Supply Pap. *Geol. Surv. GB Hydrogeol. Rep. No. 1*, 38 pp.

LASAGA, A. C. 1984. Chemical kinetics of water–rock interactions. *J. geophys. Res.* **89**, 4009–25.

LOVELOCK, P. E. R. 1977. Aquifer properties of the Permo–Triassic sandstones of the United Kingdom. *Bull. Geol. Surv. GB No. 56*, 50 pp.

MCDONALD, D. A. & SURDAM, R. C. (eds) 1984. *Clastic Diagenesis*, American Association of Petroleum Geologists, Tulsa, OK, 440 pp.

MCHARDY, W. J., WILSON, M. J. & TAIT, J. M. 1982. Electron microscope and X-ray diffraction studies of filamentous illitic clay from sandstones of the Magnus Field. *Clay Miner.* **17**, 23–39.

PACES, T. 1973. Steady-state kinetics and equilibrium between ground water and granitic rock. *Geochim. cosmochim. Acta*, **37**, 2641–63.

PLUMMER, L. N. & BACK, W. 1980. The mass balance approach: application to interpreting the chemical evolution of hydrologic systems. *Am. J. Sci.* **280**, 130–42.

—— & WIGLEY, T. M. L. 1976. The dissolution of calcite in CO_2-saturated solutions at 25°C and 1 atmosphere total pressure. *Geochim. cosmochim. Acta*, **40**, 191–202.

——, JONES, B. F. & TRUESDELL, A. H. 1976. WATEQF—A Fortran IV version of WATEQ, a computer program for calculating chemical equilibrium of natural waters. *Geol. Surv. Water-Resour. Invest. No. 76–13*, 61 pp.

——, WIGLEY, T. M. L. & PARKHURST, D. L. 1978. The kinetics of calcite dissolution in CO_2–water systems at 5°C to 60°C and 0.0 to 1.0 atm CO_2. *Am. J. Sci.* **278**, 179–216.

RAUCH, H. W. & WHITE, W. B. 1977. Dissolution kinetics of carbonate rocks. 1. Effects of lithology on dissolution rate. *Water resour. Res.* **13**, 381–94.

REARDON, E. J. & FRITZ, P. 1978. Computer modelling of groundwater ^{13}C and ^{14}C isotope compositions. *J. Hydrol.* **36**, 201–24.

RIMSTIDT, J. D. & BARNES, H. L. 1980. The kinetics of silica–water reactions. *Geochim. cosmochim. Acta*, **44**, 1683–99.
WILLIAMS, B. P. J., DOWNING, R. A. & LOVELOCK, P. E. R. 1972. Aquifer properties of the Bunter Sandstone in Nottinghamshire, England. *Proc. 24th Int. Geology Cong., Montreal*, Section 11, pp. 169–76.

A. H. BATH & A. E. MILODOWSKI, Fluid Processes Research Group, British Geological Survey, Keyworth, Nottingham NG12 5GG, U.K.
G. E. STRONG, Stratigraphy and Sedimentology Research Group, British Geological Survey, Keyworth, Nottingham NG12 5GG, U.K.

Fluid flow in the Chalk of England

Michael Price

SUMMARY: The Chalk aquifer of England can be thought of as a multi-porosity medium. The matrix is a fine-grained limestone which generally has high porosity but small pore throats, so that its permeability is typically only 0.1 to 10 millidarcys (10^{-4} to 10^{-2} m day^{-1}). A fairly uniform fracture system imparts a secondary permeability, which appears to be about 100 to 1000 mD (0.1 to 1 m day^{-1}). Where the Chalk forms a major aquifer, most of the transmissivity results from the enlargement of fractures, by solution, to form a few highly permeable zones. Over much of the outcrop, weathering leads to the development of shallow permeable layers to form an additional component. Each of these permeability systems influences some aspect of subsurface water movement, with implications for resources, quality and construction.

The Chalk of England is an unusual aquifer in that it can have several superimposed components of porosity and permeability. The initial component is the intergranular space within the matrix, which contributes high porosity but little permeability; if the Chalk possessed only this permeability, it would not be an aquifer. Superimposed upon this is a fracture component, which for the remainder of this paper will be referred to as the primary-fissure component. This fracture component has low porosity, but typically causes the permeability of the Chalk to increase by one to three orders of magnitude, thus making it a double-porosity aquifer. However, even this permeability would not be sufficient to explain the high transmissivities and well yields associated with the Chalk in many parts of England.

Two further permeability components can exist, which developed in different ways from the primary-fissure component. At shallow depths (less than about 10 m) the primary-fissure component may be developed by weathering, sometimes to the extent that in the near-surface layer the fissure–block system is destroyed. This weathering generally contributes an additional component of permeability, although it frequently lies within the unsaturated zone. Finally, within the top few tens of metres of the saturated zone, particularly at or near the Chalk outcrop, selected fissures of the primary-fissure component may become enlarged by solution; the permeability resulting from this enlargement, which will be referred to as the secondary-fissure component, accounts for the majority of groundwater movement in the highly transmissive parts of the Chalk.

In the following sections each of these components will be considered in terms of its origin, its contribution to permeability and storage, its dependence on the other components and its influence on water movement within the Chalk. In this paper I do not attempt to provide a comprehensive summary of the variations of the physical properties of the Chalk aquifer, but to give an indication of the variations that exist, how they may have arisen and how they affect sub-surface water flow.

The nature of chalk permeability

The intergranular component

The Chalk consists predominantly of two main groups of particles: a finer fraction of coccoliths and coccolith debris (Fig. 1), and a coarser fraction of foraminifera and other shell debris (Hancock 1975). Coccoliths are ring-shaped collections of calcite crystals (laths) from the skeletons of marine algae; they are typically 3 or 4 μm across but they often disintegrate into individual laths, about 0.5–1 μm across. The arrangement of these fragments during deposition leads to high porosities, but means that the pores and their interconnecting throats are small (Fig. 1) so that permeability is generally low. In England from Norfolk southwards much of the porosity has been preserved, but in parts of Yorkshire and Lincolnshire the initial intergranular porosity has been reduced by recrystallization of calcite, leading to an increase in the hardness of the chalks (Hancock & Kennedy 1967; Hancock 1975).

Table 1 shows values of porosity and intergranular permeability for chalks from several locations. These data have been chosen to give a reasonable geographical and stratigraphical spread, but otherwise more or less at random, from a few thousand such values measured by the British Geological Survey (BGS) and held on file. The values are similar to those reported by other workers (Carter & Mallard 1974; Hancock 1975; Bell 1977). The porosity values were determined by liquid saturation, and the permeability values

From Goff, J. C. & Williams, B. P. J. (eds), 1987, *Fluid Flow in Sedimentary Basins and Aquifers*, Geological Society Special Publication No. 34, pp. 141–156.

FIG. 1. Scanning electron micrograph of chalk from 103 m below ground level in the Upper Chalk, Faircross BH, Berkshire. The porosity is 46%, the vertical permeability is 6.5 mD and the density is 2.70 g/cm^3. Coccoliths, intact or nearly so, can be seen for example at A. Examples of pores are shown at B and, in contrast, D shows examples of the sub-micrometre spaces, between laths and rhombs, which form the interconnecting pore throats.

were measured on a gas permeameter and empirically corrected for gas slippage (Klinkenberg 1941).

Several features of the Chalk matrix are exemplified by the data in Table 1. These include the generally high porosity and low permeability, and the fact that the chalks are generally isotropic with respect to permeability. The chalks are also remarkably uniform in their physical properties, both at particular locations—evidenced for example by the small standard deviations of the porosities—and across the country. Generally, the Lower Chalk has the lowest porosities and permeabilities, and the Upper Chalk has the highest. The data from Bishop Burton exemplify the reduced porosity and permeability of chalks from Northern England; however, as indicated by the data from Cherry Burton, this reduction is not universal. Carter & Mallard (1974) presented data which indicate that the correlation of lower porosity with lower stratigraphical horizon results, at least in part, from the greater depth of burial (and hence greater compaction) which the lower horizons have experienced.

The small sizes of the interconnecting pore throats (Fig. 1) explain why unfractured chalks, despite their high porosities, have such low permeabilities. The effective pore-throat diameters can be measured using mercury-injection techniques (Price *et al.* 1976). Such measurements (Fig. 2) indicate that effective throat diameters are smaller in chalks from the N of England than in those from the S, presumably because of the increased recrystallization in the N. The diame-

TABLE 1

	Permeability (mD)*		Porosity (%)		No. of samples
	Geometric mean k_h	Geometric mean k_v	Arithmetic mean	Standard deviation	
Cherry Burton, Yorks Upper Chalk, 1.1–28.9 m	—	2.15	39.5	1.44	30
Bishop Burton, Yorks Upper Chalk, 1.9–24.9 m	—	0.65	23.5	2.27	25
Totford, Hants		5.07			13
Upper Chalk, 3.9–66.2 m	5.20		39.3	4.46	67
Totford, Hants		1.97			3
Middle Chalk, 69.0–79.9 m	1.70		33.4	4.73	19
Anmer, Norfolk		1.85			37
Middle Chalk (may include some Upper Chalk), 6.7–46.7 m	2.04		33.3	3.92	38
Winterbourne Kingston, Dorset Lower Chalk, 242–279 m	0.14	—	24.7	3.42	19
Faircross, Hants	3.40				65
Upper Chalk, 95.6–205.0 m		3.06			73
			40.2	4.91	81
Faircross, Hants	0.77				48
Middle Chalk,		0.61			44
205.[illegible]–263.4 m			31.0	5.47	55
Faircross, Hants	0.27				50
Lower Chalk,		0.25			51
265.[illegible]–321.5 m			25.9	7.69	55

* k_h = horizontal permeability; k_v = vertical permeability

ters also correlate approximately with stratigraphy, the largest values being found in the Upper Chalk; again, this may be due to the increased depth of burial of the lower horizons.

Study of typical mercury-injection curves from the Upper Chalk (Fig. 2) indicates that there is virtually no intrusion below a pressure of about 300 kPa, which corresponds to a pore-throat diameter of 1 µm. (The low-pressure 'tail' may be caused, at least in part, by entry of mercury into depressions on the surface of the sample.) Water-filled pores with throat sizes less than 1 µm will not begin to drain until a suction of 30 m is achieved.

Mercury intrusion into a sample, like water drainage, involves replacement of a wetting fluid by a non-wetting fluid; a mercury-injection curve can be treated as equivalent to a water-drying curve because both are controlled by pore-throat sizes. Similarly, drainage of mercury from a sample is qualitatively equivalent to entry of water into a sample; both are controlled by the largest diameter of the pore rather than by the diameter at the narrowest neck. Therefore a complete water-wetting and water-drying curve, or a complete mercury-injection and mercury-drainage curve (Fig. 3), displays hysteresis. Figure 3 shows that, although the pore throats in this sample have median diameters of the order of 0.5 µm, the pores have median diameters of about 3–4 µm, corresponding approximately to the size of a coccolith. Comparison with Fig. 1 shows that these sizes are realistic.

The primary-fissure component

Examination of almost any section in the English Chalk reveals the presence of fractures. Generally there are three more or less orthogonal sets, one set being approximately parallel to the bedding. Ward *et al.* (1968) provided a detailed description of a site on Middle Chalk at Mundford, Norfolk, where in the unweathered material these joints are more than 0.2 m apart and are apparently closed; as the effects of weathering increase, the joint spacing decreases to about 0.01 m and the chalk may eventually become a structureless mélange. At Mundford, widely spaced steeply-dipping major joints and major separation planes associated with horizontal bedding features are also present.

Few published data are available on the

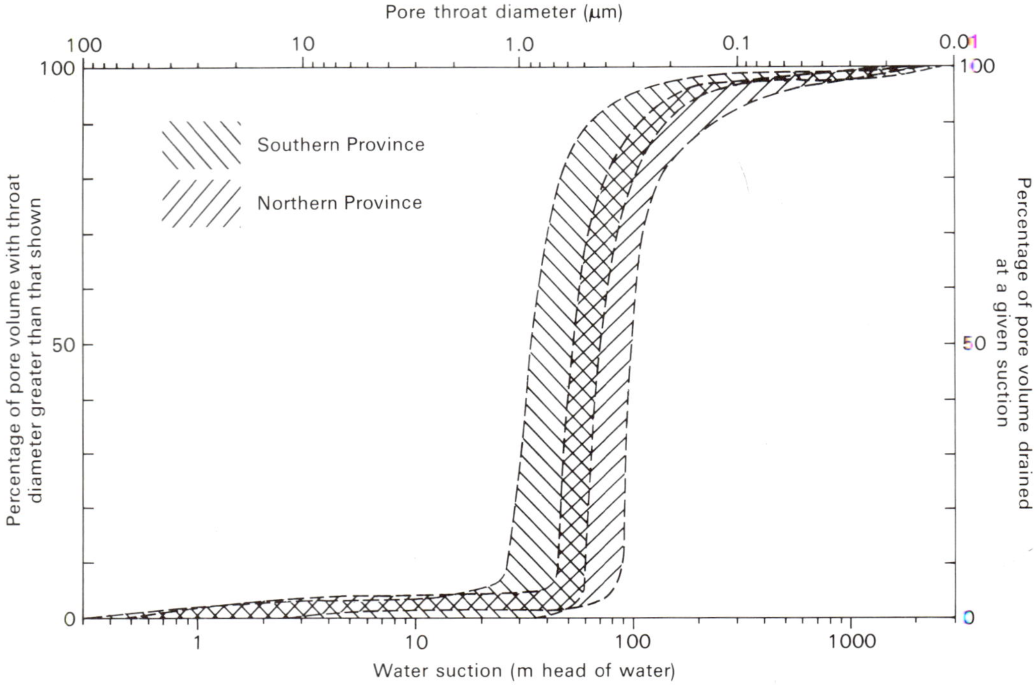

FIG. 2. Envelopes of mercury-intrusion pore-throat size distribution curves from the Upper Chalk. The Southern Province envelope is based on determinations on 21 samples from six localities, and the Northern Province envelope on seven samples from seven localities. (After Price *et al.* 1976.)

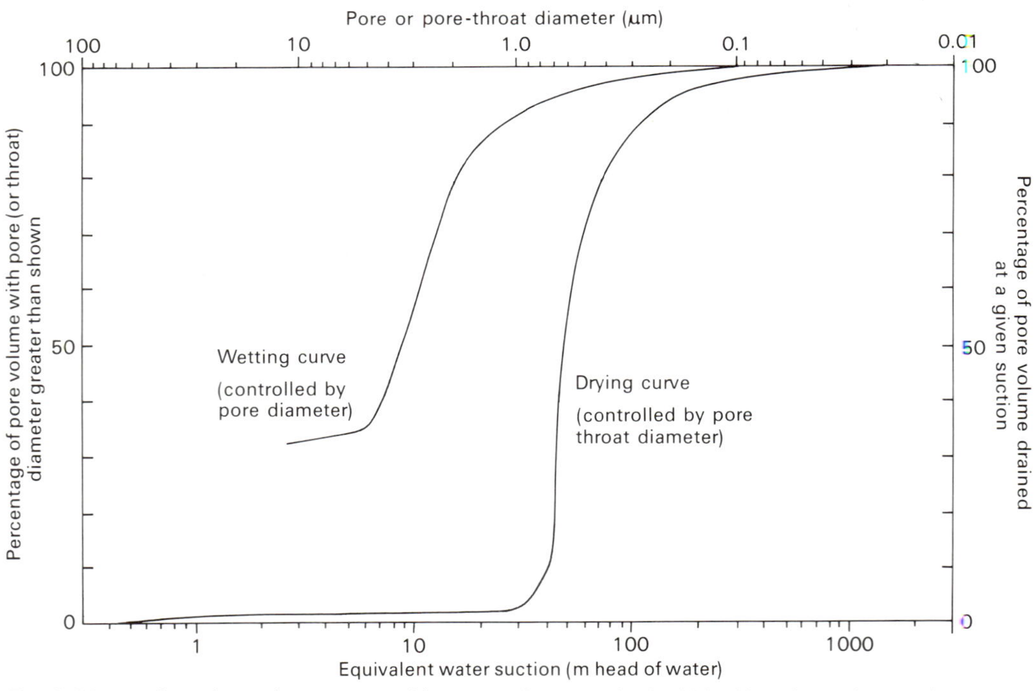

FIG. 3. Mercury-intrusion and mercury-expulsion curves for a sample of Middle Chalk from Fleam Dyke, Cambs. This sample has a porosity of 42%.

permeability of the primary-fissure component. Measurements made in the Candover Valley, Hants, using packer systems (Price *et al.* 1977, 1982) indicate hydraulic conductivity values ranging from 0.02 m day^{-1} (corresponding to 30 mD) to more than 2 m day^{-1} (3000 mD), some of the highest permeabilities occurring in the Chalk Rock. Limited data from piezometer tests in Norfolk (Price *et al.* 1979; Allen & Price 1983) imply similar values.

Measurements made during investigations for the proposed Channel Tunnel, mainly in Lower Chalk, ranged from less than 10^{-3} m day^{-1} (about 1 mD) to more than 4 m day^{-1} (more than 6000 mD) (Channel Tunnel Study Group 1965). Similar investigations during tunnelling trials at Chinnor, also in the Lower Chalk (Priest *et al.* 1976), gave conductivity values ranging from 0.07 m day^{-1} (100 mD) to 9.5 m day^{-1} (15 000 mD) with a mean value of 1.3 m day^{-1}. The Chinnor investigations were at depths of less than 30 m below ground level.

At significantly greater depths there is evidence that the primary-fissure component decreases in importance. Investigations at S Killingholme, S Humberside, for the construction of caverns to store liquefied petroleum gas (LPG), showed that the chalk at cavern level (180–190 m below ground level) is largely intact and the minor joints are closed by secondary calcite. Obvious near-vertical open joints are generally spaced several metres apart. At this depth the bulk of the chalk (which is in the lower part of the Middle Chalk and the top of the Lower Chalk) has *in situ* permeabilities of the order of 0.1 mD (less than 10^{-4} m day^{-1}). In zones where joints are concentrated, the permeability is of the order of 1000 mD (about 0.6 m day^{-1}) (Trotter *et al.* 1985).

At depths more typical of water-abstraction boreholes, open joints are probably more common. If the primary fissures were spaced 0.2 m apart in three orthogonal sets, horizontal flow parallel to one of the vertical sets would be taking place through 10 joints (five horizontal and five vertical) per square metre of rock. If the joints had smooth-sided plane parallel openings, apertures of about 55–60 μm would theoretically produce hydraulic conductivities of about 0.1 m day^{-1} with water at 10 °C. A cubic metre of chalk would contain 15 such openings, which would contribute a porosity of about 0.001 or 0.1%. In reality, the surface roughness and irregularities in such features probably mean that their average openings (and therefore their contribution to porosity) must be greater than this theoretical value.

The harder or Grade I chalks (Ward *et al.* 1968), such as the Melbourn Rock, the Chalk Rock and the more localized hardgrounds, probably fracture more cleanly than do the softer chalks. Because of this their joints are likely to possess more uniform openings, a fact that explains why these harder chalks (e.g. the Chalk Rock in the Candover Valley referred to above) are frequently more permeable than underlying or overlying strata. Their intergranular porosity is generally lower than that of softer chalks at similar horizons, and there is no evidence that the higher fracture permeability is associated with above-normal fracture porosity.

The shallow (weathered-layer) component

Over parts of the Chalk outcrop, weathering may have affected the properties of the chalk to depths of several metres. Moving up the profile, the effects are first characterized by an increased frequency of joints and an increased separation of the joint surfaces. On excavation the chalk frequently breaks up into cuboids along the joint planes, which may show straining. Further up the profile the material frequently becomes friable and rubbly. In the most extreme cases the chalk becomes structureless, with angular blocks set in a matrix of weathered chalk; when wet the matrix may take on a paste-like consistency. Ward *et al.* (1968) gave an excellent description of weathered-chalk profiles at the Mundford site.

Few data are available on the hydraulic properties of this material. At the Totford site in Hampshire (Price *et al.* 1982) it was necessary to line the upper 15 m of the borehole to support the weathered chalk; hence packer-testing of this interval was not undertaken. Television inspection revealed rubbly chalk, visible through the perforations in the lining. Flow logging indicated significant inflow into the borehole over this interval, suggesting enhanced permeability.

Measurements at a site in N Norfolk (Price *et al.* 1979; Allen & Price 1983), using piezometers installed at various depths, indicated that the hydraulic conductivity varied from more than 25 m day^{-1} at depths of 1–2 m below ground level to about 3 m day^{-1} at depths of 5–6 m. At 16 m below ground level the hydraulic conductivity was less than 1 m day^{-1}. These measurements were made in the unsaturated zone, but studies in the zone of fluctuation at a nearby site indicated that the values measured by these injection tests were approximate saturated hydraulic conductivity values.

The enhancement of the primary-fissure component of permeability at shallow depths will almost certainly vary from site to site, depending on factors such as cryoturbation. In many areas this enhanced permeability, if present, will occur in the unsaturated zone.

The secondary-fissure component

Table 2 shows that if the Chalk possessed only the intergranular and primary-fissure components of permeability, its transmissivity would be only of the order of 20 m^2 day^{-1} or less. There are areas where the transmissivity of the Chalk is this low or even lower, e.g. parts of the confined aquifer in the London Basin (Water Resources Board 1972) and in southern E Anglia away from river valleys (Lloyd *et al.* 1981), but these are not the areas where the Chalk is of major importance as an aquifer. Some of the larger well yields may be associated with Chalk transmissivities in excess of 2000 m^2 day^{-1}.

Even in these areas, however, there is evidence that most of the thickness of the Chalk may have low hydraulic conductivities with the bulk of water movement occurring at a few horizons where fissures have been enlarged by solution (Headworth 1978; Headworth *et al.* 1982). By inference, the few fissures that have been enlarged in this way must have high transmissivities. At a site in E Yorkshire, for example, Foster & Milton (1974) used data from pumping tests at high and low water tables and from geophysical logging to infer that most flow takes place in two zones, with the upper one corresponding to the zone of water-table fluctuation. They estimated that this upper zone contributes 1200 m^2 day^{-1} of a total transmissivity of 2200 m^2 day^{-1}.

During the BGS–Southern Water Authority investigations in the Candover Valley, a double-packer injection test was carried out on a borehole interval containing a major fissure which, on the basis of television inspection, appeared to have been enlarged by solution. This fissure had a transmissivity of about 700 m^2 day^{-1} (Price *et al.* 1982); this is apparently the only published value for the transmissivity of a Chalk fissure that has been obtained by direct measurement.

An interesting indirect study was provided by Atkinson & Smith (1974). They introduced Rhodamine WT into a swallow hole in S Hampshire and observed its appearance at Bedhampton springs. The straight-line distance between the two points is 5.75 km, and the time between the introduction of the dye and its occurrence at the springs at peak concentration was 62.5 h, corresponding to a flow speed of 2.2 km day^{-1}. If the conclusions of Atkinson & Smith are accepted as to prevailing conditions, the hydraulic gradient between the input and discharge points was of the order of 2 m km^{-1}.

The mean interstitial flow speed $\bar{v}_i$ through an aquifer is given by

$$\bar{v}_i = \frac{K}{n_d}\frac{dh}{dl} \tag{1}$$

where K is the hydraulic conductivity, dh/dl is the hydraulic gradient and n_d is the dynamic porosity (the porosity through which flow is occurring). If the fissure connection between the swallow hole and the springs is treated as a single aquifer with a dynamic porosity of 100% and if Darcian flow is assumed, it is possible to calculate the hydraulic conductivity of the fissure connection from Equation 1 by substituting 2.2 km day^{-1} for $\bar{v}_i$. This value works out as 1.1×10^6 m day^{-1} (equivalent to 1.7×10^6 D), and it must be emphasized that this corresponds to the hydraulic conductivity of the fissure and not of the chalk as a whole.

The hydraulic conductivity K_f of a smooth plane parallel opening is given by

$$K_f = \frac{gb^2}{12\,v}. \tag{2}$$

where g is the gravitational acceleration, b is the aperture and v is the kinematic viscosity of the fluid; for groundwater at 10 °C v is about 1.31×10^{-6} m^2 s^{-1}, and so a K_f value of 1.1×10^6 m day^{-1} corresponds to a value for b of 4.5 mm. A plane parallel opening with a hydraulic conductivity of 1.1×10^6 m day^{-1} and an opening of

TABLE 2. *Summary of typical Chalk permeability components*

Component	Typical horizontal hydraulic conductivity K_h (m day^{-1})	Typical effective thickness b (m)	Transmissivity T (m^2 day^{-1})
Matrix	0.001	200	0.2
Primary fissures	0.1	200	20
Weathered layer	10	5	50
Individual secondary fissure*	25 000	0.02	500

* Hypothetical value derived from assumed fissure transmissivity and observed fissure opening (based on conservative values for a fissure at Totford).

0.0045 m possesses a transmissivity of 5000 m^2 day^{-1}.

Given the assumptions involved, the above calculation should be regarded as entertaining rather than of direct use, but it is interesting that the calculated transmissivity, although high, is not impossible for a restricted part of the Chalk aquifer Also, although this example may be a special case with solution favoured by the relatively acid water discharged from the Tertiary cover (Foster 1975b), it is worth noting that the calculated fissure opening is modest by comparison with some fissures observed during borehole television inspections (e.g. Price *et al.* 1982, Plate Ia). Clearly, the surface roughness and tortuosity of a real fissure will reduce its permeability below that of an artificial parallel-plate opening with the same aperture.

Implications for water movement

The unsaturated zone

The unsaturated zone of the Chalk aquifer contains chalk which, for the most part, approaches saturation. Measurements made by the Institute of Hydrology at an experimental site on the Upper Chalk in Hampshire (Wellings 1984) indicate that, below about 5 m, pore-water suctions usually range from 2 to 15 m below atmospheric pressure. Study of Fig. 2 indicates that, over this range of pore-water suctions, there will be little or no change in the water content of chalk. its intergranular porosity will remain essentially fully saturated. It is therefore to be expected that the intergranular or matrix component of chalk permeability remains unchanged over most of the naturally occurring range of pore-water conditions.

The fractures of the primary-fissure component and those in the weathered zone drain much more readily than do the pores. Earlier (p. 145) it was stated that a hydraulic conductivity of 0.1 m day^{-1} would be imparted by three orthogonal sets of smooth-sided planar fractures spaced 0.2 m apart and with openings of about 60 μm. Such fractures would drain at suctions of about 0.25 m of water and would be empty at higher suctions. Combining this conclusion with that from the mercury-injection curves, it would be expected that a plot of hydraulic conductivity versus pore-water suction would show a marked fall in conductivity at a suction value of about 0.25 m from a value typical of the primary-fissure component to one typical of the intergranular component; as suction increases the conductivity could be expected to remain more or less unchanged up to a suction of the order of 30 m, when it would be expected to fall below the saturated value for the matrix.

Actual curves of unsaturated hydraulic conductivity against suction, derived *in situ* using irrigation techniques (Wellings & Cooper 1983, Fig. 2; Wellings 1984, Fig. 7), are remarkably consistent with this prediction (Fig. 4). For the Fleam Dyke site on Middle Chalk in Cambridgeshire, conductivity values for undisturbed chalk declined from more than 0.1 m day^{-1} to 10^{-3} m day^{-1} or less as suctions increased from about 0.2 m to about 0.5 m. Thereafter, conductivity remained constant to suction values of about 3 m. At the Bridgets site, on Upper Chalk in Hampshire, there were reductions in conductivity values from more than 0.1 m day^{-1} to between 10^{-3} and 10^{-2} m day^{-1} (typical saturated intergranular values for this area) as the suction value increased beyond about 0.25 m. Thereafter the conductivity stayed relatively constant until the suction reached about 10 m, when there was another sharp decline. This decline at 10 m (rather than the 30 m or more which might be expected from the evidence of mercury-injection curves) should not be interpreted too strictly; the measurement of the high suctions is less reliable than measurements in the 0–3 m range (J. D. Cooper, personal communication, 1985).

At the Bridgets site, Wellings (1984) reported the almost continuous presence of pore-water suctions in excess of 2 m, and concluded that recharge must occur through the matrix at this site as the fissures would be drained at these suctions. At the Fleam Dyke site, Wellings & Cooper (1983) reported suctions in winter which were frequently less than 0.5 m, implying a water-filled fissure system.

At Bridgets, the unsaturated matrix conductivity is high enough to permit almost all recharge to take place through the matrix—the so-called 'piston flow' process. However, some recharge does occur through fissures at other Chalk sites in southern England. Downing *et al.* (1978) reported that some wells in Chalk near Brighton may become contaminated by bacteria and suspended solids within 24 h after heavy rain, and that the tritium content of the water in some wells increased after periods of intense or prolonged rainfall to values greater than those occurring in the rain. They reasoned that when the tritium levels in rainfall were high, following the main series of thermonuclear tests, tritium entering the ground diffused from fissures into the relatively immobile chalk pore water in the unsaturated zone in the manner suggested by Foster (1975a). Later, as the concentration of tritium in the infiltrating water declined, the

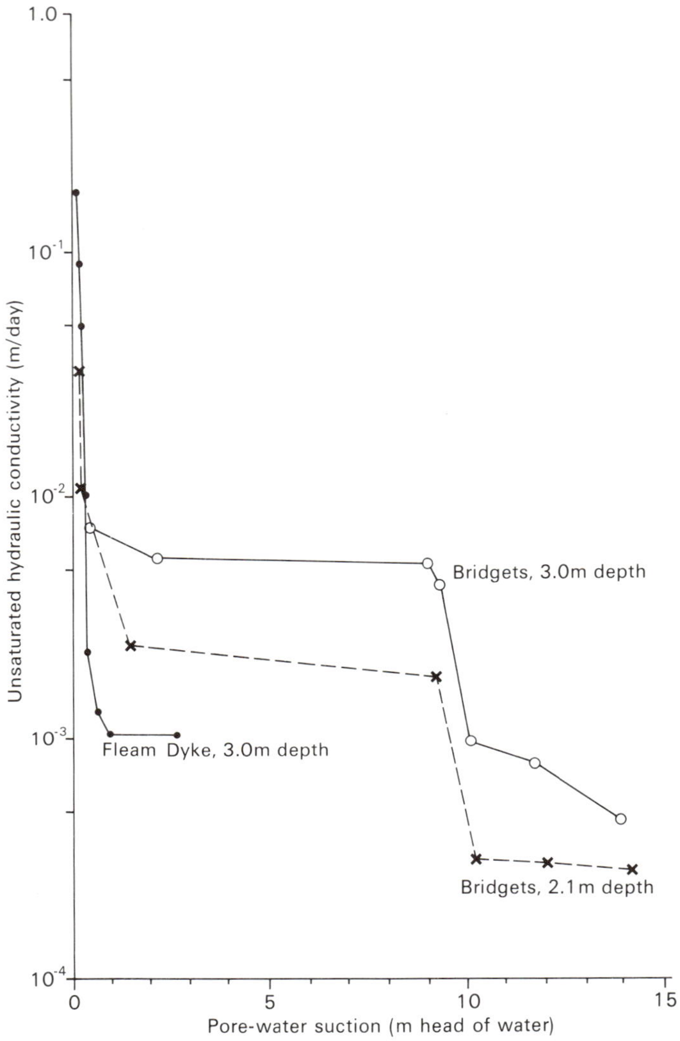

FIG. 4. The relationship between unsaturated hydraulic conductivity and pore-water suction for chalk at Fleam Dyke and Bridgets (data supplied by the Institute of Hydrology).

gradient of tritium concentration in the matrix blocks became reversed at a particular depth; tritium then moved from the pore space to the fissures, moved downwards in the fissures to a level where the tritium concentration of the pore water was lower and diffused into the matrix again. (The possibility of such a mechanism was confirmed theoretically by Barker & Foster 1981.) In this way the high-concentration pulse could move downwards by repeated interchange between the fissure water and the pore water. After periods of intense rain, infiltrating water travelling down fissures may 'collect' tritium from the unsaturated zone and carry it down to the water table, thereby arriving with a tritium concentration higher than that present in the rainfall and causing the anomalously high tritium values in sampled groundwater.

As far as groundwater resources are concerned, the mechanism—fissure flow or intergranular flow—by which infiltrating water reaches the saturated zone is of little significance. The recharge mechanism becomes very significant, however, when pollutant movement is considered. Flow through a fissure system will be much more dispersive than flow through the pore space in chalk. Study of the distribution of stable isotopes (Wellings & Cooper 1983) suggests that, under the intergranular-flow 'piston-displacement' process, seasonal variation in solute concentration in infiltrating water will be preserved as stratification in the chalk pore water for years. By implication, a pollutant entering a chalk unsaturated zone will remain relatively undispersed and undiluted if the fissure system is non-conducting. It will move slowly down to the water table—rates of about 1 m a^{-1} seem typical (Smith *et al.* 1970)—and arrive there as a concentration peak, perhaps many years after it first infiltrated the soil. Only then will it begin to be widely and rapidly dispersed.

Conversely, a pollutant entering an unsaturated zone where fissure flow is occurring will be quickly and widely dispersed, and may reach the water table within days or even hours of infiltration even where the unsaturated zone is relatively thick (Barker & Foster 1981). Furthermore, bacteria and some suspended solids can be transported through a fissure system, whereas they cannot be carried through the pores in chalk.

In summary, a pollutant entering a thick chalk unsaturated zone dominated by matrix flow may be relatively undispersed and undiluted for many years; the same pollutant entering a fissure-flow system may be widely dispersed and probably highly diluted, perhaps within a few days.

Within the shallow highly-weathered layer, where present, the possibilities of rapid dispersion are greatly increased. The high saturated hydraulic conductivity of this zone offers the possibility of absorption of water by soakway or similar drainage at high rates, but equally the presence of this high conductivity means that polluted water should not be allowed to enter such soakways because of the possibility of direct and rapid communication with the water table.

The saturated zone

In the saturated zone it has been shown that the matrix conductivity generally makes a negligible contribution to the transmissivity of the Chalk aquifer. The primary-fissure component is of importance but still does not account for the transmissivity of the most productive parts of the Chalk. There is abundant evidence that the bulk of this transmissivity is contributed at a few levels where the fractures of the primary-fissure component have been enlarged by solution (Foster & Milton 1974; Price *et al.* 1977, 1982; Owen & Robinson 1978; Headworth *et al.* 1982).

Woodland (1946) drew attention to the fact that the highest well yields in the Chalk of southern E Anglia were associated with river valleys. Subsequently, the association between high transmissivity and valleys (both flowing and dry) has been confirmed generally for both unconfined and confined situations (Ineson 1962; Water Resources Board 1972; Owen & Robinson 1978; Lloyd *et al.* 1981).

There is also considerable evidence that secondary (solution) fissures are usually concentrated at the water table or within the zone of seasonal water-table fluctuation (Foster & Milton 1974; Owen & Robinson 1978; Headworth *et al.* 1982). This has led some workers to propose that solution occurs preferentially within and near the zone of water-table fluctuation, possibly because of the presence of increased concentrations of CO_2 in the water at this level (e.g. Ineson 1962). It seems to me that such an explanation ignores the positive correlation between high transmissivity and river valleys; valleys are groundwater *discharge* areas, where water is leaving the aquifer and can be expected to be largely depleted of any dissolved CO_2. More progress can be made by considering the Chalk in terms of classical carbonate-aquifer theory.

As far as valleys are concerned, several factors probably contribute to their relationship with high transmissivity. In the first place many valleys probably follow lines of structural weakness, where primary fissures are likely to be well developed. Second, the reduction in the thickness of overlying material will lead to a reduction in effective stress, thereby favouring the opening up of near-horizontal discontinuities (Ineson 1962).

The pattern of groundwater flow to valleys is probably a significant factor in the development of permeability. Rhoades & Sinacori (1941) put forward a mechanism by which flow through an initially homogeneous isotropic carbonate aquifer would lead to the enhancement of permeability near the area of groundwater discharge. The principle is shown in Fig. 5. Flow from the recharge areas to the discharge area takes place through the full thickness of the aquifer, but flow lines are more closely spaced at shallower depths and with decreasing distance from the discharge area. The same volume rate of flow of water therefore occurs through successively smaller areas of rock—with a corresponding increase in flow speed—as the river is approached. Further, the distance from the recharge area to the discharge area will be much less for one of the shallower flow paths than for one of the deeper ones.

In combination these effects mean that the shallower waters, originating from nearer the river, will have much less time to become saturated with bicarbonate and will be much more aggressive than those originating from the interfluves. In effect, the concentration of flow towards the river leads to more rapid solution of the carbonate in the valley and to the development of what Rhoades & Sinacori (1941) termed a

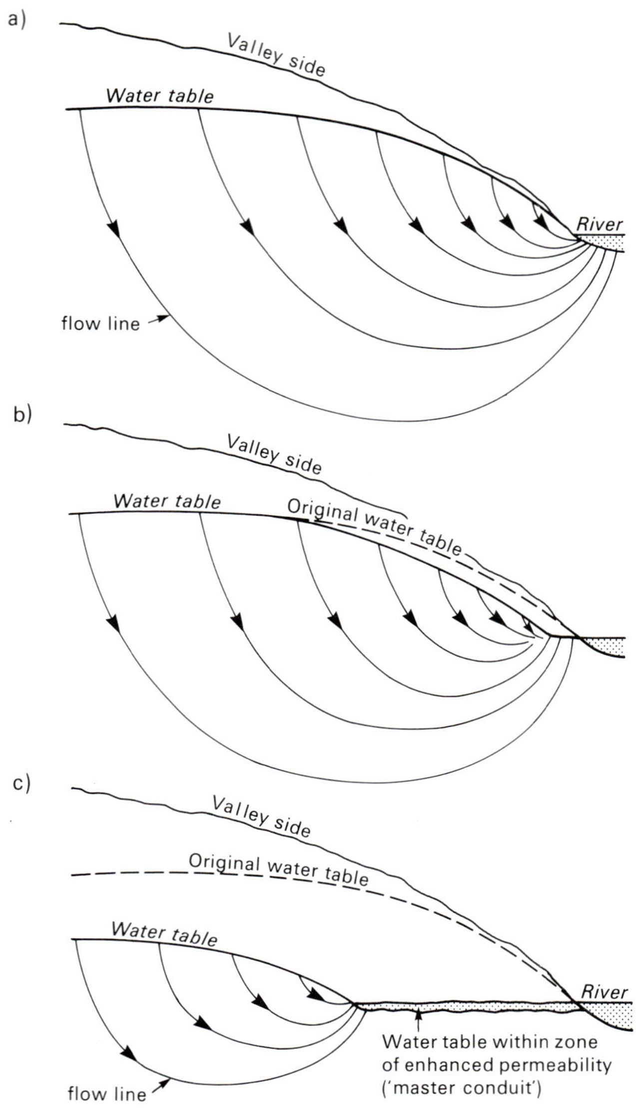

FIG. 5. Development of enhanced permeability in a carbonate aquifer, controlled by topography: (a) flow pattern to a river in a homogeneous isotropic aquifer; (b) concentration of flow near the river leads to preferential solution at shallow depths along the valley which enhances permeability and leads to further concentration of flow; (c) eventually a zone of high permeability (or a single well-developed fissure or 'master conduit') develops, and the water table is constrained to stay within this zone. (After Rhoades and Sinacori 1941.)

'master conduit' (Fig. 5b). Once this conduit (a secondary fissure in the terminology of this paper) has begun to develop, the flow pattern is modified, with flow tending to be concentrated towards and along the secondary fissure, leading to its further enlargement and growth away from the valley (Fig. 5c).

Other workers (Swinnerton 1949; Legrand & Stringfield 1971) have agreed with this mechanism of development, in which it is assumed that the aquifer is initially homogeneous and isotropic. Uniformly fractured chalk approximates tolerably well to this ideal and possesses many of the characteristics of the 'fine-textured carbonate aquifer' of Legrand & Stringfield (1971). Clearly, however, minor natural variations in the initial primary fissure spacings and apertures—including any variations related to the presence of the valley—will cause departures from the ideal case of Fig. 5(c), perhaps leading to the development of a zone of solution-enhanced fissures rather than of a single major fissure.

Once such a secondary fissure or zone has developed, the water table will be constrained to stay within or near it; if the water table tended to rise above the fissure, the increased hydraulic gradient would lead to increased flow through the fissure, conveying groundwater more rapidly to the river so that the water table would fall again. If the water table were to fall below the fissure, the contribution of the fissure to transmissivity would be lost so that the hydraulic gradient would have to steepen, causing the water table to intersect the secondary fissure again. Thus it is probably nearer the truth to say that, in many areas, the water table lies within a zone of enhanced permeability than to say that a zone of enhanced permeability develops in the zone of water table fluctuation. Evidence for such a zone, controlling the water table in the manner described above, is available from the Candover catchment in Hampshire (Headworth *et al.* 1982).

The classical explanation of permeability enhancement along river valleys by solution can also be applied to areas where Chalk is overlain by less permeable materials. Valleys in these areas may follow lines of structural weakness, and fracture development may be enhanced by stress relief caused by erosion, leading to some preferential flow and solution. More important may be the fact that valleys incised in overlying materials will offer the easiest pathway for discharge from the confined Chalk because (a) the thickness (and therefore the hydraulic resistance) of the confining bed will probably be at its least and (b) the vertical hydraulic gradient will be at a maximum (Fig. 6). Groundwater flow and solution will therefore be concentrated beneath valleys even in confined areas (Water Resources Board 1972).

In reality, the Chalk would not have been perfectly homogeneous and isotropic to begin with. Minor variations in lithology and hardness would have led to similar variations in the frequency and nature of jointing and fracturing, so that some of the primary fissures would be more open and more permeable than others. Thus preferential pathways, and hence solution and enlargement, would have been selected on the basis of lithology or stratigraphy as well as of topography. At the edge of the Tertiary cover the concentration of acidic runoff from the generally less permeable Tertiary deposits frequently leads to the development of enhanced permeability. As noted above, this may have been a significant factor in the development of the rapid-flow system at Bedhampton (Atkinson & Smith 1974; Foster 1975b). In these, and probably other, areas the secondary-fissure systems may develop in a directional fashion with serpentine conduits (like flattened tubes) concentrated at particular fissured horizons. These tubes probably branch and rejoin. Such a system would explain the apparent directional nature of the flow system at Bedhampton.

Nevertheless, the basic principle of Rhoades & Sinacori probably applies in many parts of the Chalk. Headworth (1978) drew attention to the fact that a major fissure developed in the Alresford area corresponds to the level of the floor of the Itchen valley several kilometres downstream.

As permeability increases, the water table becomes flatter and moves downwards (Legrand & Stringfield 1971). Progressive erosion and lowering of the valley floor will similarly lead to a lowering of the water table. Enhanced permeability will develop at levels related to the new position of the valley floor; the high permeability which corresponded to earlier positions of the valley floor and higher levels of the water table will then become part of the unsaturated zone. Thus enlarged fissures may be found above the present water table as well as near or below it; for example, such dry fissures have been seen,

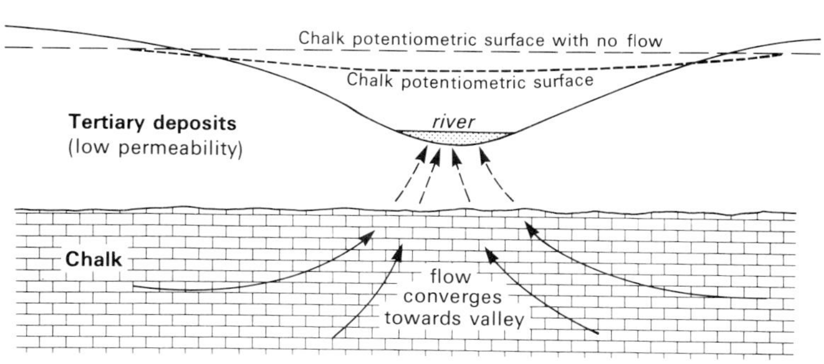

FIG. 6. The reduced thickness of impermeable material and the larger vertical hydraulic gradients along valleys mean that water will flow from confined Chalk preferentially along such valleys. The convergence of flow beneath the valley may lead to enhancement of the Chalk permeability there in the manner of Fig. 5.

using borehole television, in the unsaturated chalk of the Candover Valley, Hampshire.

A slightly different explanation for the development of Chalk transmissivity was put forward by Morel (1980), who provided a detailed account of the Chalk aquifer in the Upper Thames Valley. He accepted the classical explanation of permeability development following Rhoades & Sinacori but, in a detailed argument, reasoned that much of the present distribution of transmissivity may have become established before the Eocene strata were laid down. He suggested, for example, that the high permeability found beneath the confining strata in the Lee Valley may have developed along the line of an earlier river before the confining strata were present. Morel argued that the period of time since the last glacial recession (about 12 000 years) would have been too short for the necessary solution and fissure development to have occurred.

At present, there is on average at least 100 mm of infiltration annually over the Chalk outcrop; this corresponds to 100 l m^{-2}. 100 l of water could be expected to remove about 30 g of $CaCO_3$ in solution; if this material were removed from beneath each square metre of area of a 300 m thick chalk aquifer with a porosity of 40%, it would add about 18 ml or 6×10^{-6}% to the pore space. Thus, if all the material were removed from the saturated zone, about 16 000 years would be needed to create a fissure porosity of 0.1% evenly distributed throughout the profile. However, much of the ability of the infiltrating water to dissolve chalk will be used up in the unsaturated zone; if only 10% of the infiltration occurred as rapid flow through fissures, the period required for the necessary solution would be increased tenfold to 160 000 years. Set against this, however, are the following facts.

1 The enhanced fissure permeability does not occur throughout the full thickness of the aquifer; in combination with the patterns of concentrated flow proposed by Rhoades & Sinacori this could reduce the time required by an order of magnitude, back to 16 000 years.
2 In the colder wetter periods following the last glaciation, infiltration may have been well in excess of 100 mm a^{-1}, although admittedly the CO_2 content of the infiltrating water (and therefore its aggressiveness to $CaCO_3$) would probably be reduced because the activities of soil bacteria would be slowed.

In general, it appears that there would probably have been just enough time for the present distribution of secondary-fissure permeability and some more uniform solution enlargement of the primary fissures to have occurred since the last glacial recession, at least in the unconfined aquifer. Nevertheless, Morel's arguments for the areal development of transmissivity, particularly in the confined aquifer, remain attractive and may be applicable to areas other than the Thames Valley.

As in the unsaturated zone, so in the saturated zone the combination of rapid flow through fissures with much larger volumes of relatively immobile water in the matrix provides opportunities for solute diffusion. Downing *et al.* (1979) drew attention to the fact that the fissure volume of the confined Chalk beneath the Lee Valley, in the London Basin, must have been displaced several times since 1954, yet the water pumped from wells in this area still contained 'old' (i.e. pre-1954) water with a low tritium content. They reasoned that flow in the Chalk consists of two components—rapid flow through the main fissure system and slower movement through the matrix pore space and smaller fissures. However, the water in the intergranular pore space and smaller fissures will still be involved in the total transport mechanism of groundwater and solutes, because solutes entering the aquifer in relatively high concentrations and flowing through the major fissures will diffuse into the smaller fissures and pore spaces (provided that their molecular size is small enough to permit this diffusion). Thus, as in the unsaturated zone, a pollutant may be delayed by being 'stripped' from the fissure water and retained in the pore space. Later, as concentrations in the fissures decline, diffusion may take place in the opposite direction so that the pollutant re-enters the fissures from the pores. The end result is to delay the movement of the pollutant and spread its arrival over a longer time period. Once a pollutant has entered the Chalk, it may take a considerable time before it is completely removed.

Downing *et al.* (1979) also postulated that water molecules will move from the pore space to the fissures under the influence of a hydraulic gradient; this gradient would result from pumping that lowered the head in the fissure system. It will be shown in the following section of this paper that although such movement does occur it is probably of secondary importance, as a storage mechanism, to the reduction in fissure aperture resulting from the compaction of the aquifer.

Storage

If the matrix pore space of chalk makes a negligible contribution to the bulk permeability of the aquifer, does it make any contribution to useful storage? Study of Fig. 2 implies that even at pore-water suctions of about 10 m (the maximum that seems to occur under natural gravity drainage (e.g. Wellings 1984, Fig. 2)) only about 3% of porosity or about 1% of bulk volume will have drained. This is comparable with specific yield values measured by centrifuge drainage tests (Bird 1976). The fact that specific yield values for unconfined chalk are typically only about 1–3% implies that the water released from intergranular storage, although representing only a small fraction of the total pore water, may nevertheless be a significant part of the specific yield.

A much larger proportion of pore water is available to plants. Plant roots can exert suctions in excess of 100 m of water before wilting occurs; Wellings (1984) reported suctions approaching 40 m at the Bridgets site. At suctions of 100 m, 75–95% of Upper Chalk pore space (about 30–40% of bulk volume) will drain. The development of these high suctions at shallow depths means that steep upward gradients can develop (Wellings 1984, Fig. 2); with more than 75% of pore space drained, however, hydraulic conductivity will be very low so that the upward flow rates of water will be very small.

Under *confined* conditions, water will be released from elastic storage—by expansion of the water and compaction of the chalk—as the head (and therefore the water pressure) is reduced. The volume of water released from elastic storage in unit volume of an aquifer in response to a unit decline in head is termed the specific storage S_s. It is related to the storage coefficient S of a confined homogeneous aquifer by the expression

$$S = S_s b \tag{3}$$

where b is the aquifer thickness. Specific storage can be expressed in terms of the properties of the water and of the aquifer matrix by the equation

$$S_s = \rho g \alpha + \rho g n \beta \tag{4}$$

where ρ is the density of the contained water, g is gravitational acceleration, n is effective porosity, α is the compressibility of the rock (measured under drained conditions) and β is the compressibility of the water (approximately 4.9×10^{-10} Pa^{-1}.

The compressibility α can be calculated from values of Young's modulus Y and Poisson's ratio σ using the expression

$$\alpha = \frac{3(1-2\sigma)}{Y} \tag{5}$$

The first term on the right-hand side of Equation 4 represents the water released from (or taken into) elastic storage by the compaction (or expansion) of the aquifer framework; the second term represents the water released from (or taken into) storage as a result of the expansion (or compression) of the water itself. In the case of a fractured aquifer, Equation 4 can be expanded to

$$S_s = \rho g(\alpha_f + \alpha_m + n_f\beta + n_m\beta) \tag{6}$$

In Equation 6, α_f represents the compressibility of the aquifer framework resulting from the presence of discontinuities, α_m represents the compressibility of the unfractured matrix, n_f is the porosity contributed by discontinuities and n_m is the matrix porosity. The relative magnitudes of the terms within the parentheses in Equation 6 indicate the relative importance of the components of elastic storage. The value of β is taken to be 4.9×10^{-10} Pa^{-1}. Values for n_m generally lie within the range 0.2–0.45 and n_f is assumed to lie between 0.001 and 0.01. Values of α_m can be derived from laboratory measurements on intact chalks; using data given by Bell (1977) typical values ranging from 8×10^{-11} Pa^{-1} (Lower Chalk, Yorkshire) to 2×10^{-10} Pa^{-1} (Upper Chalk, Kent) can be calculated. Priest *et al.* (1976) quoted laboratory values of Young's modulus for the Lower Chalk at Chinnor ranging from 6.0×10^{-10} to 3.3×10^{-10} Pa and Poisson's ratios generally in the range 0.2–0.4, yielding values for α_m of 2×10^{-10} to 3×10^{-9} Pa^{-1}; the higher values correspond to material which has been affected by weathering.

Values of α_f require determination *in situ*. A complication is that such determinations will involve both the matrix and fracture components of compressibility. A second problem is that the test method may involve disturbance of the chalk and alteration of its elastic properties. For this reason large-scale loading tests using low applied stresses, such as the tank test performed at Mundford (Ward *et al.* 1968), or indirect methods deriving Young's modulus from seismic measurements (Abbiss 1979) are probably the most useful for calculation of the elastic storage. The Mundford tank experiment yielded values for Young's modulus for the Grade II chalk of 2×10^9–5×10^9 Pa (Ward *et al.* 1968). Using the value of 0.1 for Poisson's ration quoted by Ward *et al.* for the fractured chalk at this site yields *in situ* values for α in the range 4.8×10^{-10}–1.2×10^{-9} Pa^{-1}; Abbiss quoted a value of 0.24 for Poisson's ratio at this site, leading to a value for α of 3.1×10^{-10}–8.0×10^{-10} Pa^{-1}. A value of 5×10^{-10} Pa^{-1} might be typical. This corresponds to the compressibility of the total chalk mass involving both matrix and discontinuities; however, since it is more than half an order of magnitude higher than the laboratory value (8×10^{-11} Pa^{-1}) for intact Middle Chalk samples from Norfolk derived from Bell's (1977) figures, it is reasonable to assign a value of 5×10^{-10} Pa^{-1} to α_f.

Using the Middle Chalk from Norfolk as an example, the components of elastic storage of confined chalk could therefore be expected to have the values summarized in Table 3. These values indicate that most of the water released from elastic storage in chalk is derived from the compressibility of the fissures, with the expansion of water in the matrix pore space playing a significant role. The compaction of the matrix framework is of lesser importance, and the expansion of water in the fractures themselves is responsible for a negligible component of the total specific storage.

The specific storage value of 7×10^{-6} m^{-1} calculated from the values given in Table 3 would yield a storage coefficient of 0.0014 for a 200 m thickness of Chalk. This is towards the upper end of the range of published values for the storage coefficient of confined Chalk, as derived from pumping tests. If it is considered that only the upper 50 m or so of the confined Chalk functions as the effective aquifer (e.g. Water Resources Board 1972; Owen & Robinson 1978; Morel 1980), the storage coefficient would be reduced to 0.000 35; however, water would be released from elastic storage in the lower part of the Chalk and would eventually move upwards into the permeable part of the formation, where it would probably be interpreted as leakage. Leakage effects are common in interpretations of pumping tests in confined Chalk. Another factor that could result in delay in some of the water released from elastic storage is that closure of the fissures could take a finite time in response to the relatively small increases in effective stress caused by the reduction in head in response to pumping.

Conclusions

The Chalk in England can possess several components of porosity and permeability, although not all will necessarily be present at any one locality. Universally present is a matrix or intergranular component; this can contribute porosities of more than 40%, the porosity generally being higher in the southern part of England than in the northern part and also increasing up the stratigraphic succession.

The matrix permeability is generally isotropic and shows the same regional and stratigraphical trends as porosity. It is generally low, seldom exceeding 10 mD (about 6×10^{-3} m day^{-1}); a hydraulic conductivity value of 10^{-3} m day^{-1} is more typical.

The second permeability component—termed in this paper the *primary-fissure component*—is caused by a fairly ubiquitous fracture system, usually consisting of three near-orthogonal sets of joints. The degree of openness of these joints varies from place to place, depending on factors such as tectonic history and how much solution has taken place as a result of sub-surface flow. At depths much in excess of 100 m, the joints may be effectively closed. The hydraulic conductivity imparted by these joints is generally still too low to explain the Chalk's performance as an aquifer; typically it is in the range 10^{-2}–1 m day^{-1}. The presence of the primary-fissure and matrix components of porosity and permeability means that the Chalk is a double-porosity system.

The primary-fissure component of permeability can be enhanced in two circumstances. In the upper few metres of the chalk at outcrop the fracture openings can become enlarged and the block size reduced, leading to hydraulic conductivities greater than 10 m day^{-1} when the

TABLE 3. *Compressibility characteristics of Chalk*

Property	Lower Chalk (Yorks)	Middle Chalk (Norfolk)	Upper Chalk (Kent)
n_m (fraction)*	0.24	0.29	0.46
α_m (Pa^{-1})*	8.3×10^{-11}	8.1×10^{-11}	1.9×10^{-10}
$n_m\beta$ (Pa^{-1})	1.2×10^{-10}	1.4×10^{-10}	2.3×10^{-10}
n_f (fraction)	—	0.01	—
$n_f\beta$ (Pa^{-1})	—	4.9×10^{-12}	—
α_f (Pa^{-1})	—	5×10^{-10}	—

* From Bell 1977.

For Middle Chalk from Norfolk, an approximate value of S_s would therefore be

$$\begin{aligned} S_s &= \rho g(\alpha_f + \alpha_m + n_f\beta + n_m\beta) \\ &= 1000 \times 9.81\ (5 \times 10^{-10} + 8 \times 10^{-11} + 4.9 \times 10^{-12} + 1.4 \times 10^{-10})\ m^{-1} \\ &\approx 1000 \times 9.81\ (7.2 \times 10^{-10})\ m^{-1} \\ &\approx 7 \times 10^{-6}\ m^{-1} \end{aligned}$$

material is saturated. Frequently, however, this material lies in the unsaturated zone. In the top few tens of metres of the aquifer the primary fissures may be enlarged by solution. This enlargement typically seems to occur along individual near-horizontal fractures or in discrete near-horizontal zones, rather than uniformly throughout the aquifer. The non-uniform permeability so produced is termed the *secondary-fissure component* in this paper. These secondary features appear to be related to river-valley base levels, although some may have developed in the geological past (Morel 1980).

The secondary fissures, being essentially highly permeable layers, impart heterogeneity to the Chalk where they are present; thus they cause a double-permeability behaviour to be superimposed on the double-porosity behaviour. The permeability contrast between the secondary- and primary-fissure components is so great, however, that this double-permeability behaviour can be expected to appear as another double-porosity system; the Chalk can thus be a dual double-porosity aquifer. Some progress has been made in the study of the fissure permeabilities using packers; this approach needs to be combined with some of the latest analytical techniques developed in the oil industry (Bourdet *et al.* 1983), although in the unconfined condition there are likely to be problems in obtaining unique interpretations.

In the unsaturated zone, matrix flow appears to be dominant throughout the year at some sites, with pore-water suctions nearly always too high for flow to occur in fissures. At other sites there is evidence that fissure flow occurs after heavy rainfall. Good agreement is seen between unsaturated hydraulic conductivity curves measured *in situ* and predictions based on pore-size measurements and calculated fissure openings.

The small pores in chalks mean that the matrix has low permeability despite its high porosity. Most of these small pores do not drain under gravity, so that the specific yield of chalk is low; however, much of the pore water is accessible to plants.

In the confined condition, an example from Norfolk suggests that most of the water released from elastic storage is derived from closure of fissures. Some water will be released by expansion of pore water from the matrix; some of the elastic storage may take a finite time to become apparent and may be erroneously interpreted in pumping test analyses as leakage from adjacent strata.

The variability of the permeability components throughout the Chalk is so great that exceptions will be found to almost any generalization. However, generalizations can be valuable from the practical point of view. The Chalk is England's major aquifer and its unusual properties have an important influence on the supply of water and its susceptibility to pollution. A knowledge of the properties and flow mechanisms of the unsaturated zone is essential to an understanding of the risk of pollution to the groundwater of the saturated zone. In the saturated zone the secondary fissures are the pathways for most groundwater movement, and the smaller but more numerous primary fissures and the matrix pores contribute storage. Because of the extensive and permeable secondary-fissure systems in the upper part of the saturated zone, water can travel large distances with great rapidity, meaning that pollutants can be quickly and widely dispersed; the high matrix porosity means that pollutants can diffuse into the matrix so that they may be greatly diluted but may also remain there for long periods. It is hoped that the general comments in this paper will have provided some insight into the behaviour of the Chalk aquifer.

ACKNOWLEDGMENTS: I am grateful to many colleagues at the British Geological Survey and the Institute of Hydrology for helping with the preparation of this paper and for providing stimulating discussions on the Chalk generally. In particular I am indebted to Miss A. S. Robertson, Mrs A. K. Geake, Dr R. A. Downing, Dr J. A. Barker and Dr S. S. D. Foster of the British Geological Survey and Dr J. D. Cooper and the late Dr S. R. Wellings of the Institute of Hydrology for general discussion, Dr J. A. Hudson of Imperial College for reviewing the section on elastic storage, and Dr J. A. Barker and Dr R. A. Downing for reading the entire manuscript. The paper is published by permission of the Director, British Geological Survey (NERC).

References

ABBISS, C. P. 1979. A comparison of the stiffness of the chalk at Mundford from a seismic survey and a large scale tank test. *Géotechnique,* **29** (4), 461–8.

ALLEN, D. J. & PRICE, M. 1983. The in-site measurement of hydraulic conductivity of weathered or poorly consolidated materials at shallow depths. *Proc. Int. Symp. on Methods and Instrumentation for the Investigation of Groundwater Systems*, pp. 343–53, Netherlands Organisation for Applied Scientific Research, Noordwijkerhout.

ATKINSON, T. C. & SMITH, D. I. 1974. Rapid groundwater flow in fissures in the Chalk: an example from South Hampshire. *Q. J. eng. Geol.* **7**, 197–205.

BARKER, J. A. & FOSTER, S. S. D. 1981. A diffusion exchange model for solute movement in fissured porous rock. *Q. J. Eng. Geol.* **14**, 17–24.

BELL, F. G. 1977. A note on the physical properties of the Chalk. *Eng. Geol.* **11**, 217–25.

BIRD, M. J. 1976. Core analysis results from the East Yorkshire Chalk. In: FOSTER, S. S. D. & MILTON, V. A. (eds) Hydrological Basis for Large-scale Development of Groundwater Storage Capacity in the East Yorkshire Chalk, Appendix A, *Rep. Inst. Geol. Sci. No 76/3.*

BOURDET, D., AYOUB, J. A., WHITTLE, T. M., PIRARD, Y. M. & KNIAZEFF, V. 1983. Interpreting well tests in fractured reservoirs. *World Oil*, October 1983.

CARTER, P. G. & MALLARD, D. J. 1974. A study of the strength, compressibility, and density trends within the Chalk of South East England. *Q. J. eng. Geol.* **7**, 43–55.

CHANNEL TUNNEL STUDY GROUP 1965. *Channel Tunnel Site Investigations in the Strait of Dover.*

DOWNING, R. A., PEARSON, F. J. & SMITH, D. B. 1979. The flow mechanism in the Chalk based on radio-isotope analyses of groundwater in the London Basin. *J. Hydrol.* **40**, 67–83.

——, SMITH, D. B. & WARREN, S. C. 1978. Seasonal variations of tritium and other constituents in groundwater in the Chalk near Brighton, England. *J. Inst. water Eng.* **32**, 123–36.

FOSTER, S. S. D. 1975a. The Chalk groundwater tritium anomaly—a possible explanation. *J. Hydrol.* **25**, 159–65.

—— 1975b. Discussion of Atkinson and Smith (1975). *Q. J. Eng. Geol.* **8**, 155–6.

—— & MILTON, V. A. 1974. The permeability and storage of an unconfined chalk aquifer. *Hydrol. Sci. Bull.* **19**, 485–500.

HANCOCK, J. M. 1975. The petrology of the Chalk. *Proc. geol. Assoc.* **86**, 499–535.

—— & KENNEDY, W. J. 1967. Photographs of hard and soft chalks taken with a scanning electron microscope. *Proc. geol. Soc.* **1643**, 249–52.

HEADWORTH, H. G. 1978. Hydrogeological characteristics of artesian boreholes in the Chalk of Hampshire. *Q. J. eng. Geol.* **11**, 139–44.

——, KEATING, T. & PACKMAN, M. J. 1982. Evidence for a shallow highly-permeable zone in the Chalk of Hampshire, U.K. *J. Hydrol.* **55**, 93–112.

INESON, J. 1962. A hydrogeological study of the permeability of the Chalk. *J. Inst. water Eng.* **16**, 449–63.

KLINKENBERG, L. J. 1941. The permeability of porous media to liquids and gases. *Drill. Product. Practice*, 200–13.

LEGRAND, H. E. & STRINGFIELD, V. T. 1971. Development and distribution of permeability in carbonate aquifers. *Water Resour. Res.* **7**, 1284–93.

LLOYD, J. W., HARKER, D. & BAXENDALE, R. A. 1981. Recharge mechanisms and groundwater flow in the Chalk and drift deposits of southern East Anglia. *Q. J. eng. Geol.* **14**, 87–96.

MOREL, E. H. 1980. A numerical model of the Chalk aquifer in the Upper Thames Basin. *Tech Note 35*, Central Water Planning Unit, Reading.

OWEN, M. & ROBINSON, V. K. 1978. Characteristics and yield in fissured chalk. *Proc. Conf. on Thames Groundwater Scheme*, pp. 33–49, Institution of Civil Engineers, London.

PRICE, M., BIRD, M. J. & FOSTER, S. S. D. 1976. Chalk pore-size measurements and their significance. *Water Serv.* **80**, 596–600.

——, MORRIS, B. L. & ALLEN, D. J. 1979. Hydraulic conductivity tests in the unsaturated zone of the Chalk—development and initial results from two sites in north-west Norfolk. *Rep. Inst. Geol. Sci. WD/ST/79/14* (unpublished, on open file).

——, —— & ROBERTSON, A. S. 1982. A study of intergranular and fissure permeability in Chalk and Permian aquifers, using double-packer injection testing. *J. Hydrol.* **54**, 401–23.

——, ROBERTSON, A. S. & FOSTER, S. S. D. 1977. Chalk permeability—a study of vertical variation using water injection tests and borehole logging. *Water Serv.* **81**, 603–10.

PRIEST, S. D., HUDSON, J. A. & HORNING, J. E. 1976. Site investigation for tunnelling trials in chalk. *TRRL Laboratory, Rep. 730,* Department of the Environment, Crowthorne.

RHOADES, R. & SINACORI, M. N. 1941. Pattern of ground-water flow and solution. *J. Geol.* **49**, 785–94.

SMITH, D. B., WEARN, P. L., RICHARDS, H. J. & ROWE, P. C. 1970. Water movement in the unsaturated zone of high and low permeability strata by measuring natural tritium. In: *Isotope Hydrology*, pp. 259–70, IAEA, Vienna.

SWINNERTON, A. C. 1949. Hydrology of limestone terranes. In: MEINZER, O. E. (ed.) *Physics of the Earth*, Vol. 9, *Hydrology*, Dover Publications, New York.

TROTTER, J. G., THOMPSON, D. M. & PATERSON, T. J. M. 1985. First mined hydrocarbon storage in Great Britain. *Tunnelling '85*, Paper 17, pp. 1–12, Institution of Mining and Metallurgy, Brighton.

WARD, W. H., BURLAND, J. B. & GALLOIS, R. W. 1968. Geotechnical assessment of a site at Mundford, Norfolk, for a large proton accelerator. *Geotechnique,* **18**, 399–431.

WATER RESOURCES BOARD 1972. *The Hydrogeology of the London Basin with Special Reference to Artificial Recharge*, Water Resources Board, Reading.

WELLINGS, S. R. 1984. Recharge of the Upper Chalk aquifer at a site in Hampshire, England, 1, Water balance and unsaturated flow. *J. Hydrol.* **69**, 259–73.

—— & COOPER, J. D. 1983. The variability of recharge of the English Chalk aquifer. *Agric. Water Manage.* **6**, 243–53.

WOODLAND, A. W. 1946. Water supply from underground sources of Cambridge—Ipswich District. *Wartime Pamphlet 20*, Part 10, Geological Survey of Gt. Britian.

MICHAEL PRICE, Aquifer Properties Section, Engineering Geology and Reservoir Properties Research Group, British Geological Survey, Wallingford, Oxon OX10 8BB, U.K.

Thermal aspects of the East Midlands aquifer system

N. P. Wilson & M. N. Luheshi

SUMMARY: A case study of a heat flow anomaly in the E Midlands of England is reported. The anomaly has been suggested to be an effect of water movement at depth within the E Midlands basin, with recharge to the Lower Carboniferous limestones in their outcrop, eastward movement and ascent of water up a steep faulted anticline at Eakring where the heat flow measurements were made.

Numerical modelling of heat and fluid flow has been undertaken for a section running from the Peak District through Eakring to the coast. The results indicate that, although an anomaly is expected for reasonable values of hydrological parameters, its magnitude is less than that observed. The geological structure at Eakring is such that three-dimensional flow is likely to be important, and this could easily account for the discrepancy between the modelling results and the observations.

The regional water flow regime has other effects on heat flow, notably the depression of heat flow above the Sherwood Sandstone aquifer.

Regional groundwater flow systems involve fluid velocities which may be extremely low. These velocities can be estimated using approximations for potential gradients and hydraulic conductivities or natural and anthropogenic tracers. A parameter which is sensitive to low-velocity fluid motions is terrestrial heat flow, and as a consequence sub-surface temperature measurements have been used (Stallman 1963; Cartwright 1968) to help delineate flow regimes. Conversely, the effect of such regimes on the measured heat flow has been considered in investigations of deep regional heat flow (Lachenbruch & Sass 1977; Lewis & Beck 1977).

Numerical modelling of groundwater flow systems was developed in the 1960s and early 1970s (Freeze & Witherspoon 1966, 1967) and has become routine. It is now possible to model coupled heat and fluid flow in two and three dimensions (Smith & Chapman 1983; Woodbury & Smith 1985). These methods predict the variation in heat flow as measured at the surface in a basin with a constant heat input to the base of the flow regime. The areal and vertical variation of measured heat flow should provide an important constraint for such models.

A case study, using the Smith–Chapman algorithm, of a heat flow anomaly in the E of England, which has been postulated to result from water movement, is presented in this paper.

The E Midlands

Heat flow

The distribution of heat flow measurements in the E Midlands area is shown in Fig. 1. The Eakring anomaly comprises the four values of heat flow published by Bullard & Niblett (1951) which were measured in wells in the Eakring oil field [G.R. SK68 60]. These range up to 120 mW m^{-2} and are much higher than the values measured only a few kilometres away at Caunton [SK734 602] (70 mW m^{-2}) and Kelham Hills [SK759 576] (62 mW m^{-2}).

Bullard & Niblett (1951), having considered other possible causes for such an anomaly, concluded that water ascending the steep western limb of the Eakring anticline in the Carboniferous Limestone elevated the temperature at the crest of the anticline and thus the heat flow measured above.

Subsequent heat flow determinations indicate that the anomaly may extend some way to the SE (e.g. at Long Bennington [SK806 509] (88 mW m^{-2}) Richardson & Oxburgh 1978) and to the N (e.g. at Ranby Camp [SK664 808] (85 mW m^{-2}) Mullins & Hinsley 1957) but that it is still distinct. The heat flow in the area is generally higher than the U.K. average of about 55 mW m^{-2} (Burley *et al.* 1984).

Support for Bullard & Niblett's hypothesis of water movement in the Dinantian limestones is given by the measurements in the Eyam borehole in the Peak District of Derbyshire (Richardson & Oxburgh 1978). Here a very low heat flow was measured, and the temperature gradient was negative below about 600 m. This can only be accounted for by downward water flow at these depths.

The heat flow pattern is consistent with a groundwater flow regime of recharge in the outcrop area of the Dinantian, depressing the isotherms, fluid flow to considerable depths, and ascent along the approximately N–S Eakring anticline causing the high heat flow there. Where

From GOFF, J. C. & WILLIAMS, B. P. J. (eds), 1987, *Fluid Flow in Sedimentary Basins and Aquifers*, Geological Society Special Publication No. 34, pp. 157–169.

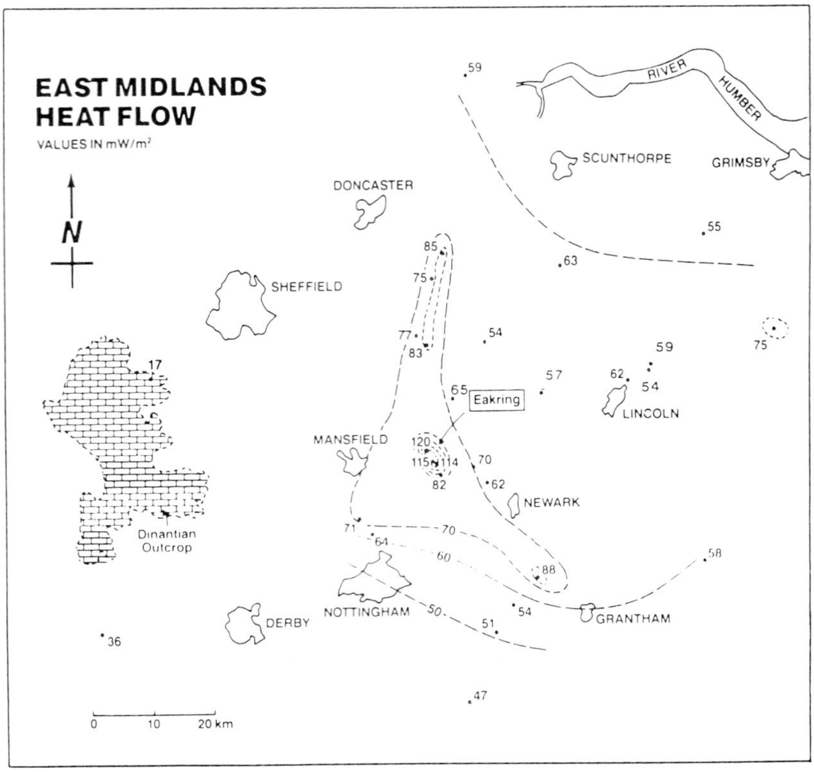

FIG. 1. Heat flow in the E Midlands of England, showing the position of the Eakring 'high'.

water movement is near horizontal heat flow would be little affected.

In boreholes penetrating the Sherwood Sandstone, a vertical discontinuity is seen in heat flow: heat flow within and slightly above this aquifer is lower than that below. This suggests a down-dip flow of water from the surface.

Hydrogeology

The geology of this area is well known, largely because of the extensive exploration for coal and, more recently, oil. A sketch geological map is shown in Fig. 2(a) and a cross-section running approximately E–W through Eakring is shown in Fig. 2(b).

Pre-Carboniferous rocks investigated by several boreholes include a variety of rock types—quartzites, phyllites and volcanics—which are all indurated and are thought to have very low permeability. For the purposes of modelling, the pre-Carboniferous is taken to be the hydrogeological basement.

The Carboniferous rocks form a section up to 2000 m thick which is most usefully discussed hydrogeologically in terms of the series Dinantian (Carboniferous limestone), Namurian (Millstone Grit) and Westphalian (Coal Measures).

The Dinantian is represented by limestones which are variably clean to argillaceous with shales which are correspondingly minor or important. The clean massive limestones were deposited over basement blocks, while the more argillaceous facies accumulated in the intervening basins (Falcon & Kent 1960). Although the intergranular permeability of even the clean limestones is very low (Gale & Holliday 1983), karstic phenomena are well developed at outcrop and may extend some distance down-dip. Fracturing is usually present at the top of the shelf limestones where these have been cored. The bulk permeability of this unit is poorly known, and this is one of the main variables in the modelling.

The Namurian consists of a mixed clastic sequence with basinal mudstones overlain by turbidite sandstones and deltaic deposits in the upper part. While some of the sandstones may be quite extensive, there is little intercommunication on a regional scale and the permeability of the Namurian as a whole is controlled by that of the argillaceous 'framework'.

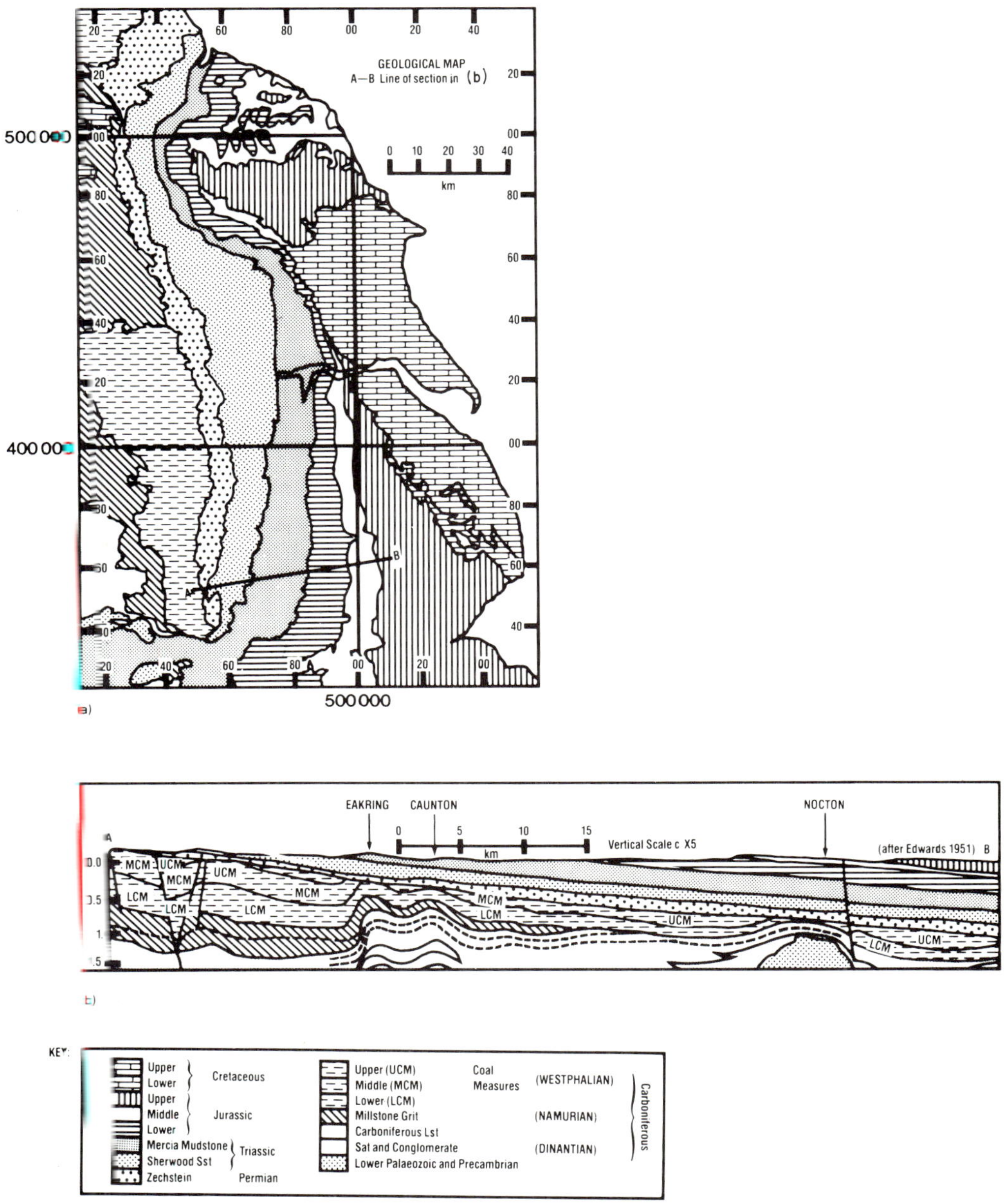

FIG. 2. (a) A sketch geological map of the E Midlands with the line of section shown in (b); (b) geological section along the line AB (largely speculative below the level of the top Dinantian). (After Edwards 1951.)

The deltaic environment of the Westphalian led to the development of laterally discontinuous sandstones in a largely argillaceous sequence, and again the regional permeability is low. Some Upper Westphalian sandstones may be more extensive in the E of the region and may have aquifer potential (Gale & Holliday 1983).

The Lower Permian is represented by patchy deposits of a basal breccia and sandstones on a peneplained Carboniferous surface. These formations become thicker and more persistent to the E, and their high permeability (up to 650 mD, Gale *et al.* 1983) makes them an important aquifer locally. The Upper Permian (Zechstein) includes the Lower and Upper Magnesian limestones, which near outcrop may exhibit important fissure flow but at depth are thought to be characterized by a low matrix permeability. To the E, in the southern North Sea basin, the Zechstein evaporites form important aquicludes.

The Triassic Sherwood Sandstone Group is largely arenaceous and the Sherwood Sandstone itself is the most important aquifer in the region. This is a thick (up to 300 m) poorly consolidated continental sandstone with porosity averaging 30% and permeabilities commonly 5–10 D (Gale *et al.* 1983). Flow velocities in this aquifer range up to 9 m a^{-1} near the outcrop (Edmunds *et al.* 1982), but away from the outcrop velocities are much lower (Bath *et al.* 1979). The Upper Triassic is represented by the thick sequence of mudstones/siltstones and minor sandstones of the Mercia Mudstone Group.

The Jurassic rocks of the E Midlands are, in general, of shallow marine origin and largely argillaceous. The mid-Jurassic Lincolnshire Limestone forms an important local aquifer near outcrop, where tritium analyses (Downing *et al.* 1977) suggest very high flow velocities. Permeabilities may be up to 30 D (Downing *et al.* 1977), but this is almost entirely due to fissures which may not persist down-dip to the E.

Lower Cretaceous rocks include shallow-water sandstones and limestones, while widespread open marine conditions are represented by the Chalk of the Upper Cretaceous. The Chalk forms an important aquifer (e.g. Foster & Milton 1976), again with fissure permeability dominant. As a unit, the Cretaceous is thought to possess low but significant permeability.

Modelling

Analytical solutions for the problem of coupled heat and fluid transfer exist only for very simple cases (e.g. Bredehoft & Papadopoulos 1965). Numerical methods are required for complex internal geometries and distributions of physical properties. In this work the finite-element algorithm developed by Smith & Chapman (1983) has been used. This method is applicable to two-dimensional geological sections with the following conditions.

1 There is no heat or fluid flow across the vertical side boundaries.

2 There is no fluid flow across the base of the model.

3 The upper boundary of the model is the water table.

4 The temperature is specified at the water table.

5 The system is considered to be in a steady state.

The first boundary condition implies a state of symmetry for the model chosen, i.e. the geometry of the model is assumed to be reflected in each side boundary. This is reasonable when modelling a clearly defined flow system, but becomes more of a restriction when only part of a large basin is being considered. Details of the method and its application are given by Smith & Chapman (1983).

The geological section shown in Fig. 2(b) has been approximated by several model sections, of which two will be considered here.

The potential used in the modelling (the 'equivalent freshwater head', Smith & Chapman 1983) is equivalent to the elevation of the water table above the datum used. This is only known in detail for some of the post-Carboniferous formations (e.g. the Magnesian Limestones (I.G.S. 1981), the Sherwood Sandstone (I.G.S. 1981; Edmunds *et al.* 1982), the Lincolnshire Limestone (Downing & Williams 1969) and the Chalk (Foster & Milton 1976; I.G.S. 1981)). For the Carboniferous rocks of the region Gale (personal communication, 1984), after studying the available pressure data, estimated a potentiometric gradient of less than 1 m km^{-1} in an easterly direction for the Dinantian. The gradients for the Upper Carboniferous sections are equally poorly defined; Gale (personal communication, 1984) suggested a slightly larger gradient for the Namurian, while the head in the Westphalian may even decrease to the W.

In model 1 the water table or fluid-potential surface is taken to be the present topography. This is an upper limit to the elevation of the water table, especially at the W end of the section where the water table, if a continuous water table exists in the karstic Dinantian limestones (see, e.g., Jennings 1971), is well below the ground surface.

In model 2 a hydraulic gradient of 1 m km^{-1} is used and the surface is assumed to rise from sea level at the coast to about 140 m at the W end of the model. Although this gradient may apply to the Dinantian and approximate to that in the lower permeability units, it is considerably lower than some of the measured hydraulic gradients in, for example, the Sherwood Sandstone at outcrop (Edmunds *et al.* 1982) and constitutes a lower limit for such units.

Each of the models shown here has an easterly extension in the form of a vertically homogeneous unit down to basement. This 'pipe' is included for the reasons mentioned above, i.e. because a complete groundwater system is not being considered. The E Midlands must be thought of as the westerly margin of the southern North Sea basin and, while the hydrodynamic situation of the whole basin is unknown, it would be incorrect to assume that the E Midlands is completely separate. Varying the permeability in the 'pipe' allows modelling of a closed system, if it has zero permeability, or an open system with free discharge of each aquifer, if it has the permeability of the most permeable unit. The real case will lie between these two extremes, closer to the closed system but with some flow through and 'discharge' offshore (Bath *et al.* 1979). The flow system near the E end of the model will therefore be mostly due to the presence of the pipe, but the rest of the model will not be locally affected.

Results

Model 1(a)

This model gives the temperature field for a section of uniform thermal conductivity and no permeability. The surface temperature is constant at 10°C and the basal heat flow is 70 mW m^{-2}. The isotherms are presented in Fig. 3, and can be seen to follow the shape of the upper surface with some damping at depth. The topography at the W end leads to an enhancement of the vertical conductive heat flow (VCHF) beneath the valley of less than 10% of the background value, and a corresponding reduction beneath the topographic highs. The VCHF across the rest of the section is within 2% of the background. This model is included as a 'baseline' against which the effects of varying thermal conductivity and water flow can be judged.

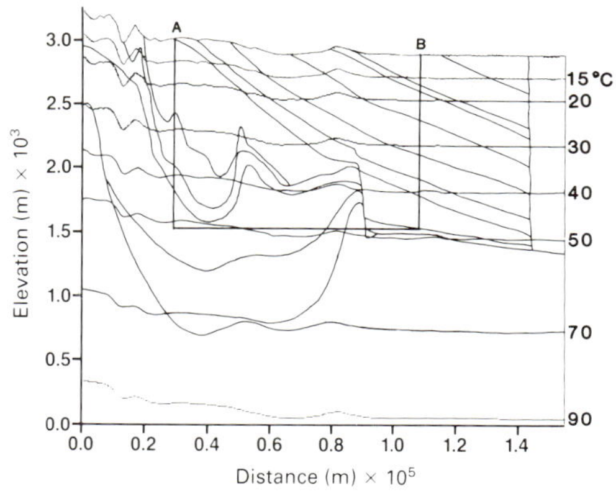

FIG. 3. Isotherms for model 1(a): uniform thermal conductivity of 2.5 W mK^{-1}. The area of the section in Fig. 2(b) is indicated. Eakring lies about 50 km from the left-hand side of the model.

Model 1(b)

Thermal conductivity contrasts between some of the units present, e.g. the basement and the Westphalian, may be important. The conductivity structure given in Table 1 is used in this model. The isotherms are shown in Fig. 4. As most of the conductivities are less than the 2.5 W mK^{-1} used in the previous run, the temperature gradients are generally greater. The enhancement of the VCHF beneath the valley at the W end of the section is increased slightly, to a maximum of 15%. The contrast between the conductivity of the basement and that of the cover produces a disturbance to the VCHF at Nocton, with a variation of between −5% and 20% from the background.

Model 1(c)

This model includes a homogeneous permeability of 1 mD for the whole section. The flow field

TABLE 1. *Values of parameters used in modelling*

Unit	Matrix thermal conductivity (W mK^{-1})	Porosity	Intrinsic permeability (m^2)
Basement	3.5	0	1.0×10^{-20}
Devonian?	3.5	0.01	1.0×10^{-17}
Dinantian	2.95	0.04	1.0×10^{-15}
Namurian	1.95	0.05	1.0×10^{-15}
Westphalian	2.10	0.05	1.0×10^{-16}
Upper Westphalian	3.25	0.10	5.0×10^{-15}
Permian	3.33	0.05	1.0×10^{-15}
Sherwood Sandstone	4.68	0.30	1.0×10^{-12}
Mercia Mudstone	2.15	0.05	1.0×10^{-17}
Lias	1.42	0.10	1.0×10^{-18}
Lincolnshire Limestone	2.93	0.10	1.0×10^{-14}
Upper Jurassic	1.66	0.10	1.0×10^{-17}
Cretaceous	3.08	0.30	1.0×10^{-14}
'Pipe'	2.0	0.05	1.0×10^{-14}

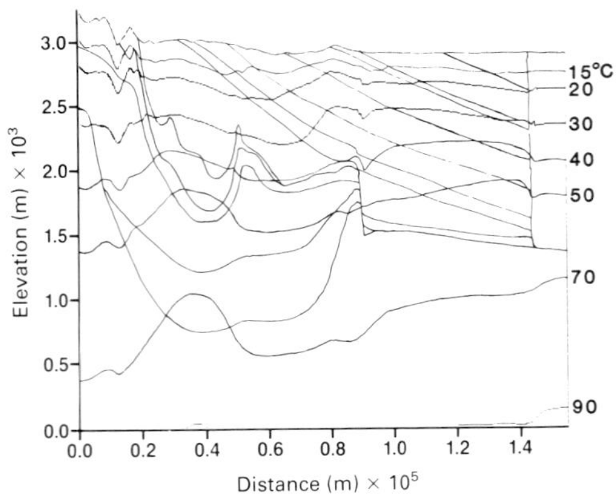

FIG. 4. Isotherms for model 1(b): thermal conductivities as given in Table 1.

(Fig. 5) shows several features. The velocities are generally very low, mostly less than 1 cm a^{-1}. Higher velocities are encountered only locally in the region of greatest relief at the W end of the model. The flow is very sensitive to the relief of the water table: very small highs produce flow systems which propagate to the base of the model with little damping. These are superimposed on a regional flow system from W to E.

Although flow velocities are very low, the effects on heat flow may be large. In the Peak District the greatest enhancement of the VCHF beneath the valley is increased from 15% to 70%. The convective component of this enhancement is a result of a localized flow system with recharge on the adjacent uplands and discharge in the valley. This valley corresponds to the position of

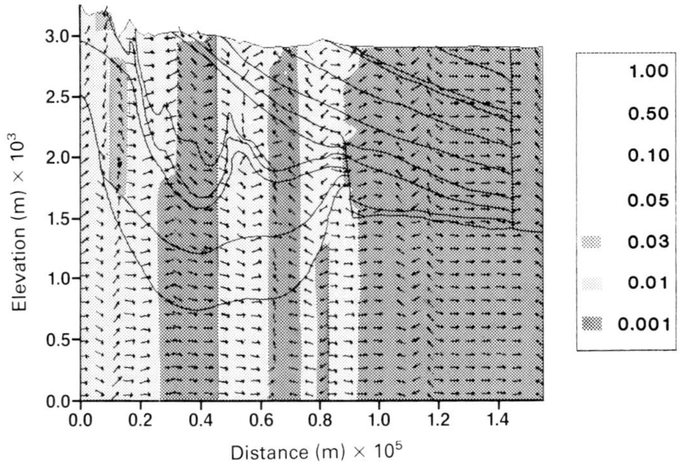

FIG. 5. The Darcy flow field for model 1(c): uniform permeability of 1 mD. The arrows show the flow directions. The shading indicates the range of Darcy velocity in m a^{-1} with the numbers in the key indicating the upper limits of the increments.

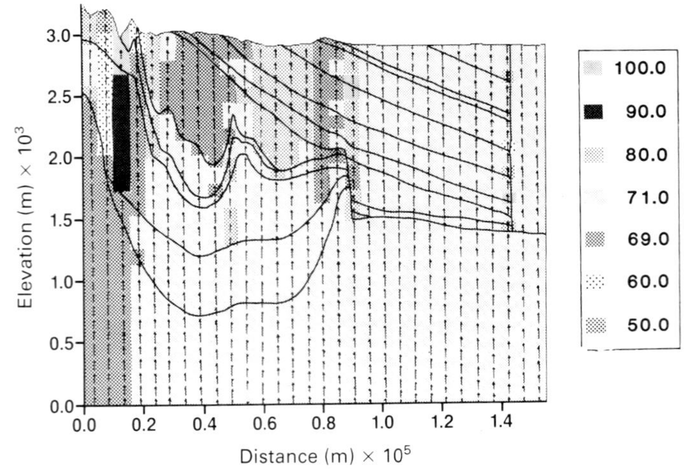

FIG. 6. The vertical conductive heat flow field for model 1(c). The shading indicates the range of heat flux in mW m^{-2}; the numbers indicate the upper limit of the increment.

Matlock on the edge of the Dinantian outcrop, where the thermal waters of the spa (at a temperature of 28°C) testify to the existence of this sort of system.

Elsewhere along the section the perturbation to heat flow correlates approximately with the topography: between the Peak District and Eakring the VCHF is slightly depressed (less than 5‰), and similarly over the outcrop of the Lincolnshire Limestone; the Trent River plain is an area of discharge and slight enhancement of surface heat flow (Fig. 6).

Model 1(d)

In this run the permeabilities given in Table 1 are used. These values are thought to be reasonable upper limits for most units. While much of the system still has permeabilities around the millidarcy level, significant differences are the introduction of a hydrogeological basement, the presence of an aquifer in the Sherwood Sandstone and the low permeability cap to this aquifer (the Mercia Mudstone and the Lias). The flow pattern is shown in Fig. 7. Velocities are still low in general, with much higher values in the Sherwood Sandstone. Local flow systems tend to be damped by the basement, and flow within the basement tends to be sub-horizontal. The aquifer forms a divide between two flow regimes: below the Sherwood Sandstone Group flow tends to be of a regional sub-horizontal form from W to E; above, the flow in the low permeability cap tends to be sub-vertical.

The heat flow field is quite different in form from the previous model (Fig. 8). Above the

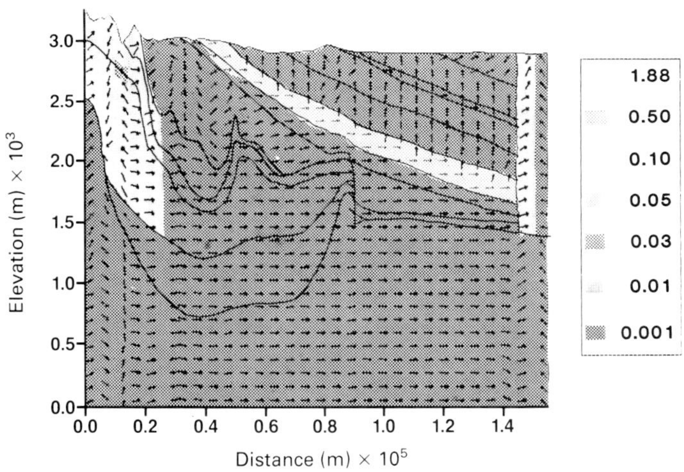

FIG. 7. The Darcy flow field for model 1(d): conductivity and permeability as given in Table 1.

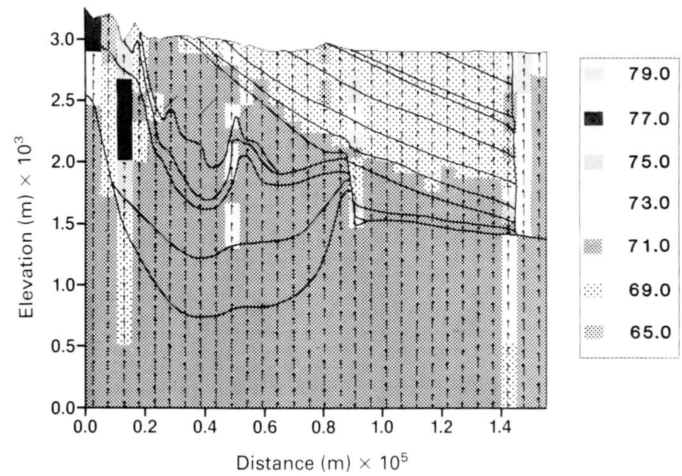

FIG. 8. The heat flow field for model 1(d). Note the slight regional depression of heat flow above the Sherwood Sandstone.

Sherwood Sandstone heat flow is uniformly depressed by less than 5%. The enhancement in the Trent River plain is removed. A slight enhancement of heat flow is seen in the Eakring region. It is, however, small in magnitude (less than 5%).

Model 2(a)

The model with a plane potential surface has been used to study the regional picture. The flow systems which depend on local gradient changes are not seen in this model and the regional trends appear. The gradient assumed for the surface is 1 m km^{-1}, which is the value estimated for the Dinantian (Gale, personal communication, 1984). It is similar to the hydraulic gradients measured in some of the confined units (e.g. 1/900 for the Sherwood Sandstone (Edmunds *et al.* 1982) and may approximate to the regional gradient.

This model has the same permeability structure as model 1(d), apart from the inclusion of an Upper Westphalian 'aquifer' with a permeability of 5 mD. However, the flow pattern (Fig. 9) is quite different: the whole system is much more dominated by easterly flow, and flow in the Mercia Mudstone unit is now down-dip rather than being dominated by vertical leakage from the Sherwood Sandstone; this in turn has induced leakage into the Mercia Mudstone from the overlying Lias.

The heat flow pattern is also different (Fig. 10): the disturbance at the W end of the section has disappeared completely and the Eakring anomaly has also disappeared, presumably because of the reduction in hydraulic gradient. The reduction of heat flow above the Sherwood Sandstone is still noticeable.

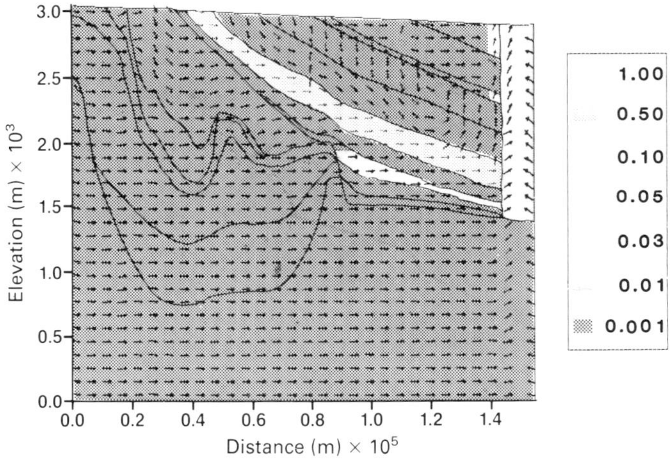

FIG. 9. The Darcy flow pattern for model 2(a) in which a plane water table with a gradient of 1/1000 is used. The conductivity and permeability are given in Table 1.

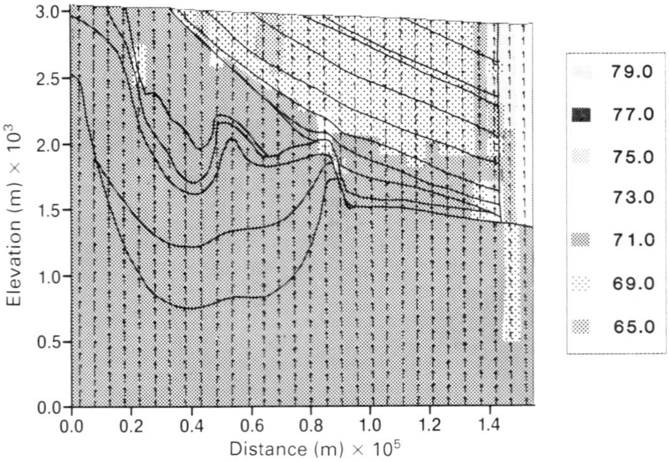

FIG. 10. The heat flow pattern for model 2(a).

Model 2(b)

We shall now investigate the effects of varying the permeability of a layer at the top of the Dinantian. This is an arbitrary layer, about 100 m thick, and is taken to represent the most permeable part of the unit. Certain other intervals may also have significant permeability, for instance dolomitized intervals in the lower Dinantian, but are unlikely to be regionally continuous.

In this model the permeability in the upper layer of the Dinantian has been increased to 50 mD. The flow pattern is very similar to that in model 2(a), but velocities in the upper Dinantian are greater (0.02–0.03 m a^{-1}). To the E of Eakring vertical leakage in the Upper Carboniferous is superimposed on the W–E flow.

In the heat flow pattern (Fig. 11) two changes can be seen: downward fluid flow reduces the heat flow above the Dinantian at the W end of the section and a small anomaly appears at the crest of the Eakring structure. The magnitude of this 'Eakring anomaly' is only 1–2 mW m^{-2}, i.e. less than 5%, but it is distinct and localized.

Model 2(c)

Here the permeability is 500 mD. Both flow fields are similar to those seen in model 2(b) but the flow magnitudes are greater. Heat flow at Eakring is enhanced above the Dinantian and below the Sherwood Sandstone by 7–9 mW m^{-2} (10%–12%) (Fig. 12). Immediately to the E and the W of the crest heat flow is enhanced by 1–3 mW m^{-2}.

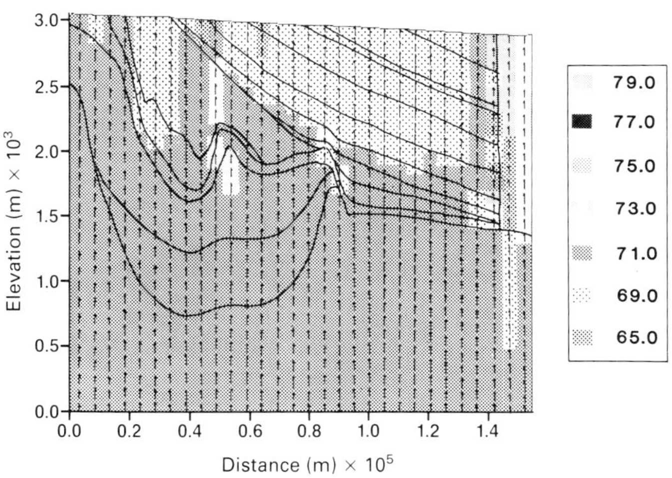

FIG. 11. The heat flow pattern for model 2(b): 50 mD in the upper Dinantian.

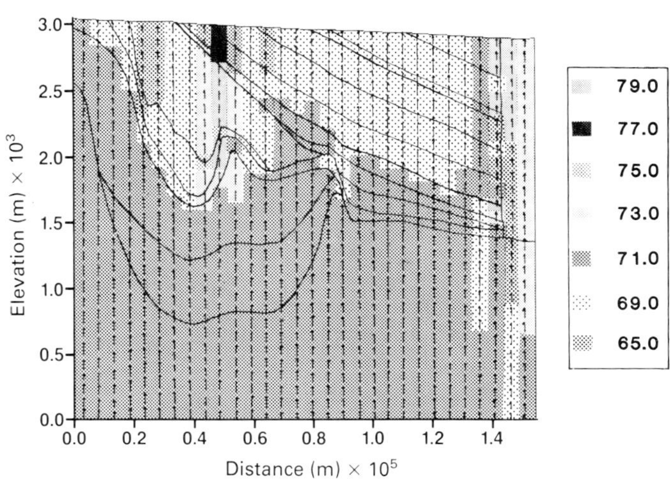

FIG. 12. The heat flow pattern for model 2(d): 500 mD in the upper Dinantian. Note the development of a heat flow 'high' above the crest of the Eakring anticline.

Model 2(d)

A permeability of 1 D in the upper Dinantian is used in this model. This must be taken as an upper limit for the permeability of the limestone. Darcy velocities in the upper Dinantian are in general between 0.1 and 0.3 m a^{-1}, but are locally up to 0.5 m a^{-1} (Fig. 13). Flow in the Namurian and Westphalian in the region from Eakring to Nocton is dominated by vertical leakage. The flow there has changed progressively from sub-horizontal in model 2(a) to vertical. Heat flow at Eakring is now between 81 and 84 mW m^{-2}, i.e. an anomaly of 15%–20%. On the flanks of the structure the VCHF is enhanced by less than 5%.

Model 2(e)

To investigate the effect of anisotropy the units which include a large proportion of mudstone have been assigned permeabilities as given in Table 2. The anisotropy ratio in each of these units is therefore 100:1. The heat flow field is very similar to that for isotropic permeability, and the anomaly at Eakring is the same size. The flow pattern is constrained, as might be expected, to be more horizontal in the anisotropic units. The fact that the heat flow field is essentially unchanged suggests that velocities in these low-permeability units are so small as to have little thermal effect.

TABLE 2. *Permeability values used in the 'anisotropy' model*

Unit	*x* permeability (m^2)	*y* permeability (m^2)
Westphalian	1.0×10^{-15}	1.0×10^{-17}
Permian	1.0×10^{-14}	1.0×10^{-16}
Mercia Mudstone	1.0×10^{-16}	1.0×10^{-18}
Lias	1.0×10^{-17}	1.0×10^{-19}
Upper Jurassic	1.0×10^{-15}	1.0×10^{-17}

Model 2(f)

In the previous models the 'pipe' at the E end of the section has been assigned a permeability of 10 mD. To ensure that this has not induced a flow system, this value has been reduced by two orders of magnitude in model 2(f). The other parameters are those of model 2(d).

Heat flow at Eakring is enhanced by 5–7 mW m^{-2} as opposed to 7–9 mW m^{-2}. Obviously, then, the situation at the E end of the model is important, but the anomaly is still distinct when the permeability there is three orders of magnitude less than that of the Dinantian 'aquifer'.

Conductive models

In addition to the models discussed above a number of others have been used to estimate the effects of conductivity and heat flow variations in the basement. A granitic body has been postulated to exist in the Newark area from the interpretation of gravity data (Gale & Holliday 1983). The western margin of this body is taken to lie just to the E of the Eakring area: the Eakring 146 borehole penetrated the Lower Carboniferous section, which contained granitic pebbles in a conglomerate, before reaching the basement of quartzites (Edwards 1967). This type of body might affect the heat flow field in two ways: high heat production in the body could lead to enhanced heat flow above it, or conductivity contrasts between the granite and the surrounding basement might give refractive effects which would alter the heat flow above. Modelling

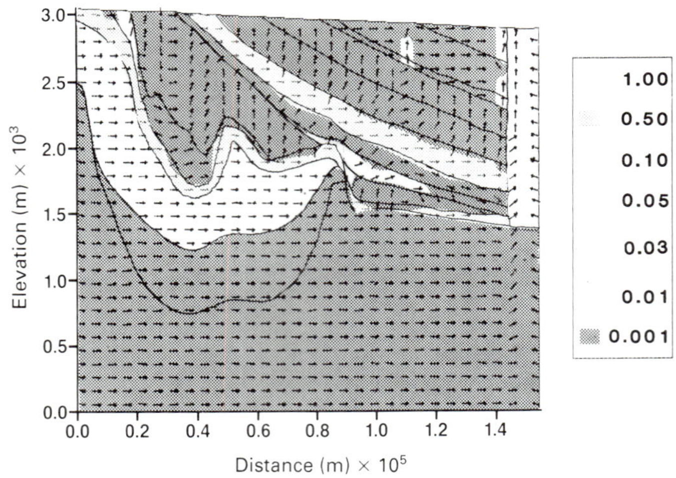

FIG. 13. The Darcy flow pattern for model 2(e): 1 D in the Upper Dinantian.

has shown that neither phenomenon is likely to explain the very localized character of the Eakring anomaly.

Comparison with observations

Several wells for which there exist measured temperature profiles lie along or close to the line of section. Comparison of the observed and calculated temperatures for model 1(d) indicated that, for wells located in a conductive regime, the conductivity of the Westphalian had been over-estimated. Figure 14 shows the comparison for a model in which the conductivity of the Westphalian has been altered. Obviously, while the modelling compares fairly well for places where convective effects are small, the anomaly at Eakring is reflected in the large discrepancy between observed and computed temperatures.

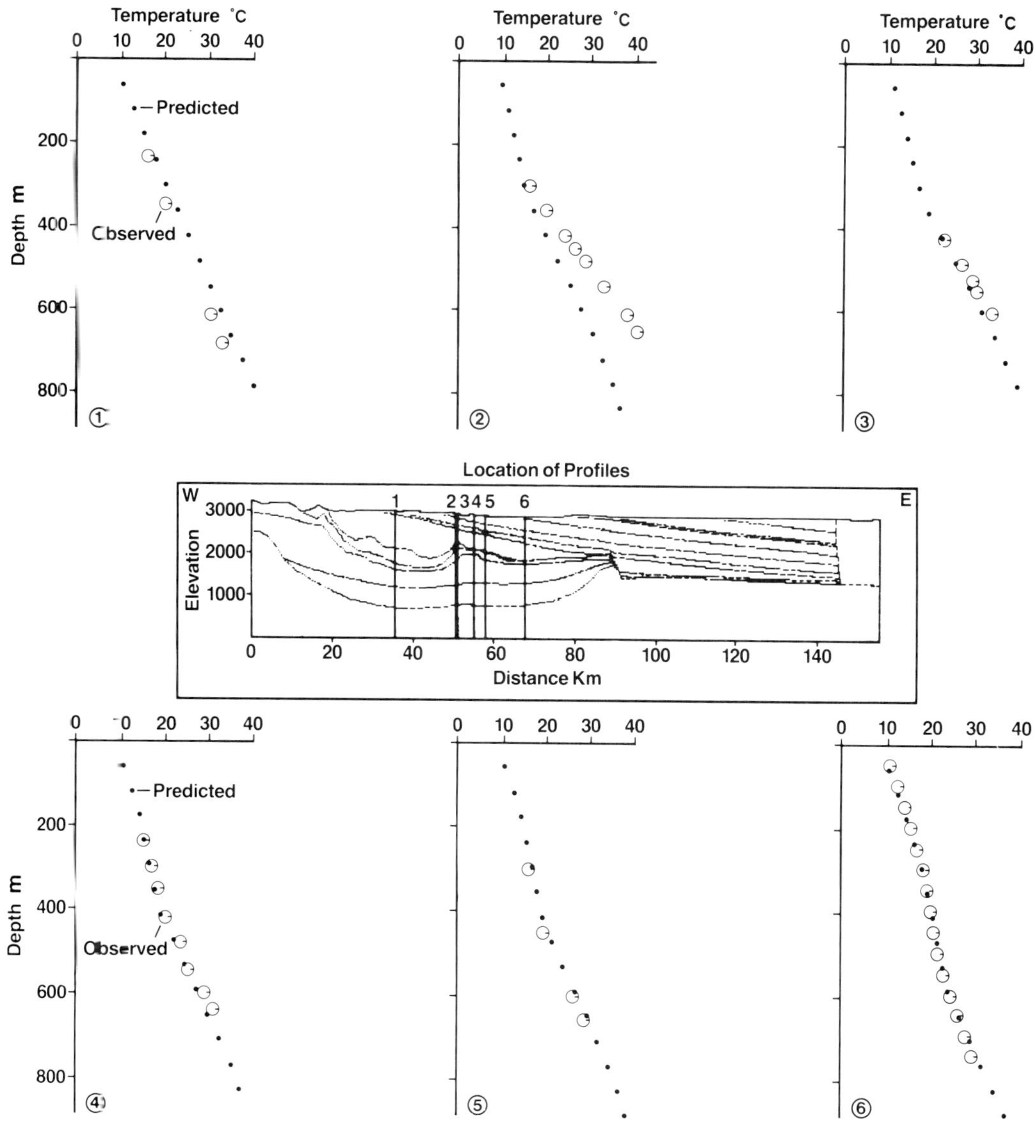

FIG. 14. Comparison of observed and predicted temperature profiles along the line of section. A surface temperature of 9°C and a heat flow at the base of the model of 65 mW m^{-2} have been used in the modelling. The profiles are from (1) Papplewick (Mullins & Hinsley 1957), (2) Eakring 6, (3) Eakring 64, (4) Caunton, (5) Kelham Hills (all from Bullard & Niblett 1951), (6) Farley's Wood (this work).

Discussion

From the results presented here it can be seen that, while a heat flow anomaly in the Eakring area can be explained with plausible values for the permeability in the upper part of the Dinantian, the magnitude of this anomaly falls well short of that observed in the field, even with a permeability which is improbably high. This discrepancy is almost certainly due to three-dimensional effects.

The Eakring oil field lies above an approximately NW–SE-trending ridge which extends from the Grantham area to Eakring, where it swings round to the N, and which we have modelled as a step with a height of about 500 m. The local structure at Eakring is more complicated. Figure 15 shows a schematic view of the Eakring structure at the top Dinantian level, which is a NNW–SSE elongated dome truncated on the W by faulting. The line of section is indicated, as are some possible water flow lines on this surface. Flow is very unlikely to be two dimensional, and channelling up the flanks of the structure will occur. It is difficult to calculate the amount of channelling that is likely to take place, but if the domal structure acts as a focal point for flow up a length L m of the fault the volume of water arriving at the crest in unit time will be greater than that moving in the line of section simply by a factor L. In this reasoning it is assumed that flow up the fault is uniform and that the channelling is 100% efficient. Therefore, for instance, channelling of 1% of the flow up a 5 km section of the fault will increase the volume flow fiftyfold at the crest. This can be no more than an 'order of magnitude' estimation, but it appears that this sort of three-dimensional effect could easily account for the heat flow anomaly observed. In detail, heat flow increases from S to N within the Eakring oil field, but the highest value (on the N flank) does not correspond to the most elevated part of the structure. This may reflect local variations of the permeability structure.

Conclusions

1 The Eakring anomaly in the E Midlands of England could be a result of water movement in the Lower Carboniferous limestones, as originally suggested. However, three-dimensional effects make the exact form of the anomaly difficult to reproduce in two-dimensional modelling and little can be said about the permeability of these limestones at depth. Flow velocities in the bulk of the Dinantian, however, are likely to be very low (of the order of 1 cm a^{-1}).

2 Modelling of regional fluid flow in sedimentary basins requires knowledge of the permeability and fluid-potential structures in the basin. Modelling of coupled heat and fluid flow, with observations of heat flow, should provide extra constraints on the flow system. However, modelling must be three dimensional for this approach to be effective.

3 Near-surface heat flow is affected by regional and local groundwater flow across the whole of the modelled section. Convective effects must be taken into account whenever the 'basement' heat flow is sought.

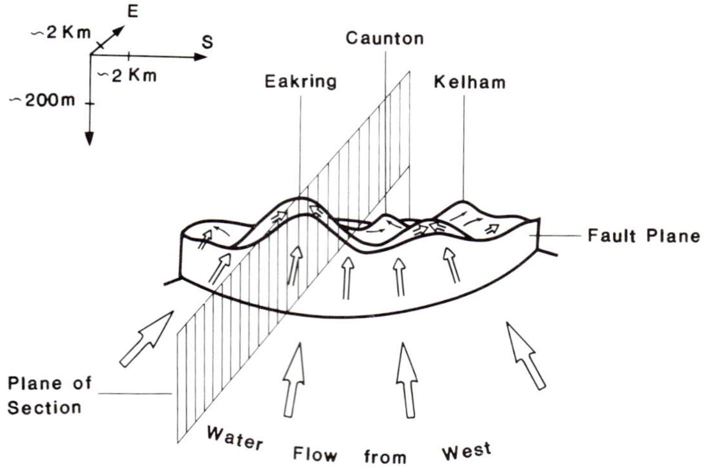

FIG. 15. A schematic perspective view of the top Dinantian surface in the Eakring area together with the orientation of the modelled section. The arrows indicate possible flow directions in the Dinantian limestones.

ACKNOWLEDGMENTS: The authors thank Dr D. S. Chapman and Dr L. Smith for kindly providing the finite-element program which was used in this study, Professor E. R. Oxburgh, Dr J. R. Bloomer, Ian Gale and Alan Smith for useful discussions, and Dr R. A. Downing for a thorough review. N.P.W. gratefully acknowledges a research studentship from the Department of Education, Northern Ireland, and financial support from BP Petroleum Development Ltd. The authors thank the Chairman and Directors of British Petroleum Co. plc for permission to publish this paper, which is contribution no. 1014 of Cambridge University Department of Earth Sciences.

References

BATH, A. W., EDMUNDS, W. M. & ANDREWS, J. N. 1979. Palaeoclimatic trends deduced from the hydrogeochemistry of a Triassic sandstone aquifer, United Kingdom. *Proc. Symp. on Isotope Hydrology, Vienna, 1978*, Vol. 2, pp. 545–68, International Atomic Energy Agency, Vienna.

BREDEHOEFT, J. D. & PAPADOPOULOS, I. S. 1965. Rates of vertical groundwater movement estimated from the Earth's thermal profile. *Water resour. Res.* **1**(2), 325–8.

BULLARD, E. C. & NIBLETT, E. R. 1951. Terrestrial heat flow in England. *Mon. Not. r. astr. Soc. geophys. Suppl.* **6**, 222–38.

BURLEY, A. J., EDMUNDS, W. M. & GALE, I. N. 1984. Catalogue of geothermal data for the land area of the United Kingdom. *Investigations of the Geothermal Potential of the U.K.*, British Geological Survey.

CARTWRIGHT, K. 1968. Thermal prospecting for groundwater. *Water resour. Res.* **4**(2), 395–401.

DOWNING, R. A. & WILLIAMS, B. P. J. 1969. *The Groundwater Hydrology of the Lincolnshire Limestone*, Water Resources Board, Reading, 160 pp.

——, SMITH, D. B., PEARSON, F. J., MONKHOUSE, R. A. & OTLET, R. L. 1977. The age of groundwater in the Lincolnshire Limestone, England, and its relevance to the flow mechanism. *J. Hydrol.* **33**, 201–16.

EDMUNDS, W. M., BATH, A. H. & MILES, D. L. 1982. Hydrochemical evolution of the East Midlands Triassic sandstone aquifer, England. *Geochim. cosmochim. Acta*, **46**, 2069–81.

EDWARDS, W. N. 1951. The concealed coalfield of Yorkshire and Nottinghamshire. *Mem. Geol. Surv. G.B.*, HMSO, London, 285 pp.

—— 1967. Geology of the country around Ollerton, 2nd edn. *Mem. Geol. Surv. G.B.*, HMSO, London, 297 pp.

FALCON, N. L. & KENT, P. E. 1960. Results of petroleum exploration in Britain 1945–1957. *Mem. Geol. Soc. Lond. No. 2*, 56 pp.

FOSTER, S. S. D. & MILTON, V. A. 1976. Hydrogeological basis for large scale development of groundwater storage development in the East Yorkshire Chalk. *Rep. Inst. geol. Sci.* **76**(3).

FREEZE, R. A. & WITHERSPOON, P. 1966. Theoretical analysis of regional groundwater flow, 1, Analytical and numerical solutions to the mathematical model. *Water resour. Res.* **2**, 641–56.

—— & —— 1967. Theoretical analysis of regional groundwater flow, 2, Effect of water table configuration and subsurface permeability variation. *Water resour. Res.* **3**, 623–34.

GALE, I. N. & HOLLIDAY, D. W. 1983. The geothermal resources of eastern England. *European Geothermal Update EEC, 3rd Int. Semin., Munich, 1983, Rep. EUR 8853 EN*, EEC, Brussels.

——, SMITH, I. F. & DOWNING, R. A. 1983. The post-Carboniferous rocks of the East Yorkshire and Lincolnshire Basin. *Investigations of the Geothermal Potential of the U.K.*, Institute of Geological Sciences, London.

I.G.S. 1981. *Hydrogeological Map of the Northern East Midlands (Scale, 1 : 100 000).*

JENNINGS, J. N. 1971. *Karst: An Introduction to Systematic Geomorphology*, MIT Press, Cambridge, MA.

LACHENBRUCH, A. H. & SASS, J. H. 1977. Heat flow in the United States. *In*: HEACOCK, J. G. (ed.) *The Earth's Crust, Geophys. Monogr. Ser.*, Vol. 20, pp. 626–75, American Geophysical Union, Washington, DC.

LEWIS, T. J. & BECK, A. E. 1977. Analysis of heat flow data—detailed observations in many holes in a small area. *Tectonophysics*, **41**, 41–59.

MULLINS, R. & HINSLEY, F. B. 1957. Measurement of geothermic gradient in boreholes. *Trans. Inst. min. Eng.* **117**, 379–93.

RICHARDSON, S. W. & OXBURGH, E. R. 1978. Heat flow, radiogenic heat production and crustal temperatures in England and Wales. *J. geol. Soc. Lond.* **135**, 323–37.

SMITH, L. & CHAPMAN, D. S. 1983. On the thermal effects of groundwater flow: 1. Regional scale systems. *J. geophys. Res.* **88**, 593–608.

STALLMAN, R. W. 1963. Computation of groundwater velocity from temperature data. *U.S. Geol. Surv. Water Supply Pap. 1554-H*, pp. 36–46.

WOODBURY, A. D. & SMITH, L. 1985. On the thermal effects of three-dimensional groundwater flow. *J. geophys. Res.* **90**, 759–67.

N. P. WILSON*, Department of Earth Sciences, University of Cambridge, Downing Street, Cambridge, U.K.

M. N. LUHESHI, BP Petroleum Development Ltd., Britannic House, Moor Lane, London EC2, U.K.

* Present address: BP Petroleum Development Ltd., Britannic House, Moor Lane, London EC2, U.K.

Section 4
Fluid Flow in Low Permeability and Fractured Media

The role of low-permeability rocks in regional flow

J. Alexander, J. H. Black & M. A. Brightman

SUMMARY: The role of low-permeability rocks in determining fluid migration within sedimentary basins has been shown to be crucial for deriving the direction and rate of fluid movement. The magnitude of the flows depends on the geometry of the sedimentary basin and on the bulk hydrogeological properties of the various strata and their relative values. Measurements of head and pore-water compositions of argillaceous formations show separate evidence for the existence of cross-formational flow resulting in groundwaters of mixed origins. Numerical and analytical models have been developed and applied to some basinal systems in England. They have been used to assess the directions of groundwater flow likely to be encountered by repository site investigation programmes and to aid in the interpretation of groundwater head and chemistry data. Such flow models still require detailed validation. Our knowledge of the flow mechanism and the physical properties of mudrocks at depth is still elementary, but the indications are that the role of low-permeability rocks in controlling fluid migration has been much underestimated.

The movement of the groundwater within low-permeability rocks has recently attracted great interest. This is due primarily to investigations in many countries (Belgium, U.S.A. and Italy), including the U.K., into the feasibility of the disposal of radioactive waste into formations of low permeability (mudrocks). The objectives in all cases have been to investigate the patterns of both local and regional groundwater movement and to assess the probable routes and speeds of subsequent radionuclide migration from any potential repositories.

These mudrocks occur within layered sedimentary sequences so that the movement of groundwater across them is directly linked to flows in intervening beds of high permeability. In water-resource assessments flows in aquifers in sedimentary basins are large and flows through adjoining mudrocks are considered to be negligible in comparison. The areal extent of a sedimentary basin is such that, even though the permeability of mudrocks is low, significant fluid movement can occur across them, given long time-scales. At depths, where flow rates are small, water found within interbedded 'aquifers' has inevitably been derived from the adjacent less-permeable rocks. In contrast with resource assessments the form and quantity of these cross-flows is of direct interest to investigations concerned with radioactive waste disposal.

In most regional flow systems low-permeability formations are as important as the more permeable strata in controlling the hydrodynamics of the system. The aim of this paper is to examine the role of low-permeability rocks in determining regional flow. This is achieved by an outline of some of the basic relationships involved illustrated by the use of simplified representative geometries. More complex geometries are examined and examples where field measurements are available are discussed.

Basic concepts of fluid movement in sedimentary basins

In any sequence of mixed high- and low-permeability strata groundwater will tend to flow along the more highly permeable layers and across the intervening less permeable layers. In sedimentary basins, a characteristic of lowland Britain, this results in sub-horizontal flow in the high-permeability rocks (throughflow) and vertical flow across intervening mudrocks (cross-flow). Geological heterogeneity has profound effects on the development of regional groundwater circulation patterns. The layering imparts a gross permeability anisotropy on the whole system, so that where a flow system would otherwise be equidimensional it is elongated horizontally and does not penetrate to as great a depth as might be expected.

The direction of groundwater movement is overwhelmingly controlled by the distribution of potential. If density contrasts and osmotic and velocity potentials are ignored, the potential distribution in a fresh groundwater system is related to topography. Thus groundwater essentially flows down and away from hills and upwards to discharge in valleys. However, when the distribution of potential is combined with layered hydrogeological properties the outcome, in terms of fluxes and flow directions, is not always obvious.

A generalized cross-section of a basin (Fig. 1a) shows some fundamental considerations. Only half the basin is shown since it is assumed to be symmetrical. Of prime importance in these

From GOFF, J. C. & WILLIAMS, B. P. J. (eds), 1987, *Fluid Flow in Sedimentary Basins and Aquifers,* Geological Society Special Publication No. 34, pp. 173–183.

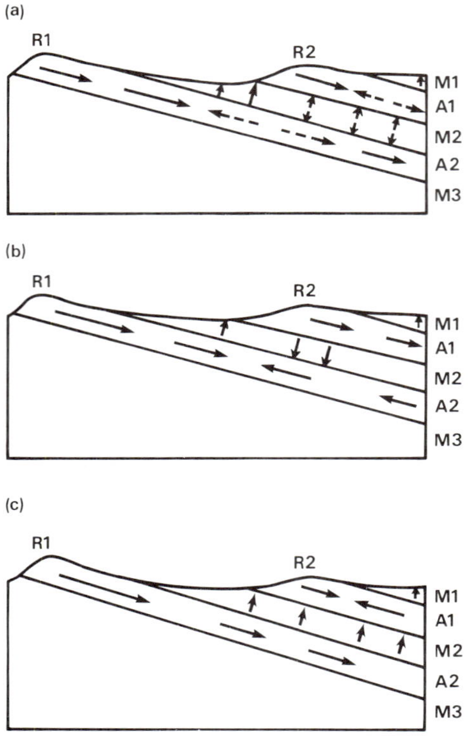

FIG. 1. Schematic cross-sections illustrating the generalized deposition of rocks of high- and low-permeability strata in a sedimentary basin and the probable flow directions.

considerations is that the mudrocks (M1, M2 and M3) have a low but measurable bulk hydraulic conductivity. Thus, since the centre of the basin is lower than elsewhere, groundwater will leak upwards and out of the system in the central zone. The flows within the basin are steady and are not sustained or modified by abstraction. In addition to the upward-pointing arrows in the centre of the basin other groundwater flow directions are predictable. These are (i) the flows down-dip away from the 'high'-altitude recharge areas (R1 and R2) within the rocks of high permeability (A1 and A2), and (ii) the upward leakage through the thinning end of M2 (the feather edge).

Of more interest are the flows which are not predictable (shown as broken double-headed arrows in Fig. 1a). If the flow beneath R2 is directed downwards it is possible for the water which has recharged at R2 to penetrate to the underlying layer of high permeability (A2) and flow up-dip as well as down-dip (Fig. 1b). In this event the layer A2 of higher permeability will contain at least one zone of zero flow (i.e. a zone of flow balance). On each side of this zone the groundwater has a very different history. The opposite set of circumstances (Fig. 1c) occurs where the head in A2 exceeds that in A1. In this case the groundwaters from A2 are more widespread within the basin. Thus it is possible for zones of zero throughflow and adjacent groundwaters of different evolution and chemistry to exist within the overlying layer of high permeability (A1).

It can be seen from this brief outline that an interpretation of the hydrochemical evolution of fluids within the sedimentary sequence will be difficult if the groundwater flow pattern is unknown. The pattern of flow clearly depends on the geometry of the system and the bulk hydrogeological properties of the rocks. The results of this interaction can be modelled numerically for any given basin, but some more general insights can be gained from consideration of the analytical 'step-and-wedge' model (Black & Barker 1981).

Critical relationships in simple configurations

The scarp-and-vale (step-and-wedge) topography of lowland Britain underlies the configuration presented in Fig. 1 and is partially a product of the lithologies involved. The simplified cross-section of Fig. 1 illustrates a basic 'unit' which can be repeated any number of times. This basic unit consists of the 'wedge' of mudrock (M2) sandwiched between two layers of higher permeability (A1 and A2); A1 is the 'step'. This geometry has the attribute of being amenable to mathematical analysis under certain boundary conditions which are summarized as follows.

1 *Assumptions concerning the head.* At the western feather edge the height of the water table coincides with the ground surface (i.e. springs should account for most excess head). From the feather edge to the base of the step the water table in the mudrock coincides with its surface. The water-table underneath the step, which in reality rises gradually (50 m km^{-1} is quite common), is assumed to rise abruptly and to remain at that altitude.

2 *Assumptions concerning the flow.* The underlying low-permeability rocks are assumed to be a no-flow boundary. The mudrock (M2) has a lower permeability than the adjacent strata (A1 and A2). The dip of the beds is so small that it can be assumed that flow in the more highly permeable layer (A2) is horizontal and that through the mudrock is vertical.

Given these conditions it is possible to vary the size of the step, the slope of the mudrock surface and the hydraulic conductivities of A2 and the mudrock (M2) and solve for head distributions. An equation can be derived to describe the variation of head along the top surface of A2 and hence delineate flow elsewhere within A2 and M2 (Black & Barker 1981).

The main conclusion of the analysis is that a zone of flow balance is produced within layer A2 if the step is at a higher altitude than the feather edge. This means that the water recharging through the step enters A2 and inhibits the down-dip passage of flow from recharge. The analysis revealed that the location of this flow balance position is a function of the ratio of the hydraulic conductivity of the high-permeability rock (A2) to that of the mudrocks (M2) (the permeability ratio). This is depicted in Fig. 2 which shows that the permeability K_{M2} of M2 decreases relative to the permeability K_{A2} of A2 so that the step governs the flow over a greater distance. This is seen as the movement of the flow balance position away from the step with increasing permeability ratio K_{A2}/K_{M2}.

The analysis reveals that the position within the basin where the cross-flow (up or down) takes longest is also a function of the permeability ratio (Fig. 2). The position of maximum vertical hydraulic gradient is underneath the step regardless of the permeability ratio.

It should be remembered that the groundwater on each side of a flow balance point may well be of different evolution and composition. On one side the groundwater may have migrated from recharge wholly within the high-permeability rock and all be of equal 'age'. On the other side it may arrive at the flow balance position having been mixed in A2 after being derived from the overlying mudrock. This water will itself have taken varying times to cross the mudrock and will therefore be of variable 'age' and evolution. This variable 'age' (the transit time) is illustrated in Fig. 3 for a chosen geometry and high

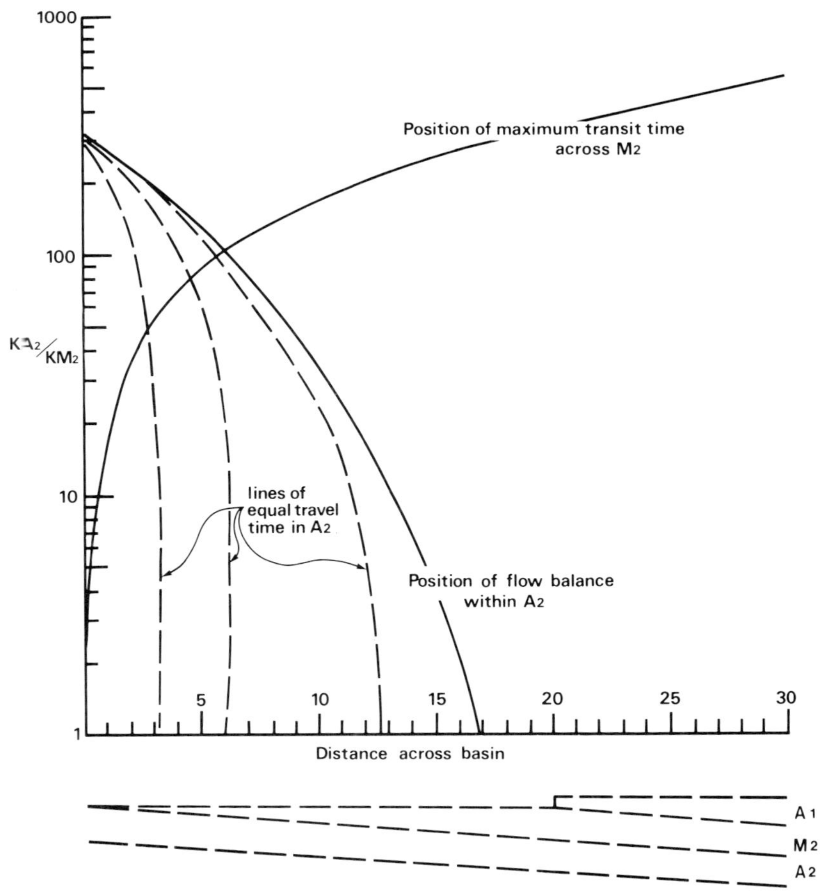

FIG. 2. The position of maximum transit time and flow balance as a function of the permeability ratio K_{A2}/K_{M2}.

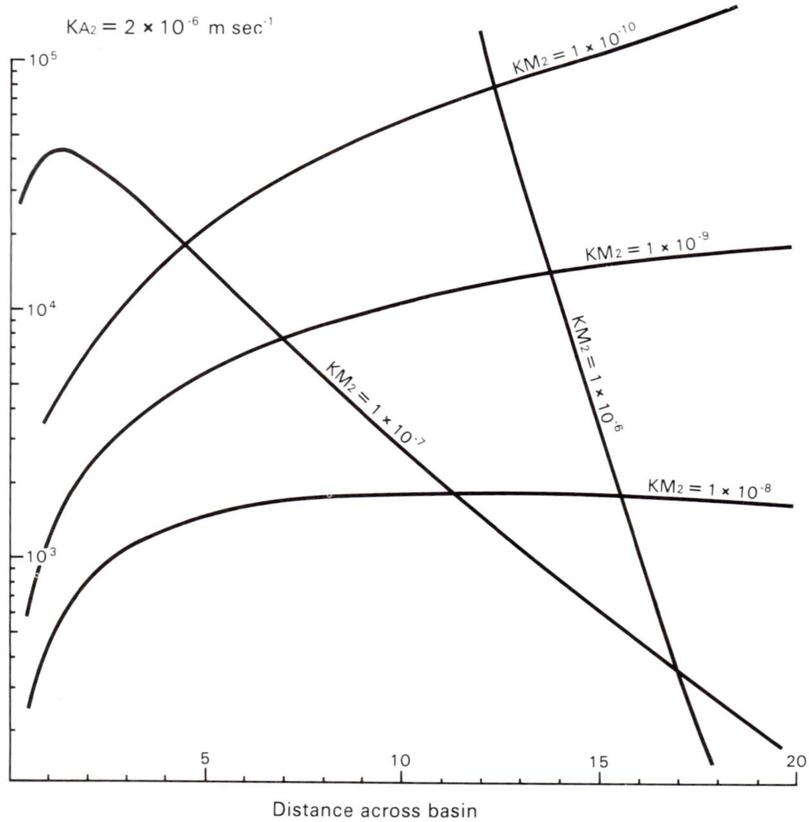

FIG. 3. Variations in the transit time across the mudrock M2 for $K_{A2}=2\times10^{-6}$ m s^{-1} and variable K_{M2}.

permeability K_{A2} in conjunction with a selection of different mudrock permeabilities K_{M2}.

Several general conclusions can be deduced from this mathematical consideration of a basin unit.

1 High points in the water-table (usually associated with high relief) will dominate the largest regions when the permeability ratio is large. This rule applies regardless of the permeability of the rock containing the watertable or that of the rock beneath.

2 The highest-velocity cross-flows occur beneath water-table highs regardless of the thickness of rocks of low permeability.

3 Groundwaters of a single origin only occur within high-permeability strata to a limited depth. Cross-flows will result in an accumulation of groundwaters in the high-permeability strata which will have had flow paths of different lengths and transit times.

4 The longest cross-flow transit times are not necessarily related to the greatest thicknesses of the low-permeability rock. However, where high ratios are involved there is a general relationship between long transit times and mudrock thickness beneath mudrock outcrop.

Some of the considerations raised by the step-and-wedge model are not altogether obvious, so it is useful to examine an actual study where direct evidence for cross-flows exists.

Evidence for cross-formational flow

The middle Thames valley area is situated on the northern limb of a broad basin similar to Fig. 1 and has been investigated in a number of ways. Several purpose-designed boreholes have been drilled into the Cretaceous and Jurassic rocks of the region to assess the occurrence and scale of cross-formational flow. Where possible, heads and permeabilities have been measured and groundwaters sampled. The most comprehensive results are from three closely spaced boreholes at the Atomic Energy Research Establishment, Harwell, Oxfordshire. Cores have been taken from all purpose-designed boreholes in this area, and the chemistry of the pore-fluids of the mudrocks has been measured (Brightman *et al.* 1985).

At the south-easterly end of the section, flow (based on head measurements in the deepest boreholes) appears to converge on the Corallian limestones (Alexander & Holmes 1983). This implies downward vertical flow through the Gault, Lower Greensand and Kimmeridge Clay lithologies overlying the Corallian, and upward vertical movement through the Oxford Clay from the underlying Great and Inferior Oolites (Fig. 4). Additionally, heads in both the Oolites and Corallian exceeded those observed towards outcrop areas giving rise to up-dip flow. This results in two-flow cells which are a particular feature of the wedge-and-step model.

Maps of heads in the three high-permeability formations (Chalk and Upper Greensand, Corallian, and Great and Inferior Oolites) were compared to deduce vertical head differences and consequent cross-formational flows (Alexander 1983). The head gradients deduced were then tested at a number of sites (three for upward flow and one for downward flow). At these sites head measurements are consistent with the predicted directions of flow within the region in general and the vertical gradients at the borehole locations in particular. The interpretation of mudrock pore-water compositions is consistent with the predicted directions of groundwater movement.

The value of using pore-water chemistry to examine cross-flow is illustrated by studies of cores from the Harwell boreholes. The chemistry of mudstone pore-waters was compared with groundwater samples abstracted from high-permeability lithologies which sandwich individual clay formations at the Harwell site. Mudrock pore-water extraction techniques have been outlined by Brightman *et al.* (1985), and the results for the Harwell site are summarized in Fig. 5. At the top of the sequence flow is certainly downwards and porewaters appear to have been 'flushed' by downward-moving fresher waters. Based on these pore-water compositions, groundwaters abstracted from the Corallian (Alexander & Holmes 1983) are fresher than they should be if derived exclusively from the adjacent mudrocks. However, the Corallian groundwater does not seem to be invading either of the adjacent mudrocks, thus corroborating the hydraulically measured convergent cross-flow. In contrast the groundwater from the underlying Great and Inferior Oolites appears to have moved into the Oxford Clay.

Rock properties used in the original wedge-and-step model (Black & Barker 1981) were based on aquifer tests and unreliable and limited measurements of mudrock hydraulic conductivity. Consequently, the real ratio of bulk rock hydraulic conductivities was not well defined but was estimated to vary between 10^2 and 10^3. The appropriateness of these ratios can be reviewed by considering the reliable measurements for the Harwell boreholes and by examining the values derived from a numerical model of the region (Brightman & Noy 1984). The field measurements indicated ratios between 10^2 and 10^5 whereas the model achieved best fits using ratios in the range 2×10^3–3×10^6 with the larger ratio being associated with greater depths.

It can be seen that investigations in the middle Thames valley present strong evidence for cross-flows and opposing flow cells within the high-permeability strata, in line with the simple wedge-and-step model. Both theoretical considerations and field observations are in agreement

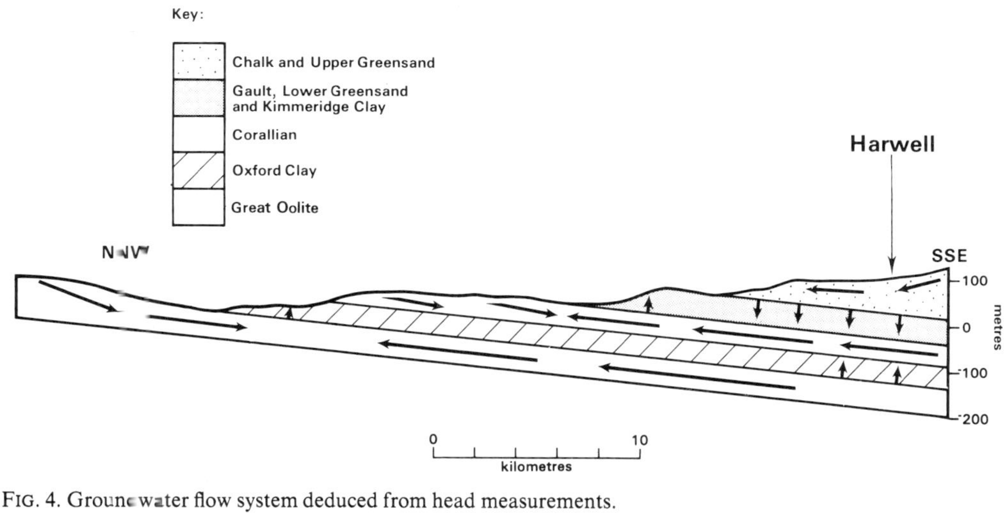

FIG. 4. Groundwater flow system deduced from head measurements.

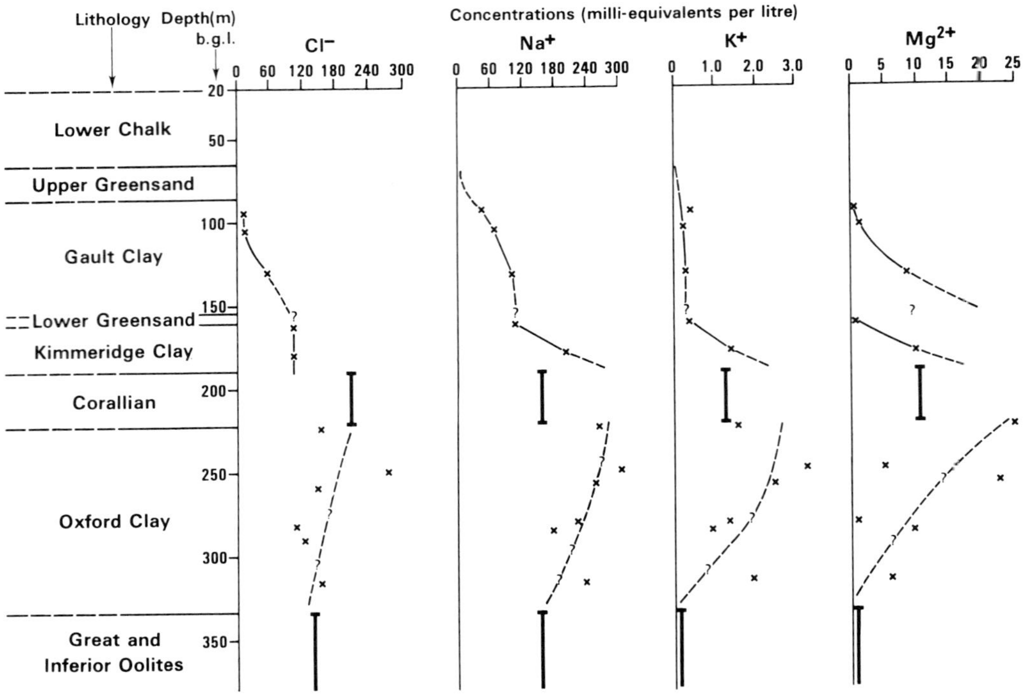

FIG. 5. Pore-water chemistry at the Harwell research site.

concerning the probable ranges of hydraulic conductivity and the resultant flow patterns.

Complex geometry

Numerical modelling

Reality is invariably more complex than the wedge-and-step model, so that flow calculations derived from areas characterized by rocks of low permeability and limited data can only be achieved using numerical models. These are site specific, and general relationships are not as easy to deduce as in analytical models. However, some insights can be gained from some two-dimensional finite-element modelling carried out using a code known as NAMMU (Rae & Robinson 1979). The system to be modelled has to be defined in terms of the average permeability and porosity applying to each element. All boundaries have to be denoted as either fixed-head or fixed-flow boundaries. The model has been used to calculate directions and rates of groundwater flow and head distributions using ratios expressing the contrasts between the permeability of lithologies (Alexander & Noy 1981) and changes in the permeabilities of compressible strata with depth (Brightman & Noy 1984). Complex geometry can be accommodated in such models, but they require greater details of the hydraulic properties of the region under study.

The NAMMU model was utilized at Harwell to evaluate how measurements from a real system with cross-formational flows compared with predictions. Since the region has been subjected to extensive studies, more specific data are available than for the reconnaissance studies such as those conducted in the Somerset and Worcester basins. For the purposes of the model the hydraulic conductivity of the mudrocks was varied according to the thickness of overburden. This produced a flow pattern that agreed closely with that already derived from field observations and shows that depth-dependent properties are to be expected in low-permeability formations. Times of travel for water to move downwards through the Gault and Kimmeridge Clay lithologies varied along the section modelled. This effect was magnified by the depth-dependent mudrock permeabilities; as a result waters of different 'ages' flow into the top of the Corallian in order to mix subsequently (Fig. 6).

Topographic variations

In the Somerset Basin an approximately NE–SW section was chosen which is perpendicular to the

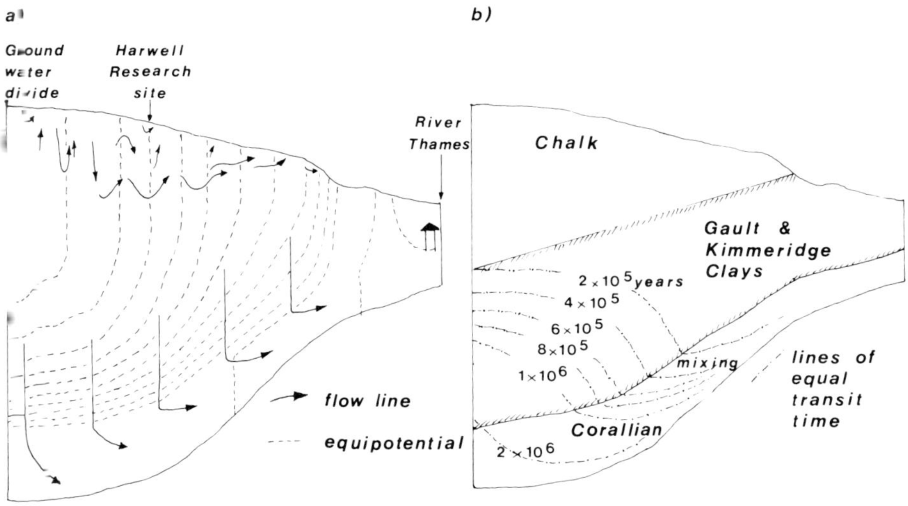

FIG. 6. Transit times for groundwater flow across the Gault and Kimmeridge Clay.

axis of the basin and includes a gently folded sandstone overlain and underlain by a thick sequence of mudrocks (Fig. 7a). The sandstones and mudrocks were assigned values of hydraulic conductivity (based on sparsely distributed data) and the flow distribution across the basin was determined.

Figures 7 (b) and 7 (c) illustrate the cellular nature of the flow system within the basin. When the ratio of sandstone hydraulic conductivity to mudstone hydraulic conductivity is low (i.e. the permeability contrasts are small), local topographic highs cause local flow cells within the regional flow system (Fig. 7b) and the influence of the marginal topographic high is reduced. A similar effect results from a decrease in the hydraulic conductivity of the sandstones with depth and increasing thickness of overburden.

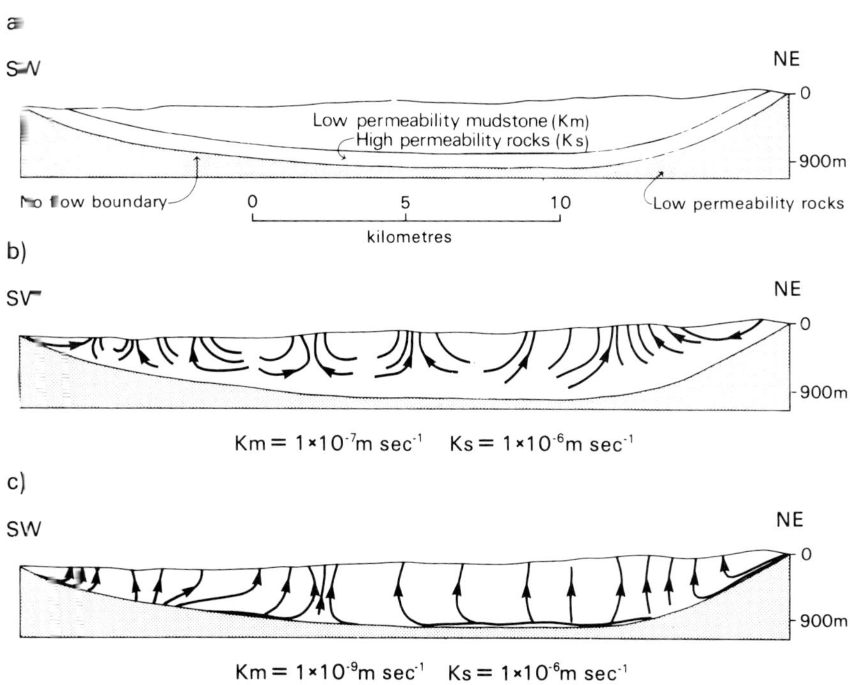

FIG. 7. Flow directions in low-permeability formations in the Somerset Basin.

When the permeability ratios increase, groundwater flows from the topographic highs on the margins of the basin progressively extend over more of the system (Fig. 7c). The location of the balance point between flows driven from the marginal hills and those driven by local topographic features is determined by the permeability ratios utilized.

Flow beneath local mounds

As downward flow beneath a topographic high is to be expected, and this is likely to oppose the upward regional movement of groundwater, a zone of stagnant conditions exists. Such a flow pattern (Fig. 8) determined by the model suggests that the stagnant zone is developed over an area not substantially greater than the topographic feature itself, and its location is highly sensitive to the lower boundary pressure used. Its size is maximized when the mudrock is anisotropic (e.g. the horizontal hydraulic conductivity is much larger than the vertical) and the contrast between the mudrock and the rocks of high permeability is small.

Faults

There are several E–W faults in the Somerset Basin, and one was included in the model to evaluate its effect in low-permeability materials. It was modelled by ascribing above-average values of hydraulic conductivity to a block of length greater than 1 km. Increases in the hydraulic conductivity ratio of the fault block reduced the southerly penetration of the water from the N of the basin. Faults as features of high permeability effectively shield the centre of the basin from the high heads associated with the basin margins and reduce overall head gradients resulting in increases in transit times in unfaulted areas. From this desk study it is apparent that faults in mudrocks may be important in controlling flow rates in deep basins far from their actual locations.

Discussion

The role of low-permeability rocks within regional flow systems is seen to be important where natural flow rates in the high-permeability rocks are small. This concept is already implicit in the regional studies performed by Drescher (1965), Tóth (1978) and Bredehoeft *et al.* (1982). Whether a flow system description needs to take account of flows across rocks of low permeability is a question of the relative properties of the high and low permeability rocks and their position within a given flow system. The theoretical and practical investigations described above seek to show that the circumstances when low-permeability rocks have to be taken into account and evaluated are commonplace. These situations occur at comparatively shallow depth, suggesting that cross-flow should be assumed in all studies. However, the notion of widespread significant cross-flow raises several issues.

The nature of the cross-flow

Low-permeability rocks within sedimentary sequences are invariably argillaceous (e.g. mudrocks, clays, siltstones), and it is through this type of rock that cross-flow has to occur. In water resources terms they are often considered impermeable but the evidence above shows that this is far from true. For the purposes of modelling regional flows they are ascribed values of bulk permeability in the range 10^{-12}–10^{-8} m s^{-1}. These values are often larger than those obtained in the laboratory on core samples, although sometimes, as is the case at Harwell, field and laboratory measurements yield the same result. Is there then any evidence to suggest that flow through mudrocks is localized to channels (or fractures) rather than diffuse flow as in 'normal porous media'?

Channelled flow is an attractive mechanism for flow through mudrocks containing evaporites, such as the Mercia Mudstones in the Somerset Basin (Alexander & Noy 1981). It serves to restrict the dissolution effects of the cross-flow on the contained evaporites. However, their persistence could be explained equally well by assuming that the cross-flowing groundwater is already saturated with respect to the evaporite salts. The major consideration against flow concentrated in fractures is the low strength of most mudrocks so

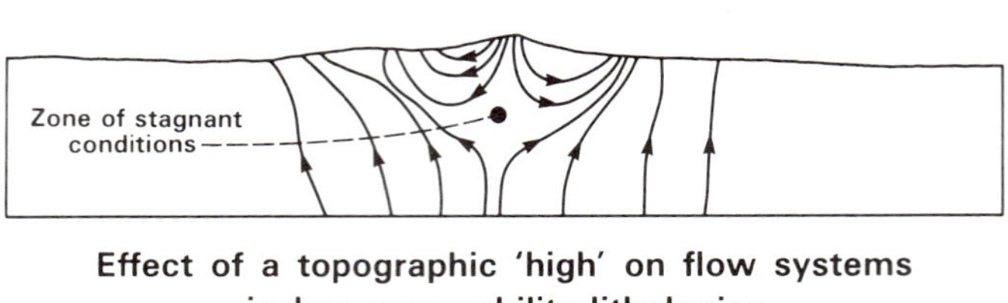

FIG. 8. Effect of a topographic 'high' on flow systems in low-permeability lithologies.

that fractures cannot be expected to remain open. This is despite the common inference that mudrocks fracture during compaction to release their excess water. It is interesting to note that, once compacted, they can sustain open fractures at higher stresses. Do these same fractures become reactivated during denudation and hence become preferred pathways for water movement? Another possible mechanism for the production of channelled flow is areal heterogeneity, i.e. the inclusion of silty and sandy lenses.

Evidence in favour of porous-medium-type flow is available from the investigations reported here. In no instance does a pore-pressure measurement approach lithostatic pressure. In all cases Darcian flow through a continuous water phase in the mudrocks yields a consistent interpretation of measured heads. It is as yet very difficult to identify the precise character of flow through mudrocks. This means that it is impossible to predict its size or its trend with depth from theoretical considerations.

Use of pore-water chemistry as an indication of flow

If groundwater commonly flows through mudrocks then the pore waters of the mudrocks should reflect this occurrence. In order to examine this feature, cores of some of the mudrocks of the middle Thames Valley have been squeezed to expel the pore waters. Although the technique is experimental and a variety of errors are possible, some general trends are evident. Firstly the water does not appear to be water of formation since it is considerably fresher than would be expected. Secondly it could be anticipated that water being expelled from the centre outwards during compaction would leave behind water of a homogeneous composition. The measured compositions vary about a general trend (see Fig. 5) suggesting that it is not water of compaction. Additionally, all the waters in the system seem rather fresh, which could be explained by a freshwater flushing event in the past. It is not known when this event occurred, but it reinforces the same conclusion based on the dissolution of $CaCO_3$ cement from the Corallian (Milodowski & Wilmot 1984).

Transit times for groundwater flow across mudrocks involve such long time-scales that detailed knowledge of the geological history of the basin is required to understand fully the chemical evolutions of the groundwater. This information needs to be combined with knowledge of the hydrogeological properties and the detailed chemistry of the clay–pore water system. Such a multicomponent data set requires the facility of a flexible interactive model to yield the complexities of the groundwater evolution. Measuring mudrock pore waters seems to be a promising technique to help understand the evolution of basins and their groundwater flow systems.

Density and osmotic effects

Density contrasts in the fluids in a given flow system have not been taken into account in any of the studies discussed here. This omission (brought about by lack of relevant data) is important since any significant density changes will result in retardation of groundwater flows through the sedimentary sequence. This may be a factor in the Somerset Basin where the predicted direction and rates of fluid migration through interbedded mudrock and evaporite sequences is such that the integrity of bedded salts could be maintained (Alexander & Noy 1981), but this requires further evaluation. Where temperature gradients are low and groundwaters are fresh, as in the middle Thames valley, it is clear that density effects will not affect flow.

Both the modelling and the interpretation of the head measurements reported in this paper are based on the assumption that 'water-table topography' is the sole cause of groundwater movement. This may often not be the case in the deeper parts of the basin systems. Osmotic effects, where water moves in response to chemical gradients, should also have limited potential to disturb the Harwell system, given the low salinities (and therefore low chemical gradients) involved. However, it is possible to envisage osmosis opposing topographically based head gradients. The application of the process to thick clays (acting as membranes) in relatively shallow groundwater systems is not too clear, but suitable conditions may be encountered deep within a basinal system (Marine & Fritz 1981). Since the magnitude of the potentials generated in a real system by osmosis are as yet unknown, it is difficult at present to evaluate the outcome of a system involving osmosis. This is because once out of the zone of highly active water the interpretation of basinal groundwater flow systems is a question of competing flow rates. When multiplied by long periods of time small errors in rates can easily result in erroneous conclusions.

Conclusions

The studies outlined in this paper have shown that the role of low-permeability rocks has to be taken into account in attempts to describe regional flow systems. In general it is surprisingly difficult to predict even the flow directions within a regional flow system involving cross-flow without detailed consideration. This is best achieved using models, both analytical and numerical. This approach has shown that the pattern of flow depends on the geometry of the system and the bulk hydrogeological properties of the rocks. The ratio of the hydraulic conductivity of high-permeability beds to that of low-permeability beds (the permeability ratio) is an important factor in determining the extent of both regional and local flow cells. If the ratio is high, regional topographic highs (e.g. marginal hills) will dominate the flow system; if the ratio is low, local topographic highs will dominate the circulation pattern. This applies regardless of what rock type contains the high point of the water table.

Downward flow beneath a topographic high is to be expected, and it will oppose upward regional movement of groundwater and a zone of stagnant conditions will result. The size of such a zone is maximized when the mudrock is anisotropic and the permeability ratio is small. The location of zones of opposed flow is a function of the permeability ratio, as is the position of the longest transit time across the mudrock within the flow system. The location of this flow path is not necessarily related to the thickness of the low-permeability beds, but is dictated by the overall geometry and permeability ratios. The cross-flows with highest velocity occur under the high point of a regional system.

Faults in mudrocks may be important in controlling flow rates in deep basins far from the location of the fault. Faults, as features of high permeability, effectively shield the centre of a basin from the high heads associated with the basin margins and reduce overall head gradients. As a consequence flow rates in unfaulted parts of a basin are reduced, resulting in increased transit times. The analytical and numerical models predict that flow cells occur within regional systems. In both cases the variation of transit time across the mudrock before mixing within high-permeability rocks leads to a groundwater chemistry that is very difficult to interpret. This effect is increased further where the permeability of the mudrocks is depth dependent.

The predictions of the modelling require testing against measurements from actual flow systems. To achieve this aim a combination of hydraulic and chemical measurements has to be made and integrated. An investigation of this type has been attempted in the middle Thames Valley area. Both heads and hydraulic properties have been measured at a number of locations. Where heads have been measured in the mudrocks they are consistent with the concept of cross-flow between the high-permeability rocks. The chemistry of mudrock pore waters provides further corroboration of the occurrence of cross-flow, even at depths of up to 300 m.

The effect of cross-flow within a regional groundwater flow regime is to produce a system which is in dynamic equilibrium throughout geological history. Although flow rates are small, when they are multiplied by the size of a basin and very long time-scales significant quantities of groundwater are involved. An adequate understanding of a regional flow system must of necessity include detailed knowledge of the role of the low-permeability rocks in the system.

ACKNOWLEDGMENTS: The authors acknowledge the help and assistance of many colleagues within the Fluid Processes Research Group, particularly D. C Holmes and D. J. Noy. Much of the work was undertaken as part of the research programme into radioactive waste management sponsored by the Department of the Environment and the C.E.C. to whom the authors are grateful. This paper is published by permission of the Director, British Geological Survey (National Engineering Research Council).

References

ALEXANDER, J. & HOLMES, D. C. 1983. The local groundwater regime at the Harwell research site. *Rep. FLPU 83-1*, Institute of Geological Sciences, London.

—— & NOY, D. J. 1981. Hydrogeological Reconnaissance study: the Somerset Basin. *Rep. ENPU 81-15*, Institute of Geological Sciences, London.

BLACK, J. H. & BARKER, J. A. 1981. Hydrogeological reconnaissance study: Worcester Basin. *Rep. ENPU 83-1*, Institute of Geological Sciences, London.

BREDEHOEFT, J. D., BACK, W. & HANSHAW, B. B. 1982. Regional groundwater flow concepts in the United States: historical perspective. *Geol. Soc. of Amer. Special paper 189*.

BRIGHTMAN, M. A. & NOY, D. J. 1984. Finite element modelling of the Harwell regional groundwater flow regime. *Rep. FLPU 84-1*, British Geological Survey, Keyworth.

——, BATH, A. H., CAVE, M. & DARLING, W. G. 1985. Pore fluids from the argillaceous rocks of the Harwell region. *Rep. FLPU 85-6*, British Geological Survey, Keyworth.

DRESCHER, W. J. 1965. Hydrology of deep-well disposal

of radioactive liquid wastes. In: YOUNG, A. & GALLEY, J. E. (eds) *Fluids in Subsurface Environments, Am. Assoc. petrol. Geol. Mem. No. 4.*

MARINE, I. W. & FRITZ, S. J. 1981. Osmotic model to explain anomalous hydraulic heads. *Water resour. Res.*, **17** (1), 73–82.

MILODOWSKI, A. E. & WILMOT, R. D. 1984. Diagenesis, porosity and permeability in the Corallian Beds (upper Oxfordian) from the Harwell research site, South Oxfordshire, UK. *Clay Miner.*, **19**, 323–41.

RAE, J. & ROBINSON, P. C. 1978. NAMMU: Finite element program for coupled heat and groundwater flow problems. *Rep. AERE R-9610*, Atomic Energy Research Establishment, Harwell.

TÓTH, J. 1978. Gravity-induced cross-formational flow of formation fluids, Red Earth Region, Alberta, Canada: analysis, patterns, and evolution. *Water resour. Res.*, **14** (5), 805–43.

J. ALEXANDER, J. H. BLACK & M. A. BRIGHTMAN, British Geological Survey, Keyworth, Nottingham, NG12 5GG, U.K.

Flow and flow mechanisms in crystalline rock

J. H. Black

SUMMARY: Recent work on the hydrogeology of crystalline rocks has evolved rapidly from basic data gathering to specific experiments for evaluating details of flow mechanisms. Single-borehole data from around the world are similar and independent of detailed geology and test techniques. It appears that most faults in crystalline rocks have an effective hydraulic conductivity of around 1×10^{-7} ms^{-1} with values for intact rock about five orders of magnitude lower. The porosity of unweathered crystalline rock measured in the laboratory usually varies between 0.1% and 1% with weathered samples attaining up to 15%. When large data sets are considered it is possible to correlate depth and hydraulic conductivity. The interpretation of hydrogeological tests has evolved from basic porous media through fissures bounded by impermeable rock and fissures bounded by permeable rock to networks and channels. Whereas most hydraulic testing is adequately interpreted using the fissured porous medium concept, tracer tests require highly complex fissure network models with channelling. There is growing evidence that groundwater flow is concentrated in a limited network of channels. This seems to apply on both the laboratory and the regional scale.

During the last decade there has been considerable research into the hydrogeology of crystalline rock stimulated by the needs of radioactive waste disposal and geothermal power exploration. For a number of reasons the work has wider relevance for the hydrogeology of aquifers, particularly where flow is associated with faults and fissures. Firstly studies for radioactive waste disposal are concerned with the migration of dissolved species rather than with how much water can be abstracted. This means that the actual mechanism of flow (i.e. within fissures, through channels or through the entire rock) is of central importance. Linked with this is a direct need to describe the interaction of the dissolved species with the rock through which they are flowing. Secondly some of the 'dual-porosity' effects which are also present in aquifers are easier to study in crystalline rocks because experimental time-scales are longer and effects are sometimes more exaggerated. Thirdly understanding the role of faults in delineating both water and hydrocarbon flow systems is assisted by the study of the hydrogeology of rocks where flow is most certainly within faults and fissures.

One of the less obvious outcomes of the study of the hydrogeology of crystalline rocks is the insight into 'standard' approaches yielded by examining their applicability to 'non-aquifers'. As a general rule studies have been of four types: (i) comparatively simple single-borehole measurements to provide input parameters for flow system (and hence safety assessment) modelling and validation; (ii) small-scale experiments to examine flow and transport mechanisms; (iii) laboratory measurements in support of the first two activities; (iv) model studies of potential nuclide or heat transport.

The aim of this paper is to provide some idea of the magnitude of the range in 'standard' hydrogeological parameters (hydraulic conductivity, porosity, head gradients etc.) together with a flavour of the concepts of flow which have emerged.

Properties relevant to flow

The very nature of crystalline rock means that flow of any significance can only be envisaged as occurring in the fissures. However, the rock matrix is not impermeable and the overwhelming proportion of water stored within the system is stored in the rock matrix. Thus, in order to describe flow through crystalline rocks accurately, four parameters are required: the flow and storage properties of both the fissures and the rock matrix. Unfortunately, when field tests are conducted they tend to measure the conductive properties of the fissures and the storage properties of the matrix. The storage properties of the fissures are extremely difficult to measure, and the hydraulic conductivity K_m of the matrix is best measured in the laboratory. The details of transport within the fissures and of diffusion into the matrix are superimposed upon this basic framework. In many instances a property of interest is defined as much by the method of determination as by its function in an equation (e.g. 'tracer' aperture and 'hydraulic' aperture).

Field tests

Single-borehole measurements

Single-borehole hydraulic tests have been carried out in every radioactive waste and geothermal

From GOFF, J. C. & WILLIAMS, B. P. J. (eds), 1987, *Fluid Flow in Sedimentary Basins and Aquifers*, Geological Society Special Publication No. 34, pp. 185–200.

programme around the world. Test methods vary from the slug and single-pulse methods used in Switzerland (Grisak *et al.* 1985b) and the U.K. (Holmes 1981) to the constant-head rate tests used in Sweden (Braester & Thunvik 1984). The methods of interpretation also vary from analysing the transient response, to being based only on the steady-state response. Almost all testing has been carried out using straddle packers with intervals ranging between 1 and 25 m. Most testing using intervals smaller than 2 m has been for special purposes such as fracture detection, whereas intervals larger than 25 m are associated with reconnaissance. In most programmes the determined transmissivity of a zone is divided by the zone length to yield an 'apparent hydraulic conductivity *K*' even though the concept is incorrect in a fissured rock. The results from some programmes in Europe and the U.K. (Fig. 1) are similar to the results reported from the U.S.A. by Brace (1980, 1984). The spread of results is very large ranging from 1×10^{-13} to 1×10^{-5} m s^{-1}. This reflects the diversity of rock conditions ranging from high values in major fault zones to low values from essentially intact rock. The magnitudes do not seem to vary greatly either with slight differences in rock type (i.e. granite versus gneiss) or test method. It would seem that a major fault zone in Britain or Sweden will inevitably yield *K* values around 1×10^{-7} m s^{-1}. Similarly the intact rock will invariably yield values around 1×10^{-12} m s^{-1}.

The variability of the results with depth is also similar from country to country. The results from two boreholes in different countries which have been tested using different test methods, straddle lengths and borehole dimensions are compared in Fig. 2. The rapidity of variation is evident from both and no clear-cut depth dependence is apparent. However, when enough results are combined, as in the work of Carlsson *et al.* (1983), a depth dependence of the form shown in Fig. 3 can be derived. The depth dependence is of the form

$$K = aZ^{-b}$$

where a and b are constants and Z is the vertical depth below ground surface.

The same form of relationship was observed in three 300 m boreholes at Altnabreac (Holmes 1981) in northern Scotland but not in a single 800 m borehole in Cornwall (Heath 1985). This may be due in part to the greater depth of weathering at Altnabreac compared with that at Carwynnen, Cornwall. As can be seen in Fig. 3, the depth dependence is more marked closer to the surface.

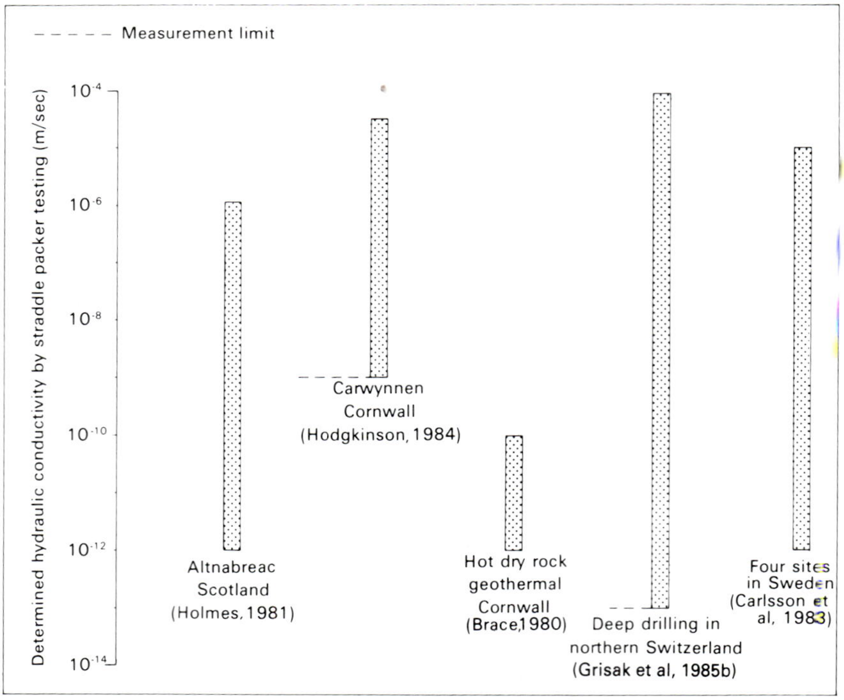

FIG. 1. Ranges of hydraulic conductivities determined by single-borehole tests in Europe and the U.K.

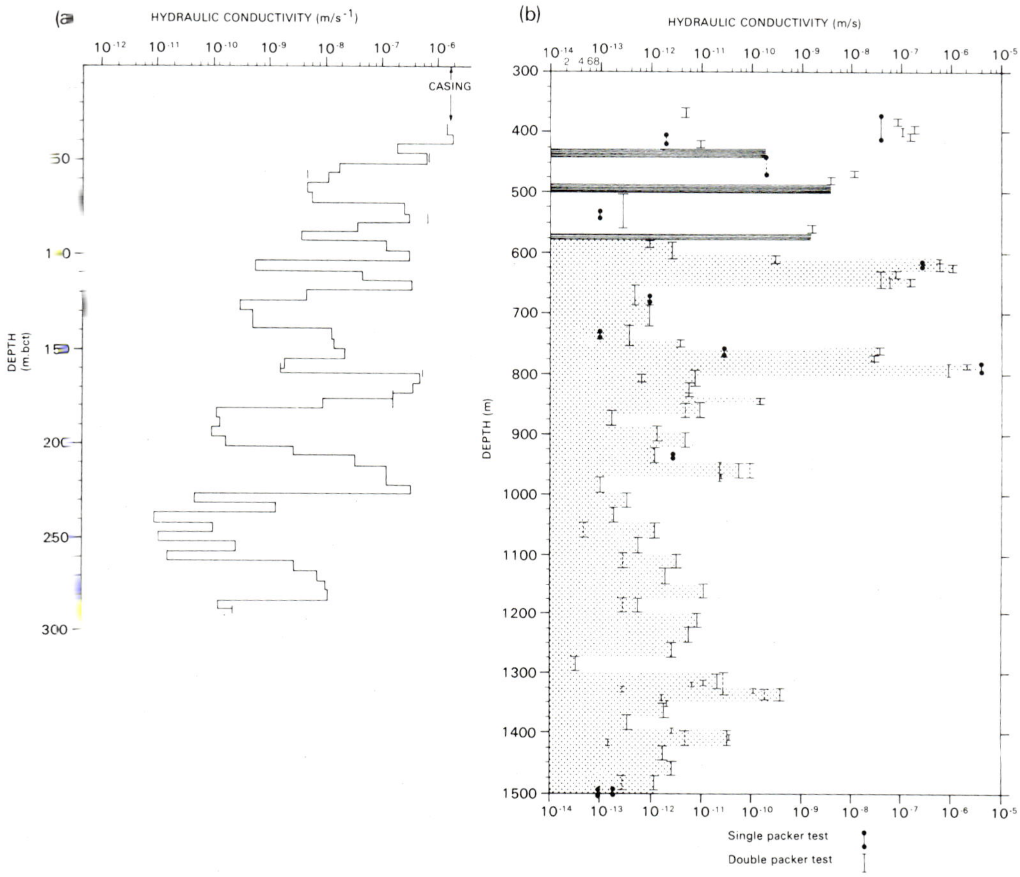

FIG. 2. Profiles of hydraulic conductivity versus depth for two boreholes: (a) Altnabreac, Scotland (Holmes 1981); (b) Böttstein, Switzerland (Grisak *et al.* 1985a).

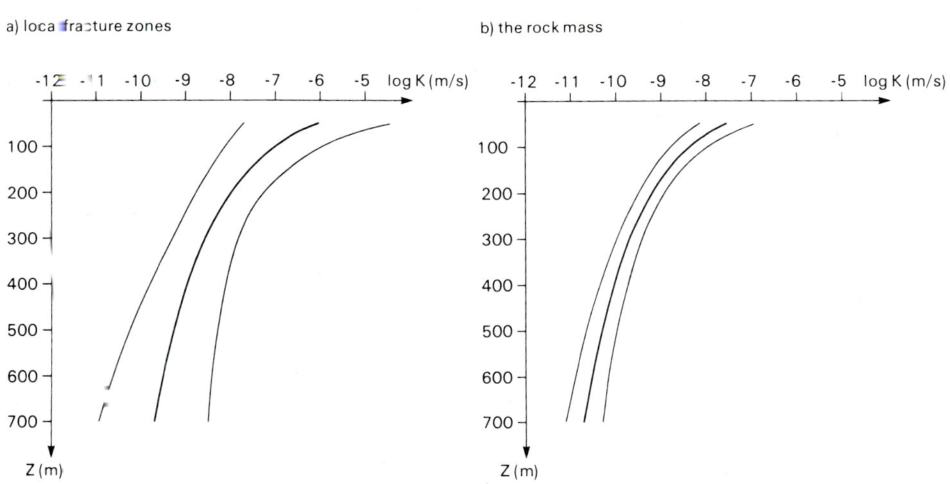

FIG. 3. Depth dependence of hydraulic conductivity for the site at Fjällveden, Sweden: (a) local fracture zones; (b) rock mass. (From Carlsson *et al.* 1983.)

Many of the more recent programmes of single-borehole measurements have also measured the head in the straddle zone. It is only an apparent equilibrium head since the borehole strongly affects the head regime locally. In general, test programmes which are based on transient test methods yield more reliable head measurements, but the form of the results seems to be common to all progammes. Often considerable lengths of time are involved in head measurement and the results are usually based on the last measurement before packer deflation. Partly as a result of this the results are often expressed as heads relative to open borehole water level (Fig. 4). This has the advantage that it is possible to identify those zones of the borehole where groundwater is entering and those where it is leaving. The result from Altnabreac (Holmes 1981) shown in Fig. 4 is typical in that the general form reflects the topographic position of the borehole. The detailed variations cannot be interpreted as is also the case for similar results from deep boreholes in Switzerland (Grisak *et al.* 1985a). In the latter case the sections tested were of granites beneath considerable thicknesses of sediments. Apart from the detailed connection and orientation of major fissures, the problem of poorly quantified groundwater density profiles is a factor. The value of such head versus depth profiles is undermined further by limited evidence that measured profiles bear little resemblance to those existing prior to drilling (Davison *et al.* 1979). To combat this tendency, rapid 'through-the-bit testing' has sometimes been proposed.

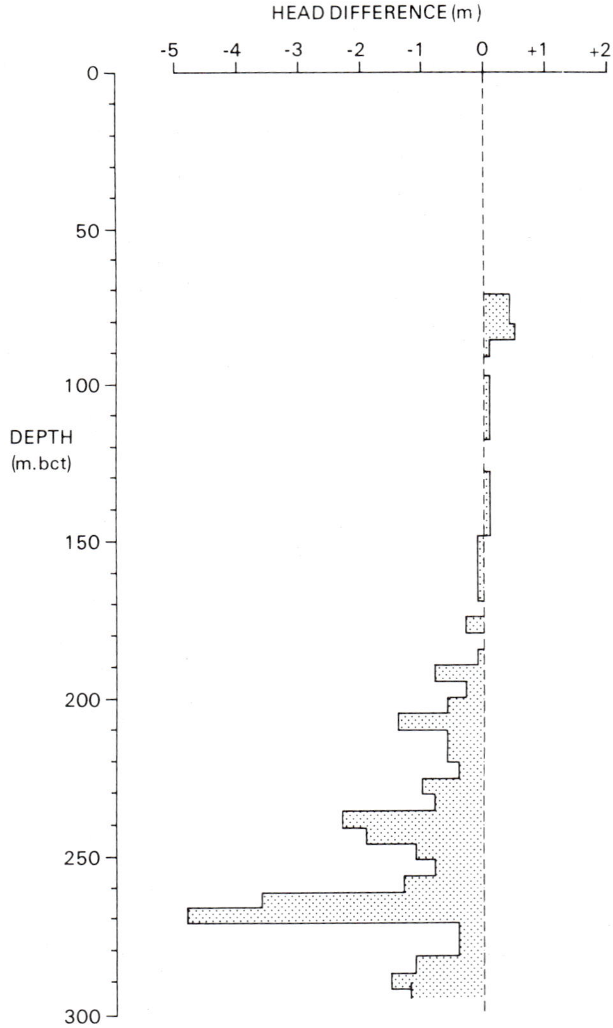

FIG. 4. Typical profile of head versus depth for a borehole in granite on the top of a hill at Altnabreac, Scotland (flow is *downwards*). (From Holmes 1981.)

Multiple-borehole tests

It is clear from a brief consideration of the value of hydraulic conductivity likely to be encountered in crystalline rocks that pumping tests on the scale of normal aquifer tests are not possible. However, a few such tests have been carried out under certain restricted circumstances. These usually involve a dense array of closely spaced boreholes with maximum separation seldom exceeding 100 m.

The problems of such a test can be illustrated using an example from the U.K. (Black 1979). The test was carried out in an array of four boreholes 300 m deep and roughly 20 m apart at the surface with water levels about 7 m down. Analysis was complicated by the inability to achieve a constant abstraction rate of 2 l min^{-1}, the variable separation between the boreholes and the vast storage in the boreholes compared with the surrounding rock. When the further complexity of significant non-horizontal flow is added it is hardly surprising that a rough-and-ready slug test was considered just as reliable as the 30 day (and commensurately expensive) multiple-borehole test.

To some extent the problems of this sort of test were addressed by the 'ventilation test' proposed by Gale & Witherspoon (1979) and carried out 310 m below ground at the Stripa Mine underground laboratory in central Sweden. In this pumping test a mine drift is used as the pumping borehole and the decline of head is observed in boreholes all around. The inflow rate was small, so a section of the drift was sealed by bulkheads and the inflow was measured by comparing the humidity of incoming and outgoing air. This proved satisfactory in operation and a whole-rock hydraulic conductivity of 1×10^{-10} m s^{-1} was determined (Gale *et al.* 1983). The head gradient away from the 'ventilation drift' was highly variable, but it was also clear that there was a near-drift skin of reduced conductivity (Fig. 5). The conductivity of this skin is about one-third of that of the rest of the rock based on average gradients. It is quite plausible, however,

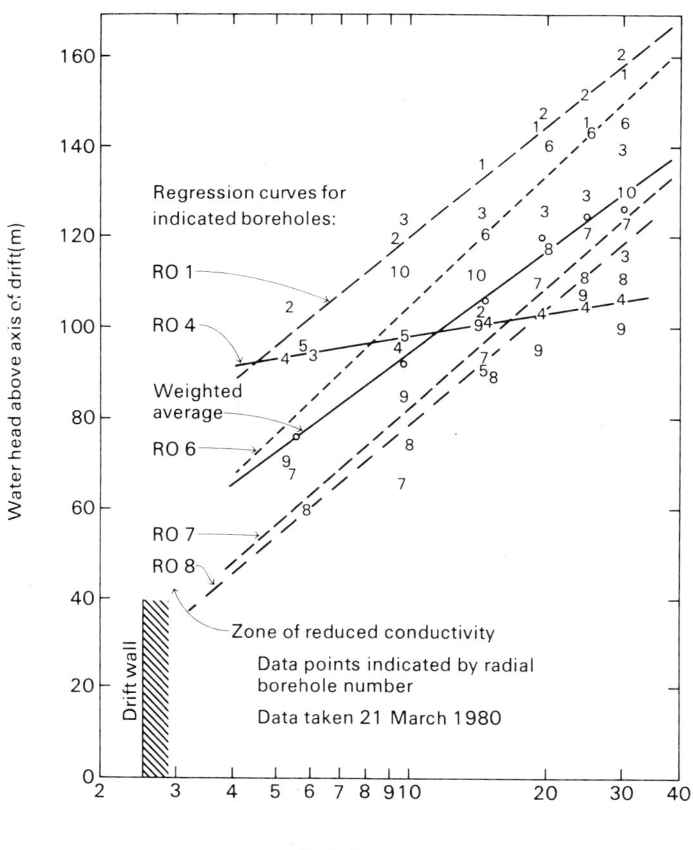

FIG. 5. Distance–drawdown plot for a ventilation drift permeability test with a nominal room temperature of 20°C. (After Gale *et al.* 1983.)

to imagine that almost all the flow is contributed by a few fractures whose pressures are not measured by the boreholes. Additionally, since the boreholes are directly along pathlines there is the possibility that they appreciably augment the natural conductivity. The basic test is currently being repeated in the Grimsel underground laboratory in Switzerland.

Other pumping tests have been conducted by Carlsson & Olsson (1985) elsewhere in the Stripa Mine. When the various specific fracture zones had been identified, the boreholes from which they were flowing were sealed and the build-up of pressure was observed. Apart from some problems with surprisingly large well-bore storage factors, dual-porosity responses were observed. However, interpretation is impeded by poor knowledge of the geometry of the system, and values of storage for the primary and secondary responses (generally assumed to be fissures and matrix) are only an order of magnitude different. Unless there is pervasive weathering of the matrix adjoining the fissures these values seem unlikely.

In a sense all these tests suffer from an ill-defined flow system geometry. This leads to doubts over the use of particular geometries which are assumed in order to interpret the test results. There has been a tendency to try to answer these doubts by devising special experiments or techniques.

Borehole-based experiments

In the investigations described previously the interpretation procedure has been chosen to fit the form of the data obtained. Dissatisfaction with the applicability of some of these 'standard aquifer' procedures has led to some experiments designed to evaluate new interpretational concepts. These investigations tend to be carried out at well-controlled experimental sites such as Carwynnen, Cornwall, or the Oracle research site near Tucson, Arizona.

One of the aspects which recurs in many tests is what form of geometry to assume for the interpretation. In standard aquifer methods a cylindrical symmetry based on testing boreholes which intercept sedimentary layers perpendicularly is assumed. No such basic geometry can be assumed in crystalline rock. In an attempt to test the ability to derive truly three-dimensional properties from an array of randomly oriented boreholes, a series of standard tests was carried out at the Oracle site near Tucson. The results are reported by Hsieh *et al.* (1983) and demonstrate how to compute all three principal values of the hydraulic conductivity tensor in three dimensions as well as the corresponding principal directions. This is an improvement on the approach of Louis (1974) in which the boreholes had to be oriented in advance perpendicular to the principal directions.

The same approach to interpretation underlies the hydraulics part of the Crosshole Project at Stripa Mine. In the Crosshole Project a fan-shaped array of boreholes 250 m long has been drilled to investigate 3×10^6 m^3 of rock about 350 m below ground (Black & Holmes, 1985). The site has been extremely well characterized geometrically by reflection radar techniques. The unique development in the hydrogeological work is the use of sinusoidal fluctuations of pressure to improve detection and exaggerate the dual-porosity response (Black & Barker 1983). An unusual aspect of the testing is the use of computer-controlled pumps so that it is fully automatic. This enables the analytical assumptions of the interpretation to be met with an accuracy hitherto unknown in hydrogeological testing. For instance, a source zone isolated by packers can be stimulated with any form of hydraulic signal (sinusoid, square wave, pulse etc.) whilst the rest of the borehole is maintained at a constant pressure. The testing also includes single-borehole tests, and these are being compared with various geophysical test techniques including radar (Fig. 6). The crosshole testing is designed to be analysed either like the Oracle data (i.e. as a uniform anisotropic porous medium) or in terms of identified discrete features. One of the useful aspects of the approach used at Oracle is that it provides a direct field indication of the degree to which rock behaviour deviates from the assumptions.

To check whether the tested rock mass behaves as an anisotropic continuum, a polar plot of the computed directional hydraulic conductivities should be examined. If they delineate a distinct ellipsoid, the continuum assumption holds; if not, its usefulness on the scale of the test should be questioned. The early results from the crosshole sinusoidal testing indicate distinct dual-porosity behaviour but the directional diffusivities do not lie on a smooth ellipsoid (Fig. 7). This points to an interpretation in terms of either a sparse network of discrete fissures or a network of probabilistically described fissures.

The network approach has developed over the past few years and seeks to describe fissured rocks in terms of a network of interconnected finite-area fissures (Robinson 1983; Long & Witherspoon 1985; de Marsily 1985). The fissure system is described in terms of probability distributions of four parameters: fissure orientation, fissure frequency, fissure transmissivity and

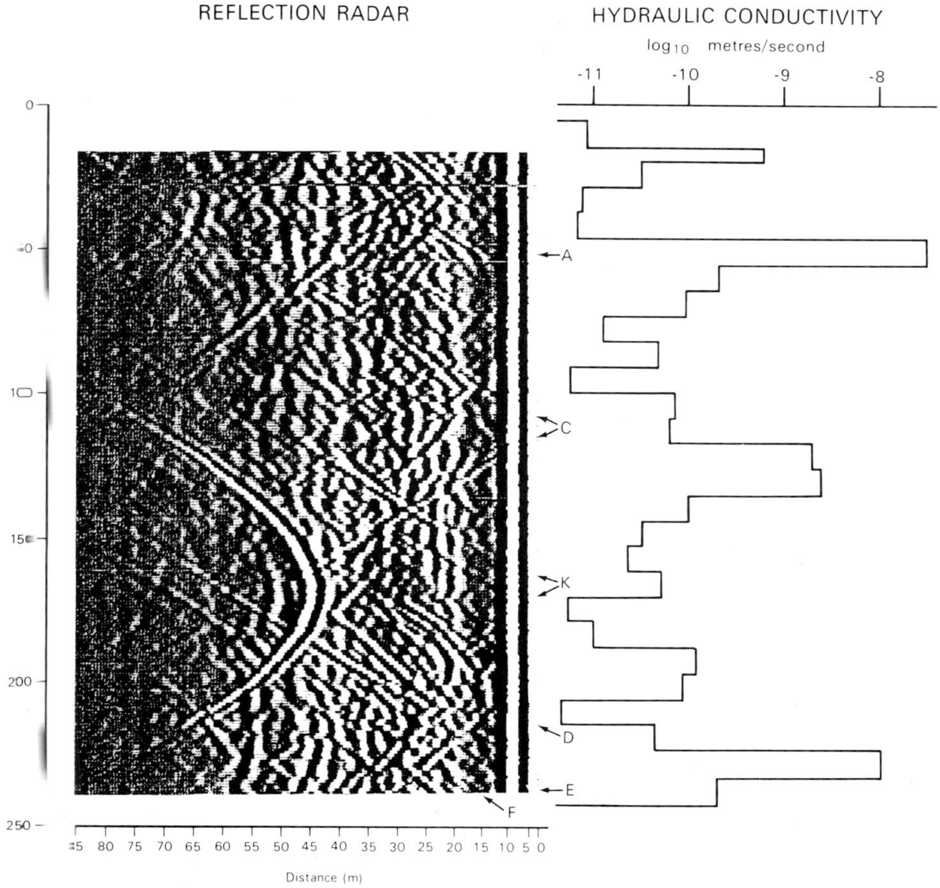

FIG. 6. Comparison of single-borehole hydraulics and radar for borehole F4 of the Stripa Project, Crosshole Programme.

fissure area (or alternatively distance between interconnections). It is then possible to construct a computer model of a particular fissure network whose properties accord with the probability distributions (Fig. 8). The permeability of this modelled network can then be calculated. By repeating this procedure a large number of times the relationships between the parameter distributions and the effective permeability can be deduced. Interesting aspects of this approach include evaluating the density of fissures required first for 'interconnectedness' to occur and second for the region to be considered to be representative (Fig. 9). In other words, if the area of a particular fissure is known, the size of the smallest volume in which fissures of that area do not need to be identified specifically can be calculated. Given the probability distributions underlying Fig. 9, the smallest volume having average properties (i.e. the 'representative elementary volume') is about 10 times the size of the mean fissure length. By including terms to describe the interaction of dissolved species with the sides of the fissures it is comparatively easy to extend this description such that it becomes a transport model. The importance of the network approach is that it tends to be conservative in operation (i.e. possible errors would lead to improvements as regards safety assessments) and it includes effects like directional hydraulic conductivity, connectivity and dispersion within its basic formulation. Another important aspect is that this modelling and interpretation approach alters the methodology of field testing. It is now important to identify clearly which fissures are hydraulically conductive and which are not in order to derive effective fissure frequency. In addition, some method must be devised to yield the parameter describing fissure area.

An attempt at this new form of testing has

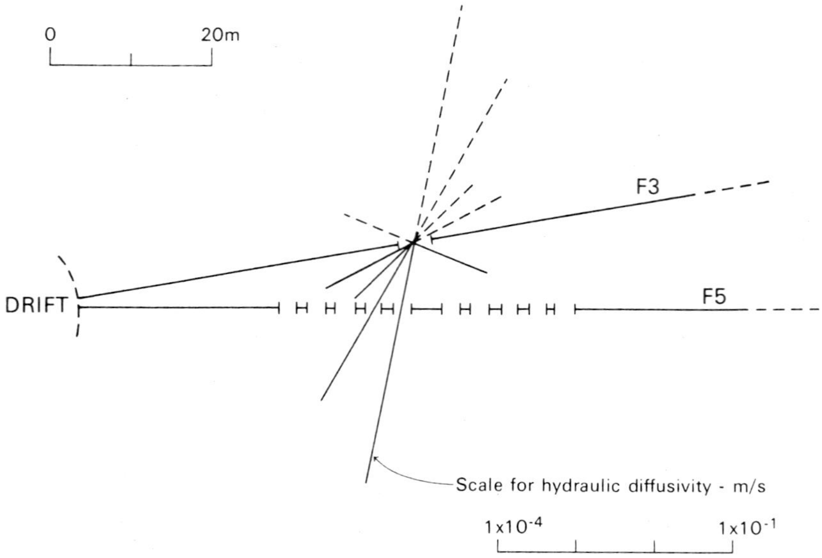

FIG. 7. Representation of the F3–F5 crosshole sinusoidal results in terms of directional hydraulic diffusivity.

FIG. 8. Typical computer-generated two-dimensional fracture network. (From Robinson 1983.)

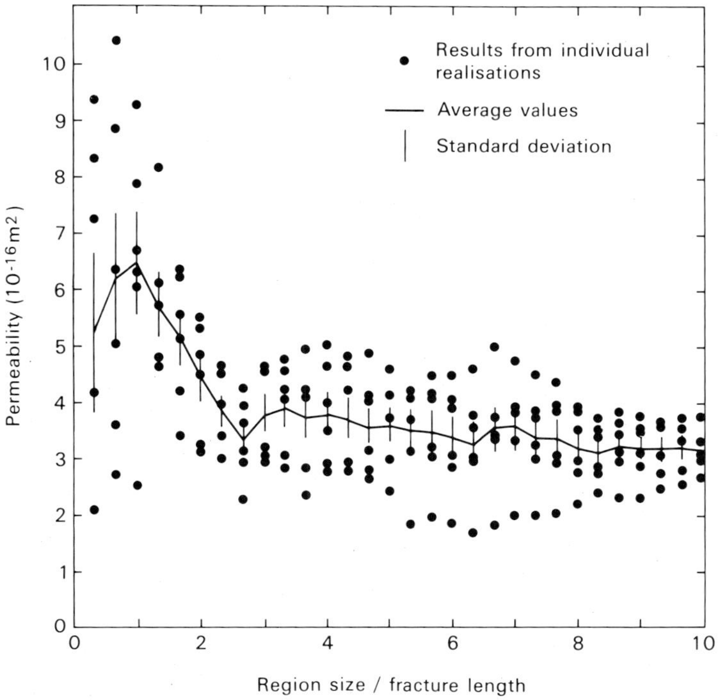

FIG. 9. The permeability of a two-dimensional fracture network as a function of region size. (From Robinson 1983.)

been made at Carwynnen test quarry in Cornwall (Hodgkinson 1984; Heath 1985). In this work about [illegible] km of boreholes was straddle-packer tested using a 1 m spacing. The fissure orientations were obtained from siesviewer logs while the transmissivities were measured by steady-state injection testing. Fissure frequency was simply the distance between each intersection of the borehole by a hydraulically active fissure. The fourth parameter concerning interconnection length was obtained by assuming that each conductive fissure had a constant head at some distance from the borehole where the tested fissure joined the rest of the fissure network. A most likely distance to this intersection was derived by iterating for each injection test result. By considering all the results an average distance of 9.8 m was obtained. This is about 1.5 times the mean fracture spacing along the borehole (Hodgkinson 1984). The techniques are a little unwieldy but at least demonstrate that a network approach is broadly practicable.

The need for a more refined concept of flow in fissures beyond laminar flow in plane-parallel fissures was also seen in tracer tests. Again in the Stripa Mine a single-fracture experiment with sorbing and non-sorbing tracers was carried out between 1981 and 1985 (Abelin *et al.* 1985). A natural fracture cutting a drift of the Stripa Mine 350 m below ground was chosen for injection with tracers. The intention was to inject the tracers into the natural flow field of the mine at a radial distance of about 5 m from the drift wall. Some hydraulic testing was performed which, if translated into plane-parallel fissure apertures, indicated tracer arrivals up to 1000 times earlier than measured. There was some difficulty over tracer recovery but more significantly there was evidence of highly non-uniform flow. This was obtained from core taken from the plane of the fracture. The surface of the fracture was examined in the laboratory to determine the distribution of sorbed tracer over the plane of the fissure. It was evident that the flow had been neither radial into the drift nor linear in any given direction (Fig. 10). This variability was ascribed to irregular channelling within the fissure. The investigation has been extended to examine how tracers migrate through small networks of fissures in an experiment known as the 3D Migration Test (Birgersson *et al.* 1985). This test is also at the Stripa Mine and consists of a 75 m long specially excavated drift with injection points into 'averagely fractured rock' up to 55 m away from it. A

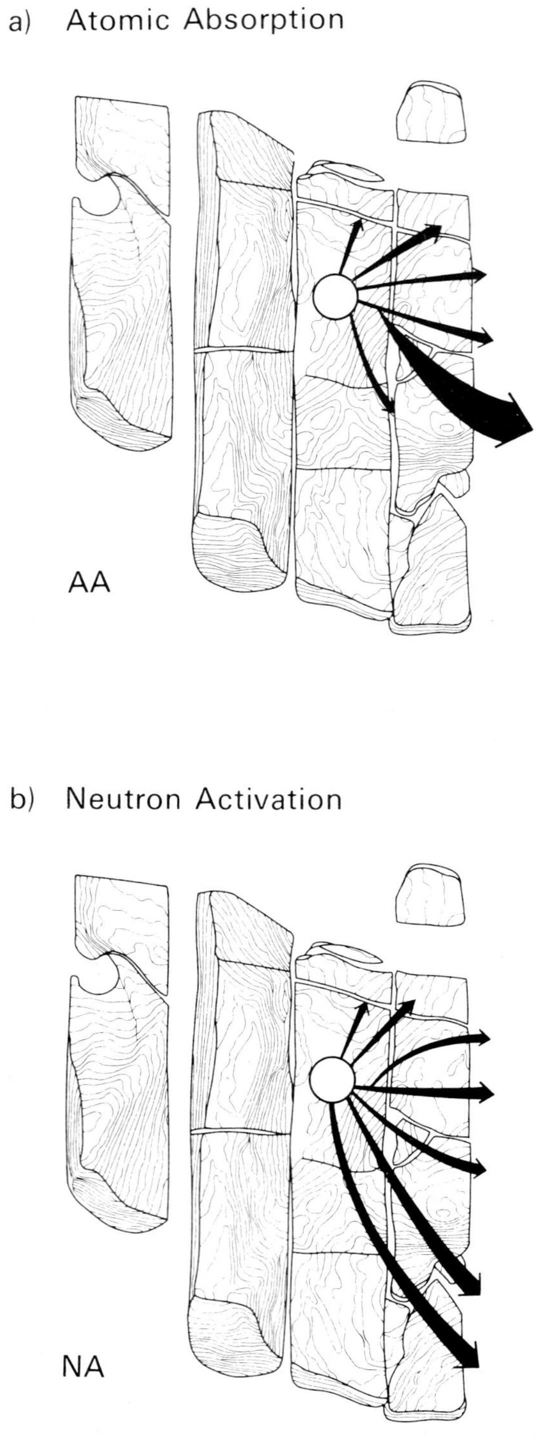

FIG. 10. Plausible channels for the flow out from the injection hole of the single-fracture experiment of the Stripa Project based on (a) atomic absorption and (b) neutron activation.

novel aspect of the experiment is the water-collection method whereby 350 sheets of poly (vinyl chloride) film of dimensions 2 m × 2 m are glued to the walls and ceiling of the test drift. In this way the distribution of natural inflow can be assessed. It was found that about 90% of the inflow came in to 5% of the sampling sheets (Fig. 11a). The tracer has so far arrived in the same fashion, restricted to just a few of the sheets (Fig. 11b). This uneven distribution of inflow is interpreted as further evidence of the importance of channelling and limited interconnections in networks of channels.

Laboratory measurements

Laboratory measurements divide into two forms of testing relevant to groundwater flow through fissured crystalline rocks: properties of the rock matrix and studies of small fissures. Fissure studies are a problem since they require restressing and disturbance is very great. In addition, a large amount of literature on the topic which shows considerable diversity of opinion has been published. For these reasons laboratory work on fissures will not be considered further here.

In fact few results of determinations of matrix hydraulic conductivity and porosity have been reported. Results for hydraulic conductivity are remarkably constant considering the range of sample origin from 2000 m depth to the surface (Heard *et al.* 1979; Bradbury 1983). The results lie between 1×10^{-13} and 1×10^{-11} m s^{-1}. Porosity measurements are more common and range from the detailed consideration of the nature of the porosity (Katsube 1983) to relatively diverse data sets for assorted samples (Alexander *et al.* 1981). In both these papers the porosity of intact crystalline rocks is reported as being between 0.1% and 1%. In some special cases associated with alteration, porosities may rise higher than about 15%, but the consistently large values reported by Norton & Knapp (1977) are seen as being a product of the unusual measurement technique employed.

Large-scale flow systems

Large-scale flow systems in crystalline rock are almost impossible given the comparatively low value of bulk rock hydraulic conductivity. However, crystalline rocks are often quite extensive laterally and in depth so that large-scale systems are possible in the absence of more conductive rocks. Also, since they are not generally used for water supply, there have been no long-term

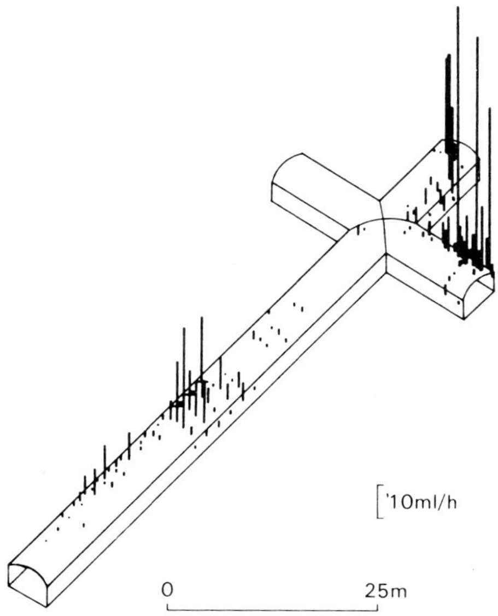

Water inflow rates into the sampling areas before drilling the injection holes

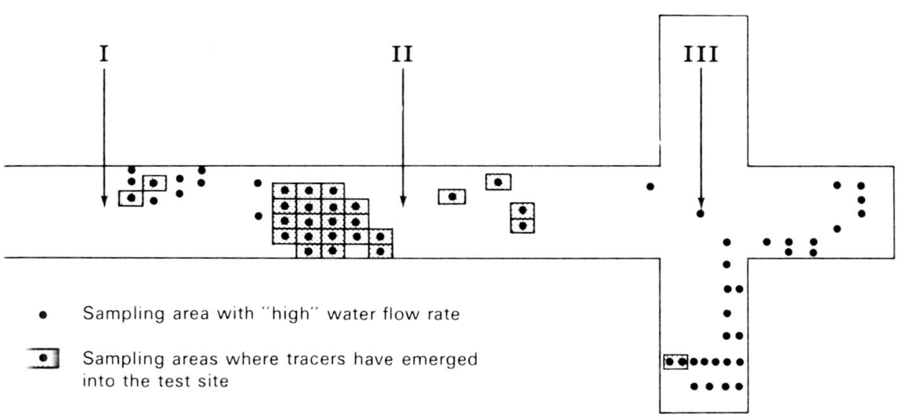

Tracer distribution (April -85)

FIG. 11. The 3D Migration Test of the Stripa Project: (a) water inflow rates into the sampling areas before drilling the injection holes; (b) tracer distribution (April 1985). (From Birgersson *et al.* 1985.)

observations during which it would be possible to monitor the evolution of a regional flow system. Instead, large-scale flow systems in crystalline rocks are largely computer model studies based on data synthesized from a wide area. These studies are of interest in that they are based on certain concepts of groundwater flow in crystalline rocks. Two such studies illustrate the similarities and differences.

The first study (Davison 1985) is being performed at the Underground Research Laboratory (URL) site on the edge of the shield in Eastern

Manitoba, Canada. At this site, which covers an area of about 25 km^2, over 120 boreholes have been drilled and many have been completed with 'West-Bay'-type monitoring casing. The boreholes are up to 1100 m deep and some groups have been aquifer tested. This has resulted in the deduced hydrogeology shown in cross-section and plan in Figs 12 and 13 respectively. The cross-section is gradually becoming more complex as more boreholes are drilled and tested but the original concept of sub-horizontal zones of dominating transmissivity can still be observed. Flows within the region are variable but pumping tests have been analysed on the basis of the rock matrix not yielding significant water during the time-scale of the pumping test. This assumption is not made in the regional modelling of the site. The modelling of the site is being 'validated' by excavating a shaft and comparing measured data with model results for such a perturbation of the system. This modelling suggests that the areal extent of the perturbation will not exceed 1 km.

The second study is one of seven investigations of various potential repository sites in Sweden (Carlsson *et al.* 1983). The area, which is known as Fjällveden and is just S of Stockholm, was first examined geologically and geophysically to identify the major structural features (Fig. 14). These are translated into a set of fracture zones for modelling purposes. They are assumed to be sub-vertical and have conductive properties defined by the single-borehole testing mentioned above. In a similar manner to the Canadian approach the fracture zones are modelled explicitly and finite-element meshes are increased in density in the fault or fracture zones. Regional flows over distances of several kilometres result from this approach. It is interesting to note in both approaches how fractures are accounted for by considering them as explicit features set within a rock mass with average properties. The major difference between the two approaches is the conviction in Canada that sub-horizontal features are dominant in large-scale flow systems. Similar features are apparently identified at all investigated sites in Canada. In Sweden, again in shield rocks, sub-vertical features are considered to be commonplace and dominant.

Discussion

The hydrogeology of crystalline rock is evolving rapidly at the moment. A few years ago test interpretations changed from being based on porous media to considering fissure flow. This was laminar flow between plane-parallel walls and it became fashionable to quote single-borehole test results determined from porous-medium analysis in terms of an effective single-fissure aperture. This situation rapidly evolved into plane-parallel fissures in porous rock and the complexity of possible analytical techniques outstripped the accuracy of the data. However,

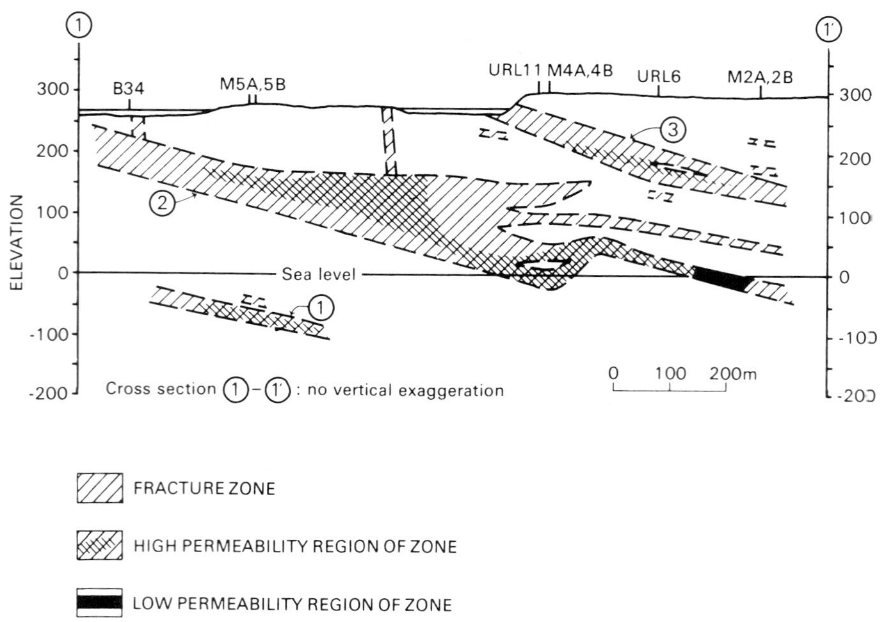

FIG. 12. Cross-section of the hydrogeology of the URL. (From Davison 1985.)

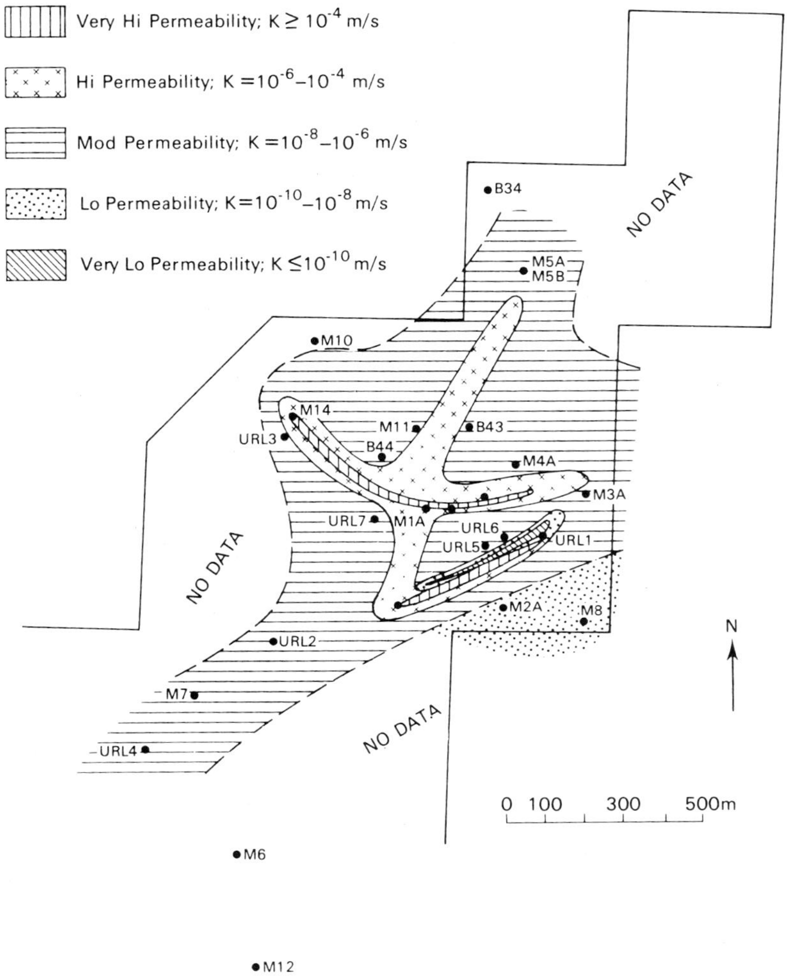

FIG. 13. Plan of the hydrogeology of the major fracture zone(s) at the URL. (From Davison 1985.)

the emergence of increasing numbers of specially designed experiments, especially those using tracers, began to redress the balance and justify some of the analytical complexities. It is perhaps salutory to bear in mind one of the results of the single-fracture experiment (Abelin *et al.* 1985): '. . . it is possible to obtain a good fit with all of the models . . . The fitting thus cannot differentiate between the mechanisms.'

It is interesting, however, to note the current trend of almost all research towards either channelling or networks. One group has even suggested networks of networks! There are some encouraging aspects such as the independence of chanelling from scale. The fracture at the URL on the scale of kilometres has strong similarities with the fracture of the single-fracture test.

Conclusions

Studies of the hydrogeology of crystalline rocks have value in forcing a re-examination of the applicability of many standard approaches to field testing. Many programmes around the world began as data-gathering exercises to delineate the range of various parameters found in crystalline rocks.

Single-borehole tests have yielded similar ranges of average hydraulic conductivity. This ranges from about 1×10^{-7} m s^{-1} in fracture or fault zones to 1×10^{-12} m s^{-1} in intact rock. The latter value is corroborated by the few laboratory measurements available. There is a general depth dependence in the hydraulic conductivity if sufficiently large data sets are considered. It

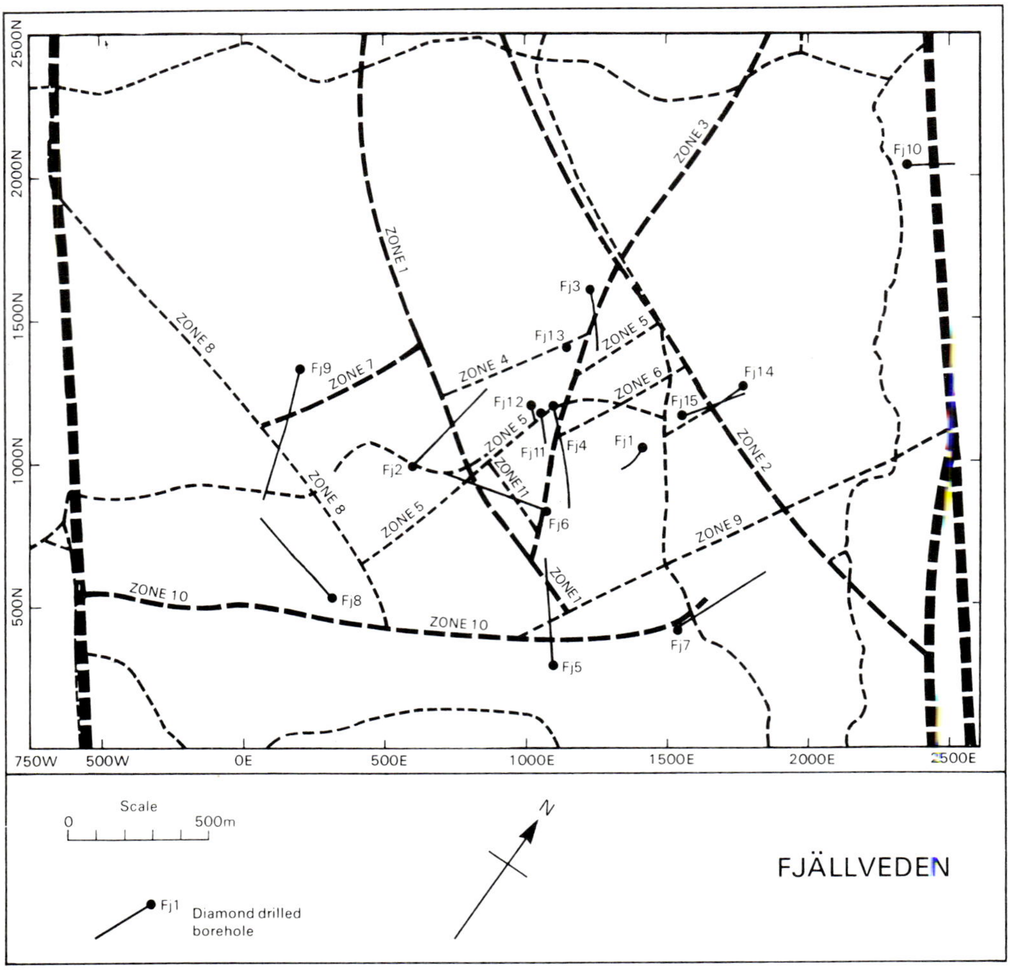

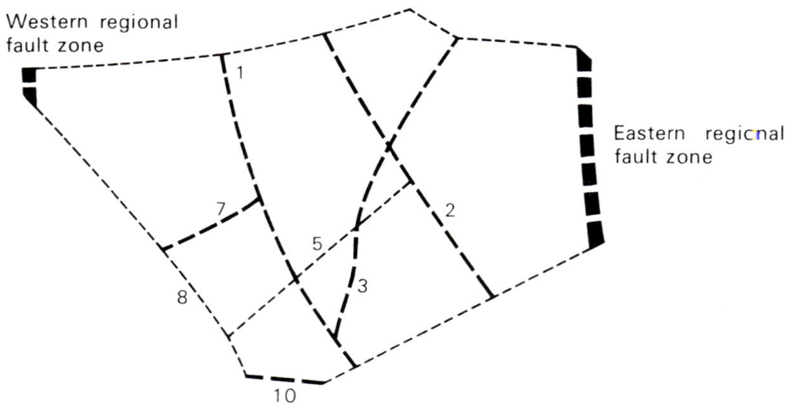

FIG. 14. Map of the fracture zones at the surface at Fjällveden, Sweden, and representation of the fractures as included in the regional groundwater modelling.

cannot be used to predict a value at a particular site. Heads determined in single-borehole tests vary depending on the orientation and connectivity of the more highly conductive fissures but reflect the general topographic position. There is limited evidence that they are corrupted by the presence of the measuring borehole.

Multiple-borehole tests have not been very successful, often as a result of poorly defined test geometry. This has been resolved by more carefully designed tests with precise objectives in mind. From another viewpoint testing has been improved by more technically advanced test equipment and by using geophysics to improve the definition of the geometry. Above all the complexity of possible analysis has been increased widely. Most detailed experimentation now points to the over-riding importance of channels and networks in understanding the flow mechanism in fissured crystalline rocks.

With this in mind it remains to be seen whether the interpretation procedure can be made simpler and therefore more usable or whether it will be necessary to refine testing procedures to suit each individual application. It is already apparent that a description, either quantitative or qualitative, depends on what it will be used for.

ACKNOWLEDGMENT: This paper is published with the permission of the Director of the British Geological Survey (Natural Environment Research Council).

References

ABELIN, H., NERETNIEKS, I., TUNBRANT, S. & MORENO, L. 1985. Final report of the migration in a single fracture—experimental results and evaluation. *Stripa Project Tech. Rep. 85-03*, SKB–KBS, Stockholm, 162 pp.

ALEXANDER, J., HALL, D. H. & STOREY, B. C. 1981. Porosity measurements of crystalline rocks by laboratory and geophysical methods. *Rep. ENPU 81- 0*, Institute of Geological Sciences, London.

BIRGERSSON, L., ABELIN, H., GIDLUND, X. & NERETNIEKS, I. 1985. Water flowrates and tracer transport to the 3D drift in Stripa. *Proc. OECD–NEA Information Symp. on in situ Experiments in Granite Associated with the Disposal of Radioactive Waste, Stockholm, June 4–6, 1985*, pp. 82–94, OECD–NEA, Paris.

BLACK, J. H. 1979. Results of a multiple borehole pumping test in low permeability granite. *Proc. Workshop on Low-flow, Low-permeability Measurements in Largely Impermeable Rocks, Paris, March 19–21, 1979*, pp. 183–95, OECD–NEA–IAEA, Paris.

—— & BARKER, J. A. 1983. Application of the sinusoidal pressure test to the measurement of hydraulic parameters. *Proc. Workshop on Geologic Disposal of Radioactive Waste—in situ Experiments in Granite, Stockholm, October 25–27, 1982*, pp. 121–30, OECD–NEA–KBS, Paris.

—— & HOLMES, D. C. 1985. Hydraulic testing within the Crosshole Investigation Programme at Stripa. *Proc. Workshop on Design and Instrumentation of in situ Experiments in Underground Laboratories for Radioactive Waste Disposal, Brussels, May 25–28, 1984*, pp 116–27, OECD–NEA–CEC, Brussels.

BRACE, W. F. 1980. Permeability of crystalline and argillaceous rocks. *Int. J. rock Mech. min. Sci. Geomech. Abstr.* **17**, 241–51.

—— 1984. Permeability of crystalline rocks: new *in-situ* measurements. *J. geophys. Res.* **89** (B6), 4327–30.

BRADBURY, M. H. 1983. Permeability measurements in crystalline rocks and sandstone. *Rep. AERE-R10836*, AERE, Harwell, 17 pp.

BRAESTER, C. & THUNVIK, R. 1984. Determination of formation permeability by double packer tests. *J. Hydrol.* **72**, 375–89.

CARLSSON, L. & OLSSON, T. 1985. Hydrogeological and hydrogeochemical investigations in boreholes—final report. *Stripa Project Tech. Rep. 85-10*, OECD–NEA–SKB, Stockholm.

——, WINBERG, A. & GRUNDFELT, B. 1983. Model calculations of the groundwater flow at Finnsjon, Fjallveden, Gidea and Kamelunge. *Tech. Rep. 83-45*, SKBF–KBS, Stockholm, 208 pp.

DAVISON, C. C., GRISAK, G. E. & WILLIAMS, D. W. 1979. Field permeability and hydraulic potential measurements in crystalline rock and solute transport through finely fractured media. *Proc. Workshop on Low-flow, Low-permeability Measurements in Largely Impermeable Rocks, Paris, March 19–21, 1979*, pp. 139–56, OECD–NEA–IAEA, Paris.

DAVISON, C. C. 1985. URL drawdown experiment and comparison with model predictions. *Proc. 20th Information Meet. of the Canadian Nuclear Fuel Waste Management Program*, Vol. 1, *AECL Tech. Rec. TR-375*, pp. 103–24.

GALE, J. E. & WITHERSPOON, P. A. 1979. An approach to the fracture hydrology at Stripa, preliminary results. *Rep. LBL-7079, SAC-15*, Lawrence Berkeley Laboratory.

——, ——, WILSON, C. R. & ROULEAU, A. 1983. Hydrogeological characterization of the Stripa site. *Proc. Workshop on Geologic Disposal of Radioactive Waste—in situ Experiments in Granite, Stockholm, October 25–27, 1982*, pp. 79–98, OECD–NEA–KBS, Paris.

GRISAK, G. E., PICKENS, J. F., BELANGER, D. W & AVIS, J. D. 1985a. Hydrogeologic testing of crystalline rocks during the NAGRA deep drilling program. *Tech. Ber. 85-08*, NAGRA, Baden, Switzerland, 194 pp.

——, —— & AVIS, J. D. 1985b. Principles of hydrogeologic investigation at depth in crystalline rock. *Mem. 17th Int. Cong. of the International Association of Hydrogeologists: Hydrogeology of Rocks of Low Permeability, Tucson, AZ, January 7–12, 1985*, pp. 52–71.

HEARD, H. C., TRIMMER, D., BONNER, B. & DUBA, A. 1979. Permeability of generic repository rocks at simulated *in situ* conditions. *Proc. Workshop on Low-flow, Low-permeability Measurements in Largely Impermeable Rocks, Paris, March 19–21, 1979*, pp. 69–84, OECD–NEA–IAEA, Paris.

HEATH, M. J. 1985. Geological control of fracture permeability in the Carnmenellis granite, Cornwall: implications for radionuclide migration. *Mineral. Mag.* **49**, 233–44.

HODGKINSON, D. P. 1984. Analysis of steady-state hydraulic tests in fractured rock. *Rep. AERE-R11287*, AERE, Harwell, 36 pp.

HOLMES, D. C. 1981. Hydraulic testing of deep boreholes at Altnabreac: development of testing system and initial results. *Rep. ENPU 81-4*, Institute of Geological Sciences, London.

HSIEH, P. A., NEUMAN, S. P. & SIMPSON, E. S. 1983. Pressure testing of fractured rocks—a methodology employing three-dimensional cross-hole tests. *Rep. NUREG/CR-3213*, U.S. Nuclear Regulatory Commission.

KATSUBE, T. J. 1983. Pore structure parameters of igneous crystalline rocks—their significance for potential radionuclide migration and storage. *Proc. Int. Conf. on Radioactive Waste Management, September 12–15, 1982*, pp. 118–23, Canadian Nuclear Society, Toronto.

LONG, J. C. S. & WITHERSPOON, P. A. 1985. The relationship of the degree of interconnection to permeability in fracture networks. *J. Geophys. Res.* **90** (B4), 3087–98.

LOUIS, C. 1974. *Introduction à l'Hydraulique des Roches.* Bureau de Recherches Géologiques et Minières, Orleans.

DE MARSILY, G. 1985. Flow and transport in fractured rocks: connectivity and scale effect. *Mem. 17th Int. Cong. of the International Association of Hydrogeologists, Tucson, AZ, January 7–12, 1985*, pp. 267–77.

NORTON, D. & KNAPP, R. 1977. Transport phenomena in hydrothermal systems: the nature of porosity. *Am. J. Sci.*, 913–36.

ROBINSON, P. C. 1983. Connectivity of fracture systems—a percolation theory approach. *J. Phys. A*, **16**, 605–14.

J. H. BLACK, Fluid Processes Research Group, British Geological Survey, Keyworth, Nottingham, U.K.

The flow of water and displacement of hydrocarbons in fractured chalk reservoirs

David A. Brown

SUMMARY: The fractured chalk reservoirs of the southern part of the Central Graben do not exhibit the flow characteristics more usually encountered in continuous sandstone formations. The Danian and Maastrichtian age formations of the Ekofisk Field are typical of these fractured chalk reservoirs. The Ekofisk Field has been extensively studied to understand better the geology and controlling fluid flow mechanisms prior to commencing enhanced oil recovery projects. During waterflooding the imbibition of water from the fracture network into the rock matrix will control the displacement of hydrocarbons. Extensive laboratory studies have been carried out to quantify the imbibition characteristics of the Danian and Maastrichtian formations. Several parameters that control the amount of water imbibed have been found. A field water-injection pilot project has been implemented to verify the mechanisms and laboratory results and to prove numerical techniques suitable for the prediction of recoveries under a full-field waterflood.

The Ekofisk area fields are located in the southern sector of the Central Graben (Fig. 1). The reservoir horizons consist of chalk formations of Danian and Maastrichtian age. Most of the commercial development in this area to date has been in the Norwegian sector; however, potential commercial fields are also present in the U.K. sector. Commercial accumulations have also been found in the Danish sector. The quality and type of hydrocarbons found in these reservoirs range from gas condensates to oils with low gas-to-oil ratios. To date most production has been by primary depletion; however, secondary and tertiary recovery techniques are now being evaluated.

The Ekofisk Field is typical of these fractured chalk reservoirs and extensive studies have been carried out to examine the potential of water-injection techniques. The field and the results of some of the studies carried out are reviewed in this paper. A brief summary of the mechanisms of production from fracture reservoirs will also be presented.

Ekofisk Field

The Ekofisk Field (Fig. 2) located in Block 2/4 of the Norwegian continental shelf is the largest of the fields in the Ekofisk area and contains more than 40% of the hydrocarbon reserves of the Central Graben chalk reservoirs. The original exploration well was drilled in 1969. Three further wells were drilled before development commenced. Further details of the appraisal and development of the field are given in papers by Byrd (1975) and Van den Bark & Thomas (1981). A detailed description of the present geological and geophysical interpretation of the Ekofisk Field has been given by Brewster *et al.* (1986).

In the Ekofisk Field the Ekofisk (Danian) Formation, which has a maximum thickness of 600 ft, is separated from the Tor (Maastrichtian) Formation, which is up to 400 ft thick, by a tight zone of low porosity which is pressure sealing over most of the areal extent of the reservoir. The thickness of this tight zone varies from 50 to 120 ft. The tight zone may be faulted out in limited areas.

The porosity distribution in each formation varies both areally and vertically, with the best porosities being found towards the centre of the field. The average porosity in the Ekofisk Formation is 32%. However, porosities of up to 48% have been encountered. The average porosity in the Tor Formation is 28%, and the maximum porosity encountered in this zone is 41%. Porosity is lost towards the flanks, and once the water zone is encountered falls to about 15%. This loss towards the flanks may be due to late diogenesis and is discussed in more detail by D'Heur (1984). Distinct porosity layers can be identified in each formation, and indeed these same layers can be correlated with all the fields in the Ekofisk area. These layers and their respective stratigraphic nomenclature are identified in Figs 3 and 4.

The water saturation in the central portion of the field can be as low as 5%–7%; however, the average saturation is 12% in the Ekofisk Formation and 10% in the Tor Formation. The base of net hydrocarbons is not flat but is higher towards the flanks. This is due to the decreasing porosity trends towards the reservoir flanks. With the possible exception of a localized area of the W flank the underlying water is not movable and

From GOFF J. C. & WILLIAMS, B. P. J. (eds), 1987, *Fluid Flow in Sedimentary Basins and Aquifers*, Geological Society Special Publication No. 34, pp. 201–218.

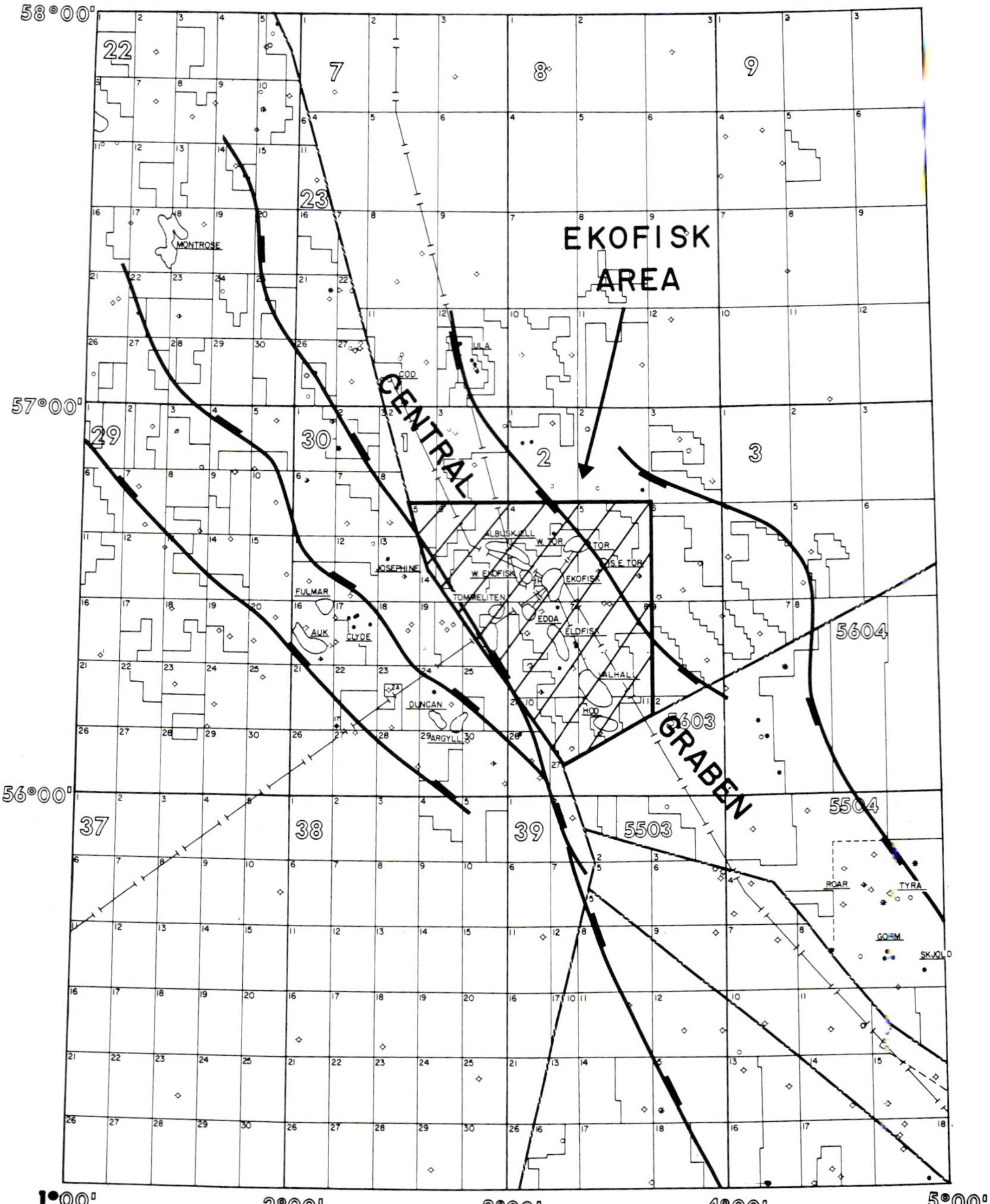

FIG. 1. Tectonic elements map—S central North Sea.

the aquifer does not contribute greatly to the hydrocarbon recovery.

The formation layering, the effect of the tight zone and the non-active aquifer highlighted above are illustrated in the vertical pressure distributions obtained from recent RFT (Repeat Formation Test) logs (Fig. 5). Distinct pressure regimes can be identified in the Ekofisk and Tor Formations. Localized differential depletion can also be seen. The rapid increase in pressure in the

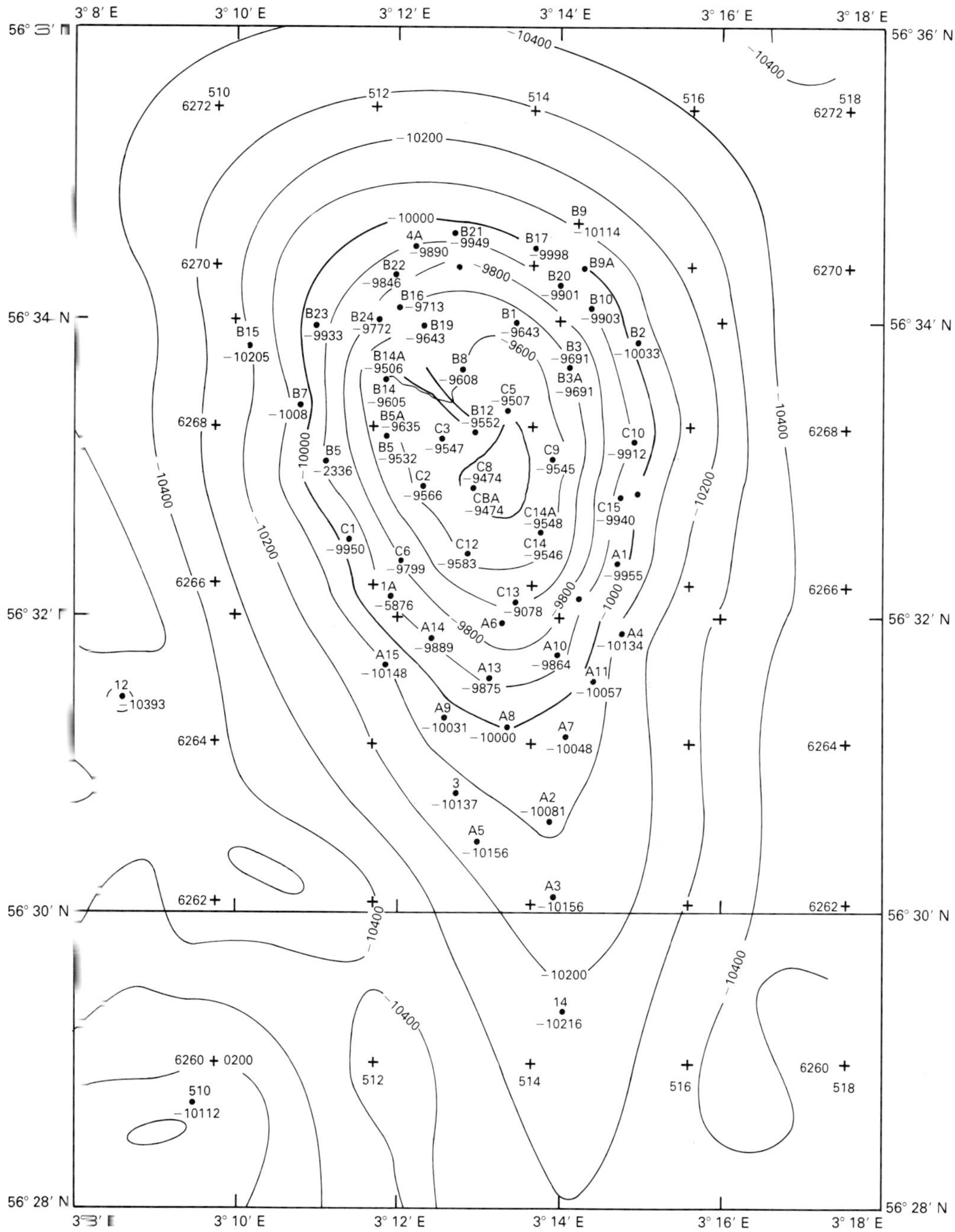

FIG. 2. Map of the top structure of the Ekofisk Formation (feet below sea level).

lower section of the reservoir is indicative of the tight underlying formations, and indeed this trend continues with depth until the original pressure gradient is achieved.

In the Ekofisk Field there is a good correlation between porosity and matrix permeability for each horizon. Figure 6 shows results of core analysis on the Upper Ekofisk, Lower Ekofisk and Tor Formations. The correlations for the Lower Ekofisk and the Tor are similar, as shown

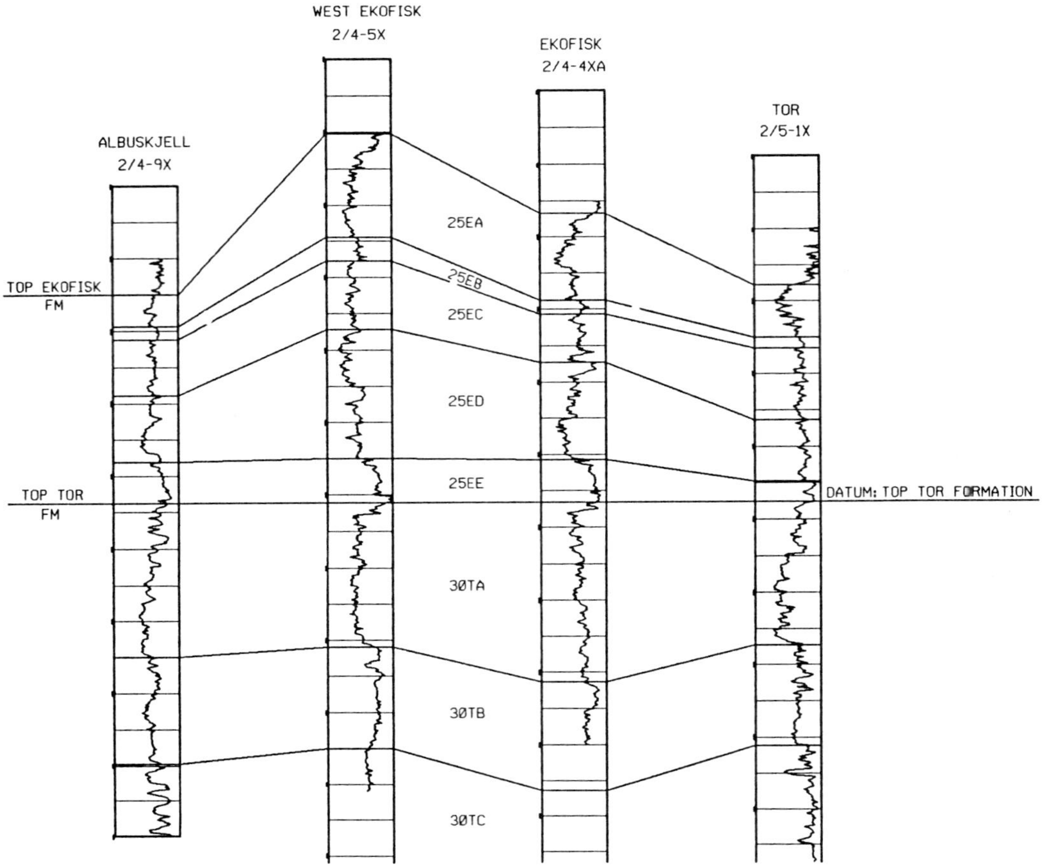

FIG. 3. Chalk zone porosity correlation in the Ekofisk area.

STAGE	GROUP	FORMATION	PHILLIPS GROUP ZONATION	
PALEOCENE	MONTROSE	MAUREEN EQ.	25 MA	MAUREEN FORMATION EQUIVALENT
PALEOCENE	CHALK	EKOFISK	25 EA	UPPER POROUS ZONE
			25 EB	TOMMELITEN TIGHT ZONE
			25 EC	REWORKED DANIAN ZONE
			25 ED	REWORKED MAASTRICHTIAN ZONE
			25 EE	EKOFISK TIGHT ZONE
MAASTRICHTIAN TO TURONIAN		TOR	30 TA	UPPER
			30 TB	MIDDLE
			30 TC	LOWER
		HOD	31HA-32HD	UPPER
			33HE	MIDDLE
			34HF	LOWER
		PLENUS MARL	34 PM	
CENOMANIAN		HIDRA	35 HA	

IN THE PHILLIPS GROUP CODE, NUMERALS ARE CHRONOSTRATIGRAPHIC BUT WITH NO ABSOLUTE AGE INFERRED AND LETTERS ARE LITHOSTRATIGRAPHIC

E.G. 25EA 25 INDICATES LOWER PALEOCENE
E INDICATES EKOFISK FORMATION
A INDICATES UPPER POROUS ZONE

FIG. 4. Stratigraphic nomenclature of the chalk group in the Greater Ekofisk area.

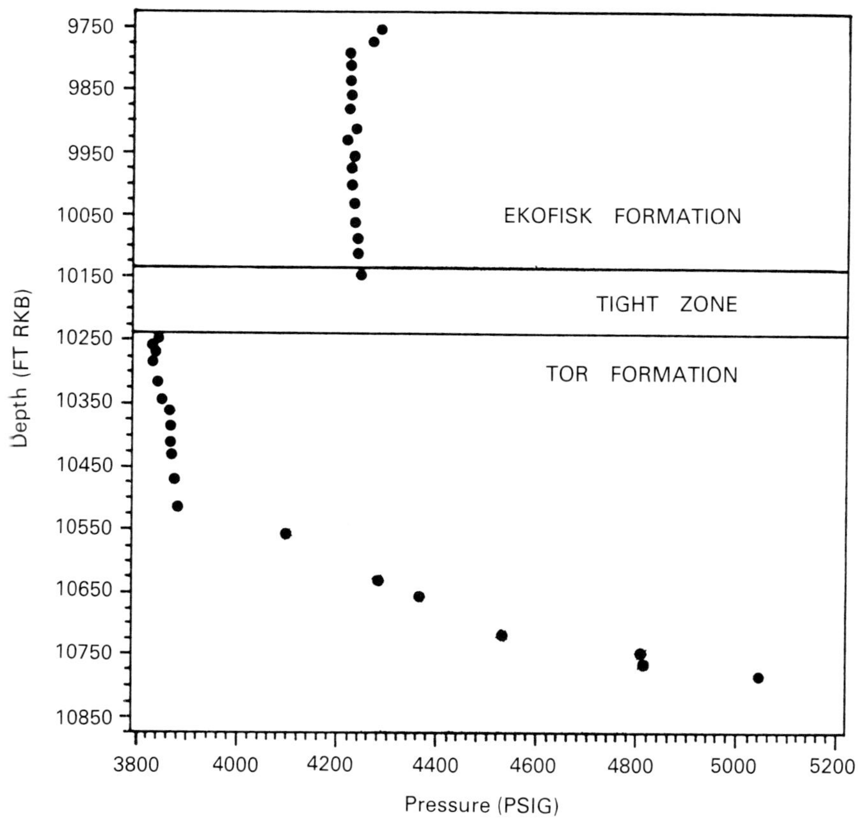

FIG. 5. Ekofisk RFT pressures logged on October 17, 1984.

in Fig. 7. However, the trend for the Upper Ekofisk is distinctly separate. From palaeontological analyses on core samples it has been found that the Lower Ekofisk Formation contains a considerable amount of reworked material of Maastrichtian age (Skovbro *et al.* 1982). This may in part account for some of this similarity.

From the porosity–matrix permeability correlations the maximum permeability expected in the Ekofisk Field is 8 mD. However, well-test results indicate effective permeabilities of up to 150 mD. This enhancement in permeability is due to fracturing of the formation. Three types of fractures, healed, tectonic and stylolite associated, have been identified from cores.

The healed fractures are filled with a base material similar to the chalk and do not provide enhancement of the permeability (Fig. 8).

The tectonic fractures are those most likely to enhance the effective permeability. It has been found from core studies that these fractures are predominantly sub-vertical (60°–75° dip). The intensity of fracturing varies both vertically and areally; fracture spacings as small as 10–15 cm have been observed (Fig. 9).

Stylolite-associated fractures form adjacent to stylolites. The usual length of such fractures is only 5 cm. This type of fracture does not have a great influence on the effective permeability (Fig. 10).

As outlined above fracturing can be detected from cores. It is also possible to detect some fractures from well logs, but this is not a reliable technique. Zones highlighted from logs as being possibly fractured have shown no fracturing in cores recovered over the same interval. The primary method for the detection of fracturing is from well testing.

It is possible to derive a map of matrix permeability from porosity maps and porosity–matrix permeability correlations. If this map is compared with a map of measured permeabilities from well tests (Fig. 11) a map of fracture intensity can be obtained (Fig. 12).

At the original reservoir conditions of 7200 lb in^{-2} absolute (psia) the Ekofisk field

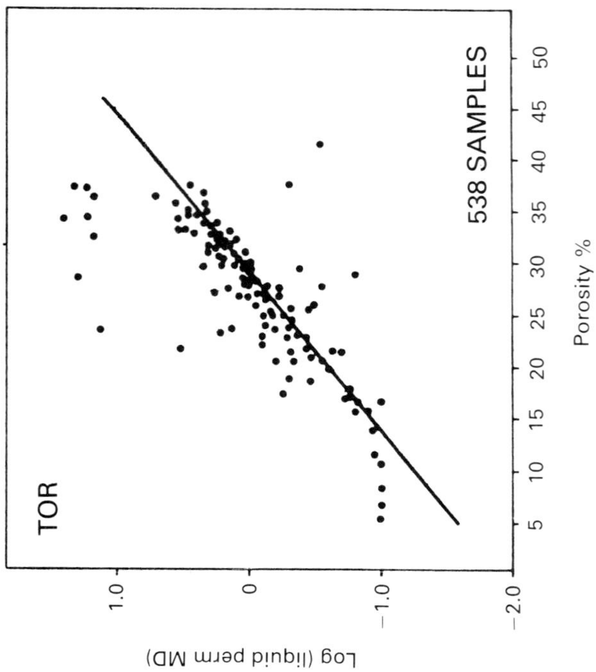

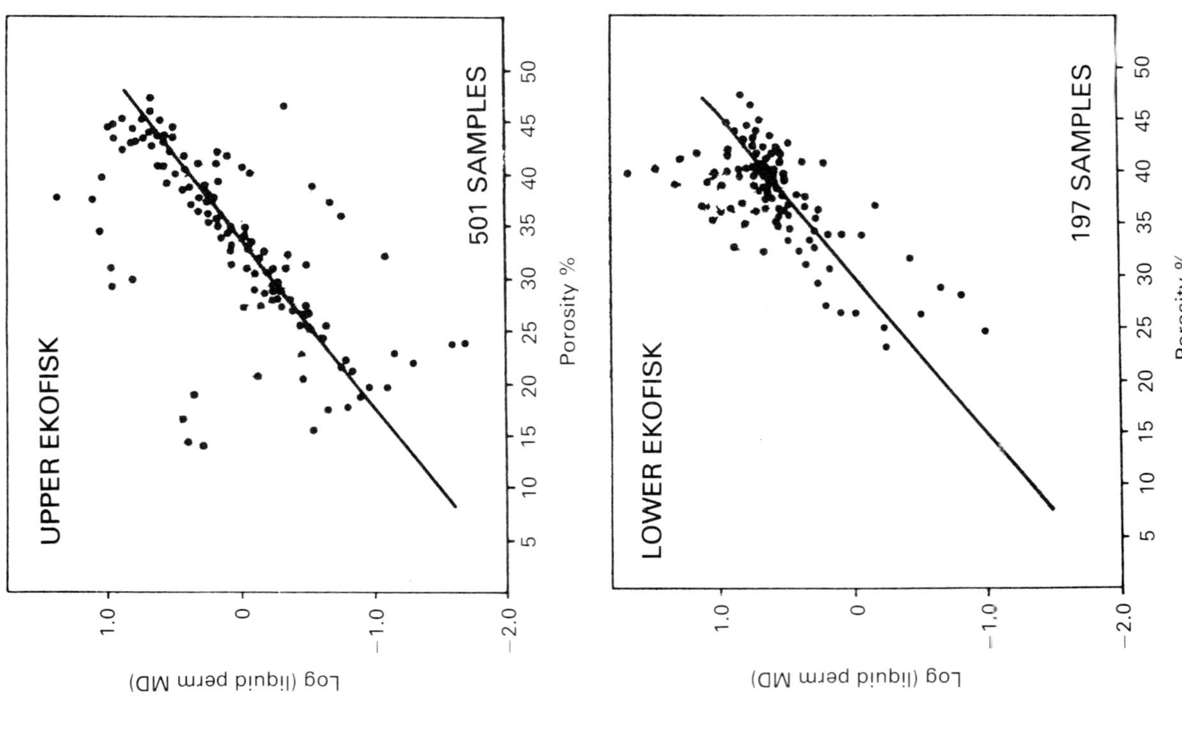

FIG. 6. Porosity versus permeability for the Ekofisk Field core samples.

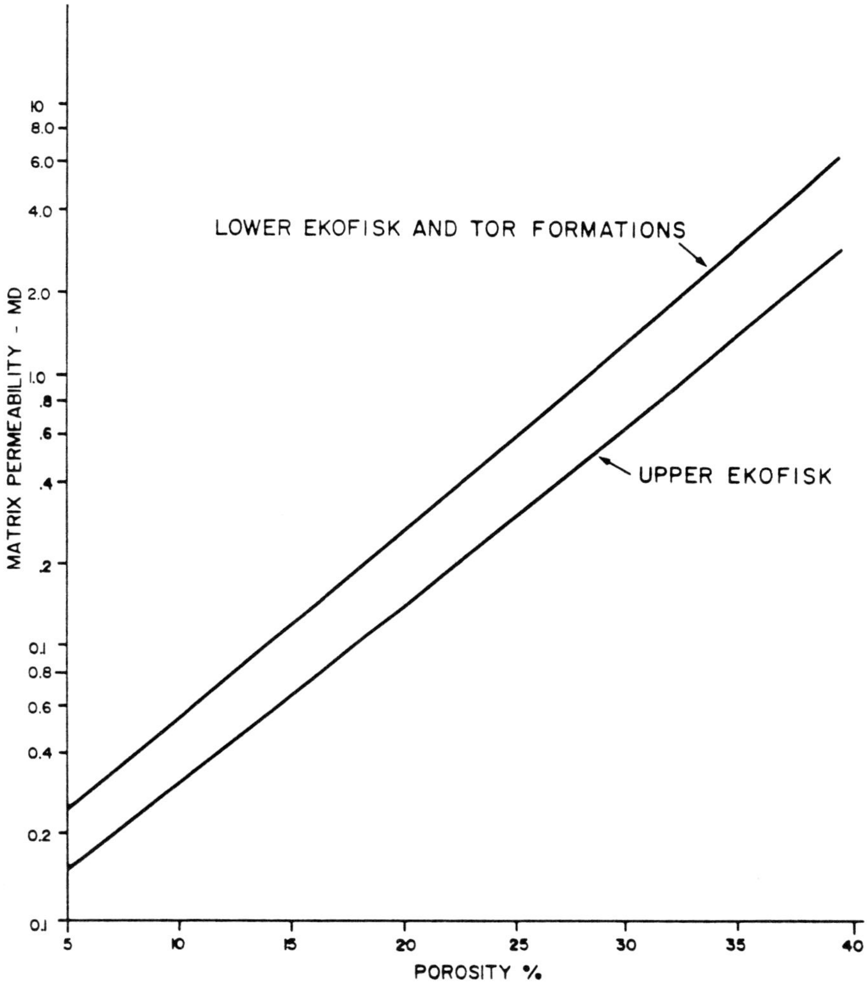

FIG. 7 Matrix permeability versus porosity for the Ekofisk Field.

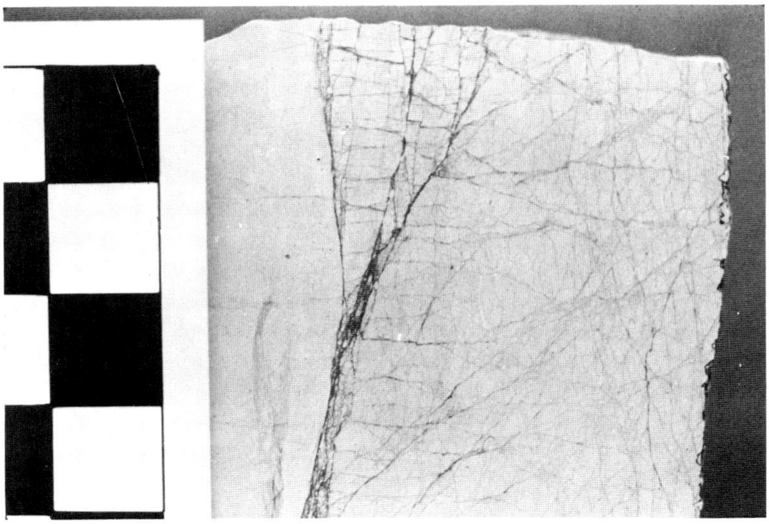

FIG. 8. Example of healed fracture network (scale bar in inches).

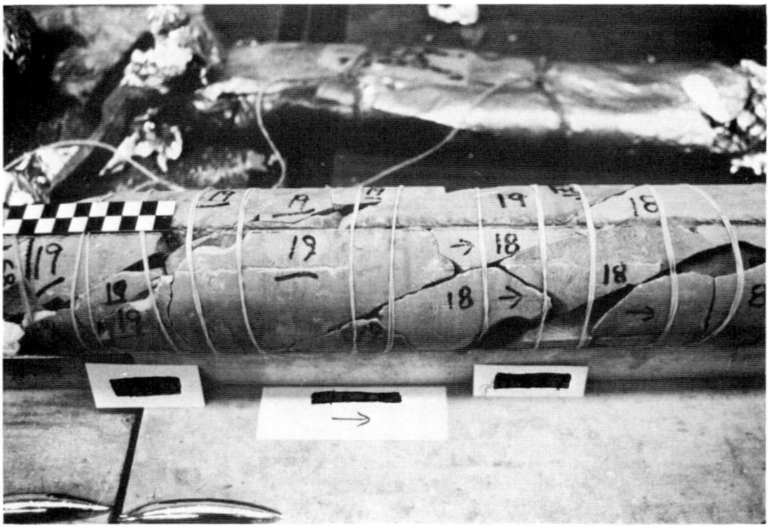

FIG. 9. Example of tectonic fractures (dips 60°–70°, scale bar in inches).

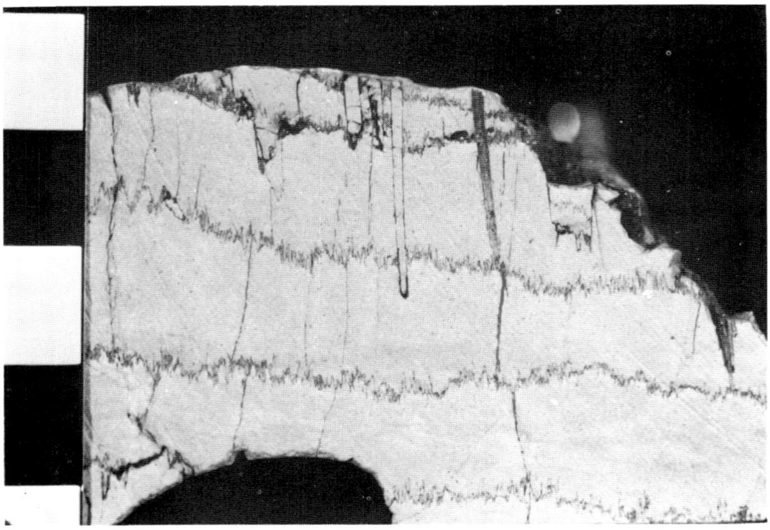

FIG. 10. Example of stylolite-associated fractures (scale bar in inches).

produced a 36° API oil with a gas-to-oil ratio of 1700 SCF bbl^{-1}. The original oil in place is estimated as 6686×10^6 bbl. Production to January 1, 1985, was 737×10^6 STB of oil and 2582×10^9 SCF of gas. Gas in excess of contract sales volumes is re-injected into the crestal area of the Upper Ekofisk Formation.

Mechanisms of recovery in fractured chalks

Fluid flow under natural depletion

Several models have been proposed to describe the flow of fluids in fractured reservoirs. The

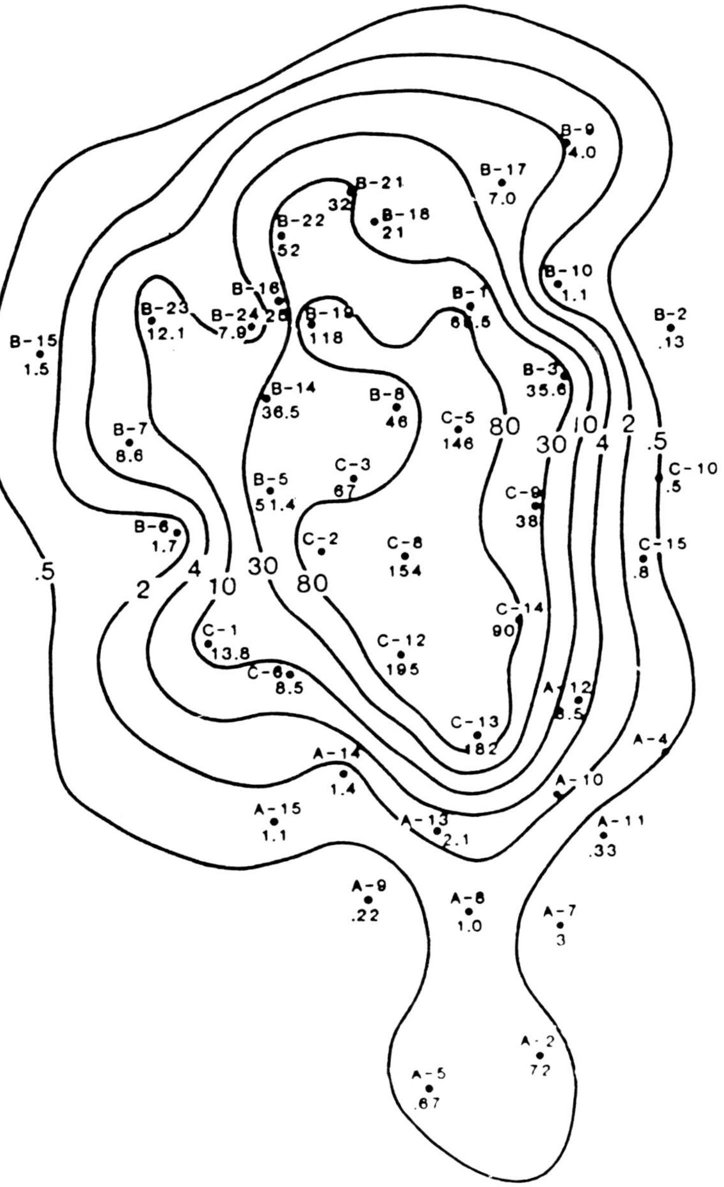

FIG. [illegible]. Isopermeability map for the Tor Formation in the Ekofisk Field (in millidarcies).

common factor of these models is that they all describe basically a system of two components. The first component, the fracture network, may or may not be continuous in any plane. The second component is the matrix rock, which also may or may not be continuous in any plane. Typically, the matrix rock contains all the pore volume of the fracture–matrix system and the fractures have a small or negligible pore volume (Fig. 13). The fractures have a high permeability value and the matrix has a low permeability value.

Under primary depletion conditions only the fracture network produces initially; however, the volume of the fracture network is quickly depleted. At this point the matrix starts to feed fluid to the fracture system. A transition period then occurs where the production rate is controlled by the conductivity and volume of the fractures and also by the rate of fill-up from the matrix. Finally,

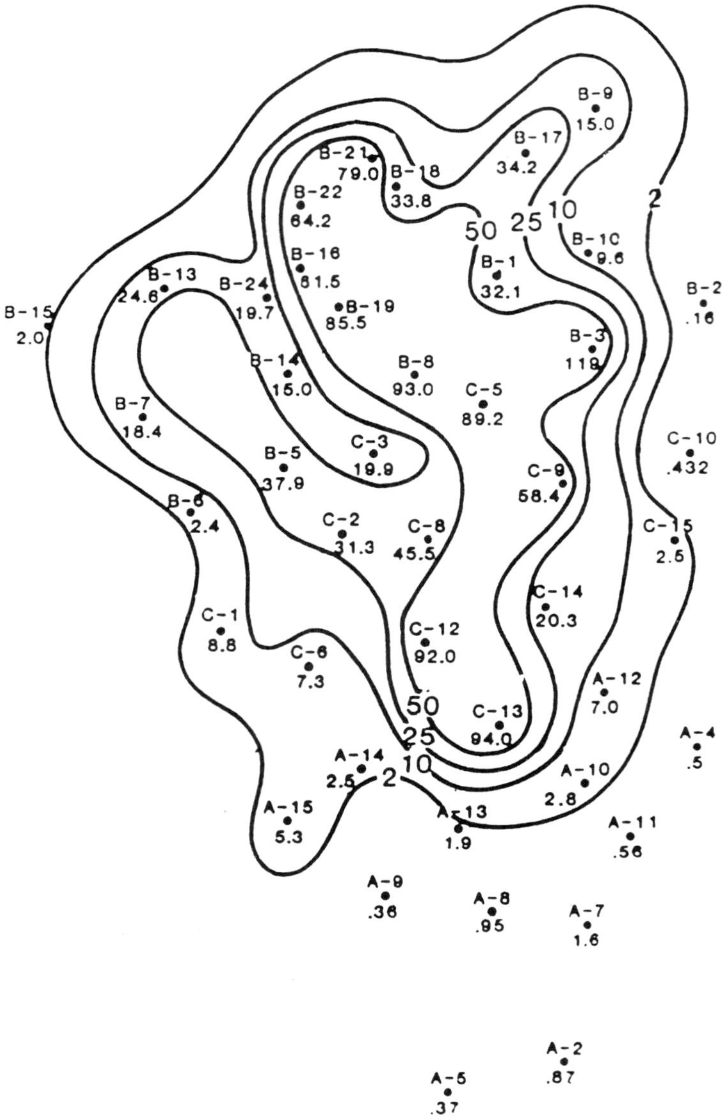

FIG. 12. Fracture intensity map for the Tor Formation in the Ekofisk Field.

a stage is reached where the production rate is controlled almost totally by the matrix system. The fracture flow is at steady state and only the fluid influx from the matrix is produced. The fracture effect at this stage is only to enhance the net flow capabilities of the chalk matrix.

Once this third phase is reached, the driving force for the production of hydrocarbons must come from the matrix rock itself and the fluids contained therein. The expansion of the pore fluids with the relief of pressure is the primary mechanism for expelling the hydrocarbons from the matrix into the fracture system. Above the bubble-point pressure only oil and water will expand. Below the bubble-point pressure the existence of a dispersed free-gas phase will provide further energy to the system. With the lowering of pressure, still more gas will be evolved from the oil to provide a further drive force. Eventually, however, a point is reached where the evolved gas begins to form a continuous phase through the matrix and, instead of expanding to expel oil into the fracture network, it will be produced itself. This results in trapping of the oil

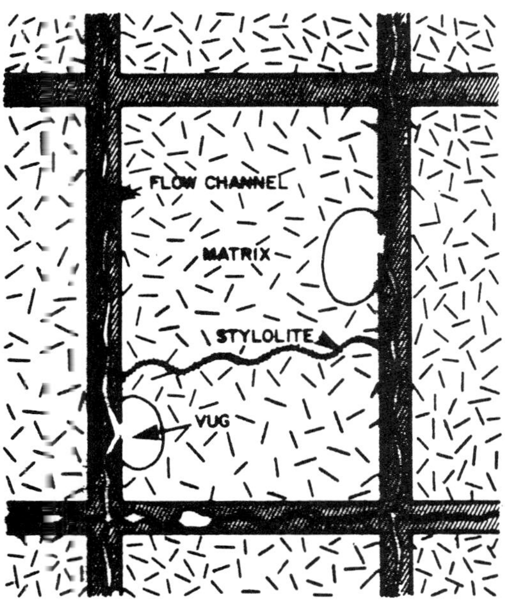

FIG. 13. An element of a fractured reservoir. (After Reiss 1980.)

in the matrix due to gas blocking. Whatever drive force is present will then follow the path of least resistance through the continuous gas phase rather than through the oil.

Some allowance has to be made for the pore-volume compressibility of the rock system with the relief of pore pressure. Dependent upon the matrix properties and the intensity of fracturing, capillary drainage may contribute to the expulsion of hydrocarbons from the matrix to the fracture system.

In homogeneous formations such as typical sandstone reservoirs the absence of fractures limits the recovery mechanisms to those described above for the matrix rock alone. There is no fracture network to enhance the permeability, and flow rates are controlled solely by the matrix rock properties. Also, there is a greater volume of continuous rock so that blockage of pockets of oil by gas will not be as significant to the recovery.

Fluid flow with water movement (injection)

Wherever a potential (pressure) gradient occurs in a reservoir, fluid will move to equalize the potentials. The fluid will flow through any permeable material placed in its path; however, it will move preferentially along the path of least resistance.

In non-fractured systems fluid will flow through the matrix. In a fractured system very little fluid will flow through the matrix; rather, it will move preferentially through the fractures. Any pressure drop established over a matrix block will be controlled by the abilities of the surrounding fracture network to pass fluid.

In a fractured reservoir it is necessary to establish a potential gradient in order for the water to migrate into the matrix from the fracture to displace hydrocarbons. If the matrix blocks are not large the controlling mechanism for the counter-current movement of water from the fractures into the block and of oil from the block into the fracture system is capillary imbibition (Fig. 14).

Classical capillary pressure concepts define imbibition as a flow resulting in an increase in wetting-phase saturation (Craig 1971). A decrease in wetting-phase saturation is referred to as drainage. The processes of drainage and imbibition are not ideally reversible, and the difference is referred to as hysteresis.

Examples of capillary pressure curves are shown in Fig. 15 for water-wet and oil-wet rock samples. For each case, only imbibition curves are presented. The limit of spontaneous imbibition, which is the point at which the water will no longer imbibe into the rock without the application of an external force, either viscous or gravitational, is of particular significance. The lower portion of the oil-wet imbibition curve has to be forced to generate the pressure gradients required.

Ekofisk waterflood studies

Laboratory investigations

The laboratory study of imbibition in its simplest form consists of immersing an oil-saturated core sample (at irreducible water saturation) in a water-bearing container, as illustrated in Fig. 16. The water imbibes into the sample, displacing oil. The displaced oil rises owing to gravity and is collected in a graduated tube (Torsaeter 1984). From this experiment the rate of imbibition with time can be obtained. Under these conditions the displacement mechanism is capillary imbibition only. To study the combined effect of imbibition and gravity forces the apparatus can be turned on its side and spun in a centrifuge.

Imbibition experiments have been run at both room and elevated temperatures. It was found that increasing the temperature of the experiment increased the amount of water imbibed. The final results of the experiments were independent of the temperature path taken (Fig. 17).

The imbibition characteristics of the Ekofisk

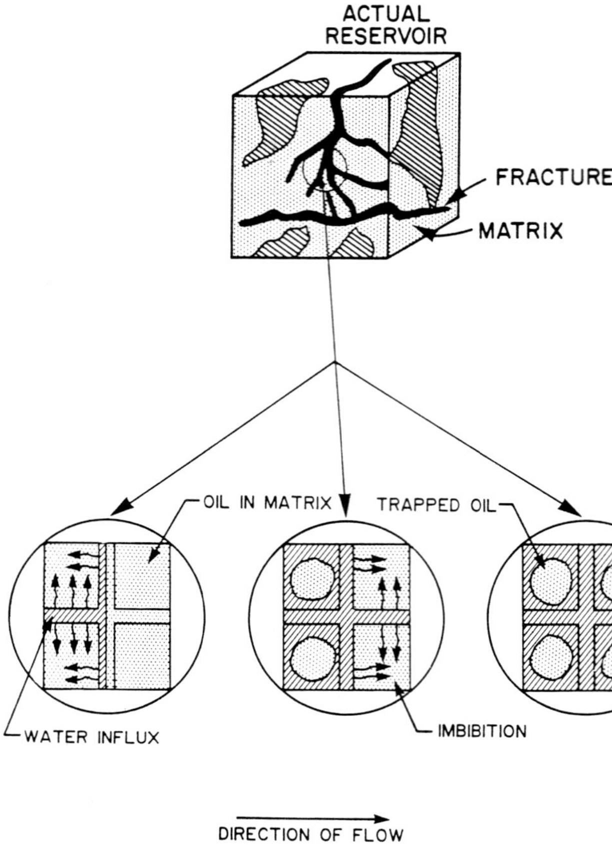

FIG. 14. Water flooding in a fractured reservoir.

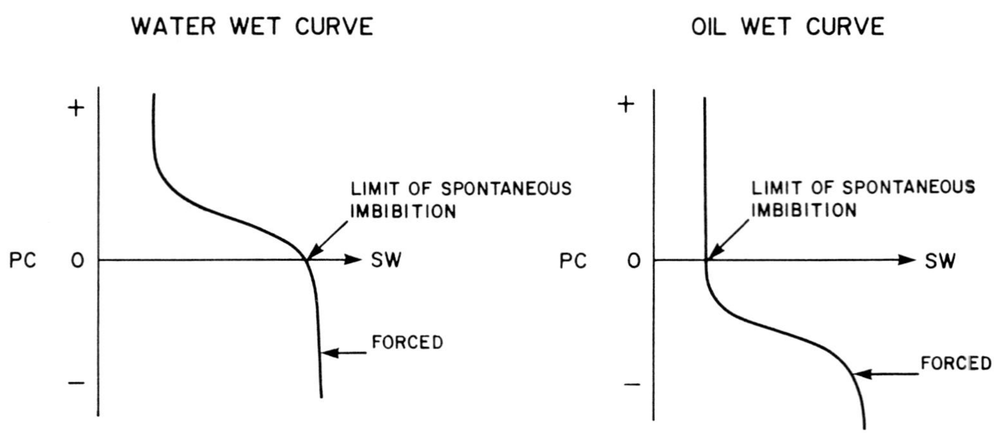

FIG. 15. Wettability as a function of the shape of the imbibition capillary pressure curve. PC = capillary pressure or oil phase pressure minus water phase pressure.

Field chalks have been examined in detail. Distinct differences were observed in the results obtained from Ekofisk Formation and Tor Formation core samples.

A summary of some of the results from the imbibition studies carried out is presented in Fig. 18. The data are correlated against sample porosity. For all samples good trends can be identified for irreducible water saturation. However, only the Tor Formation samples exhibit

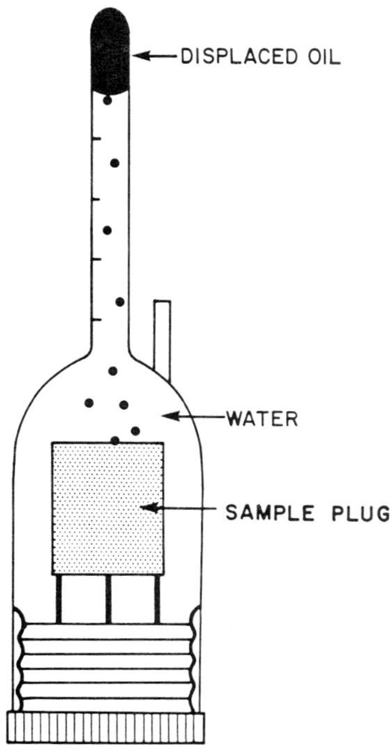

FIG. 6. Class 3D imbibition cell.

consistent trends for the amount of water imbibed and the final water saturation achieved during the experiment. The Lower and Upper Ekofisk data are presented with the Tor Formation correlations as a reference. Both show considerable scatter with only a limited amount of Lower Ekofisk samples exhibiting reasonable imbibition values.

Extensive studies have been carried out to investigate the cause of the differences in wettability and imbibition characteristics of the Tor and Ekofisk Formation samples. It was found that after cleaning using strong solvents, such as tetrahydrofuran (THF), samples from all sources exhibited high imbibition properties (Baldwin 1985).

Detailed X-ray photoelectron spectroscopy analysis of the surfaces of the chalk material revealed an organosilicate film which was present on all the samples examined irrespective of origin. This film is 20–30 nm thick. Examining the chemical nature of the materials extracted during cleaning resulted in the identification of an oxidation level in the surface film that correlated with the wettability trends of the samples. It is felt that the relative polarity of these oxidized organic films controls the hydrophillic character of the rock surface and hence the imbibition properties. The source of this

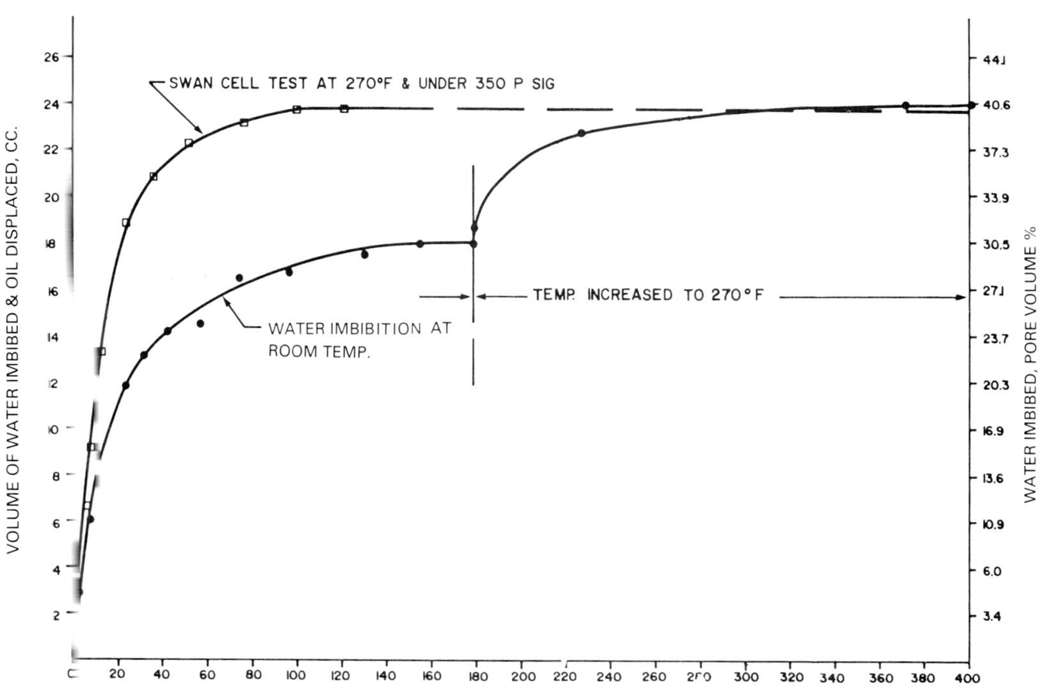

FIG. 17. Imbibition testing of chalks.

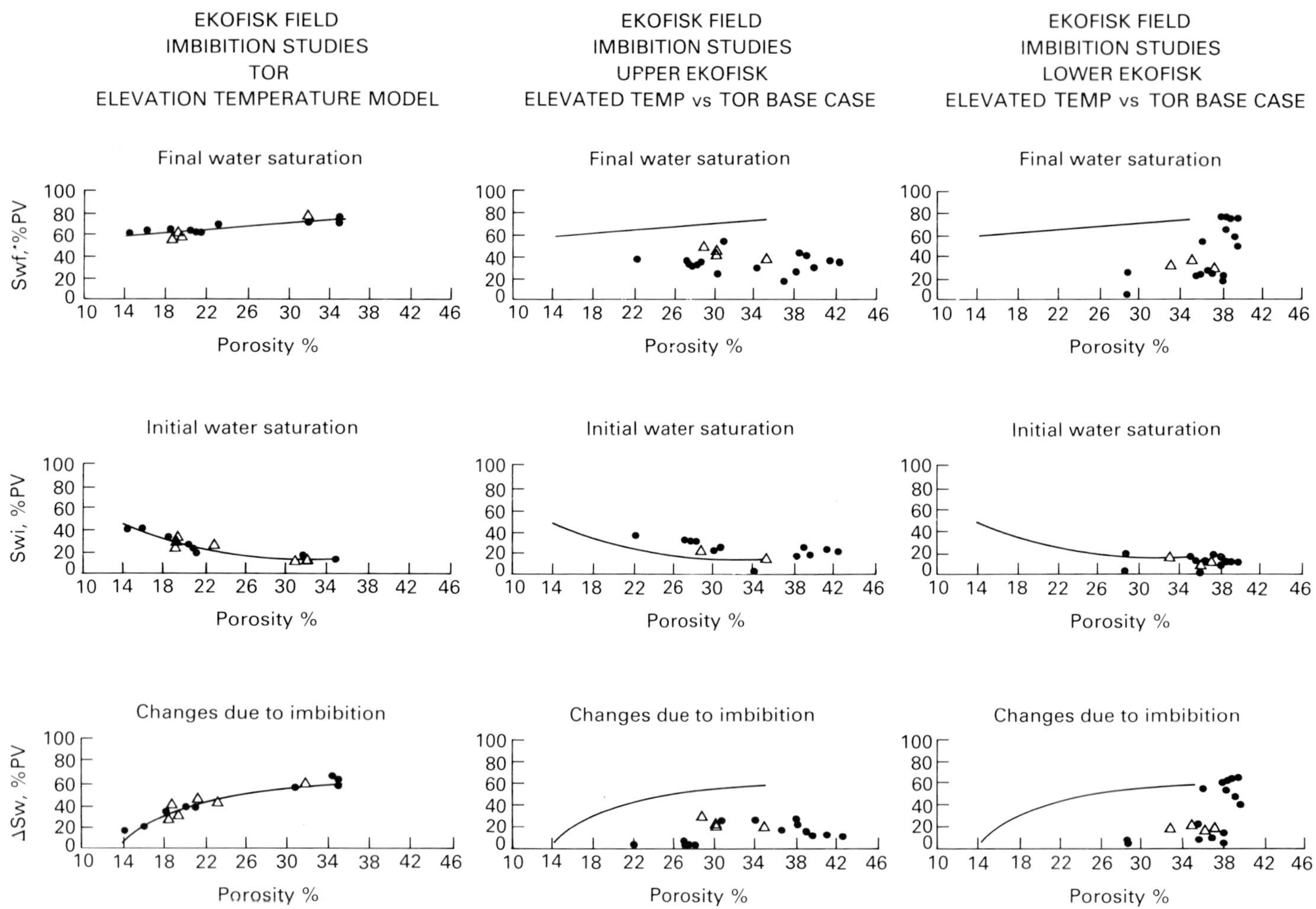

FIG. 18. Comparison of imbibition test results for the Tor, Lower Ekofisk and Upper Ekofisk Formations.

oxidized material is still being investigated, but it is believed that it may have migrated into the reservoir with the oil.

Scanning electron microscope analysis of the Ekofisk Formation samples enabled a secondary coating to be identified on the surface of the chalk material (Fig. 19). The coating is approximately 0.5 μm thick. The material of this coating is possibly a hydrocarbon residue and it can be removed with solvents such as toluene or THF. This coating is not observed on Tor Formation samples.

Pilot water-injection project

To provide information on the imbibition characteristics of the formation *in situ* rather than in the laboratory a single-injection-well pilot project was implemented. Water was injected into the Tor Formation of the Ekofisk Field in well 2/4 B-16 (Fig. 20). Water injection commenced in April 1981 and continued at rates of up to 32 000 BWPD for a period of 3.5 years. All three of the primary observation wells (2/4 B-19, 2/4 B-22 and 2/4 B-24) showed a response to the water injection by a decrease in the producing gas-to-oil ratio. Two of the wells, 2/4 B-22 and 2/4 B-24, produced appreciable volumes of water. Only one well, 2/4 B-22, showed an increase in reservoir pressure; however, the other two wells showed a decrease in the rate of pressure decline.

Figure 21 shows the water-cut of observation well 2/4 B-22. As can be seen water breakthrough did not occur until mid-1982, more than a year from the start of injection. The water-cut increased rapidly to 60% and then levelled off. When the injection into 2/4 B-16 was temporarily halted an immediate fall in produced water levels was observed at 2/4 B-22. The response of this well is typical of that of an imbibition-controlled environment. The results of numerical modelling of the pilot project utilizing the results of the laboratory work are also shown in Fig. 21. A good match was obtained. The simulator used was a dual-porosity three-phase black-oil model. A more detailed discussion of the Ekofisk pilot test has been given by Thomas *et al.* (1984).

Several conclusions can be drawn from the results of this pilot test, the most important of which are that the laboratory measurements of imbibition were confirmed and that water could be injected into the reservoir at commercial rates without rapid breakthrough to producers.

Full-field water-injection proposal

Given the imbibition characteristics summarized above and the success of the pilot injection

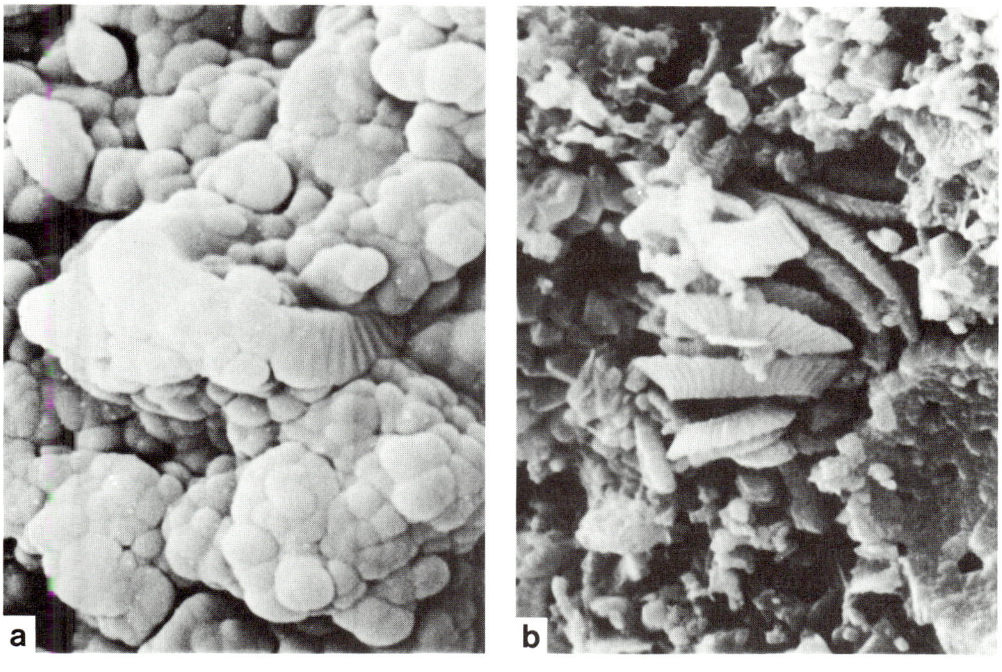

FIG. 19. Scanning electron micrograph showing the surface film on Ekofisk Formation samples: (a) as-received sample; (b) cleaned sample.

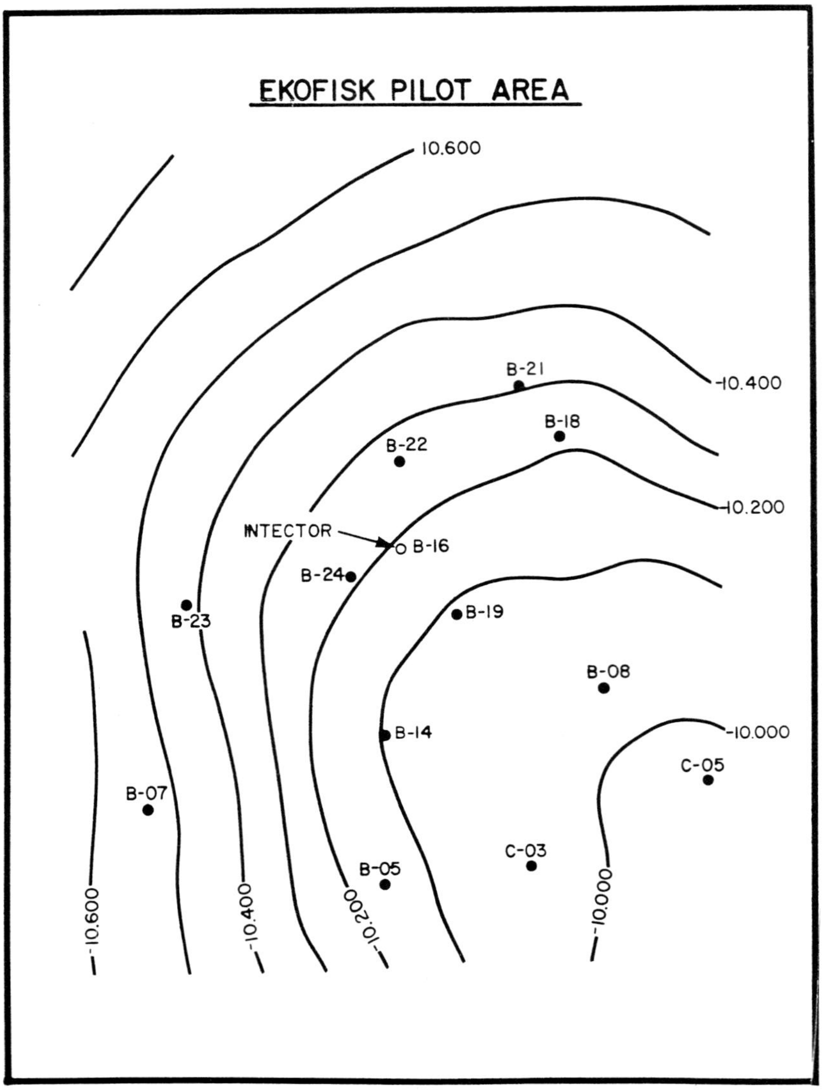

FIG. 20. Ekofisk pilot area: top of the Tor Formation (feet below sea level).

project, a commitment has been made to implement a full-field water-injection programme into the Tor Formation. Injection is scheduled to commence in 1987 and to reach a peak rate of 375 000 BWPD into 20 wells. The eventual recovery will be dependent upon the imbibition characteristics of the formation and the ability to maintain the reservoir pressures and hence the well deliverabilities at commercial levels.

A second pilot injection project is to be carried out to evaluate the characteristics of the Ekofisk Formation. Dependent upon the results of this test water injection may be extended to include the Ekofisk Formation. Gas injection into the crest of the Danian will continue.

It is expected that the Tor Formation water-injection programme will increase the recovery from the Ekofisk Field by 170 MMSTB oil or 3% of Original Oil in Place (OOIP). It is not a project that will result in a rapid increase in production rate. The benefits of the project will require years to realize and will be spread over the remaining life of the field (Fig. 22).

Conclusions

1 The imbibition properties of the chalk matrix material have been highlighted as being the

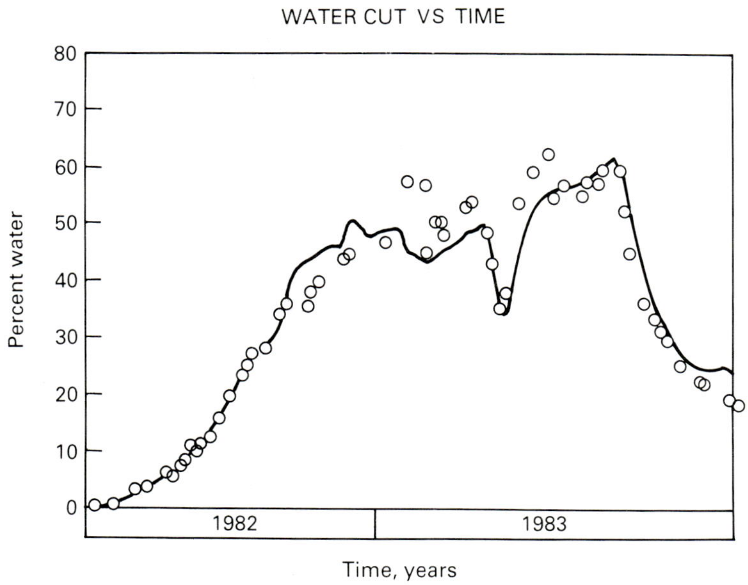

FIG. 21. Pilot history match: o, observed; —, modelled.

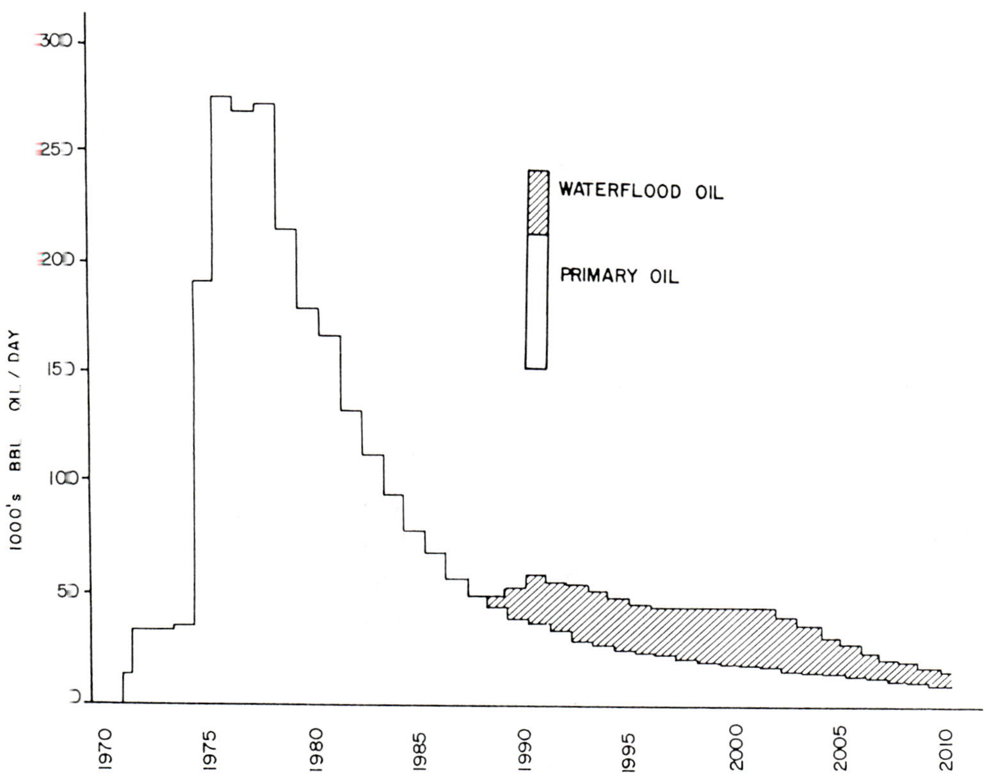

FIG. 22. Oil production in the Ekofisk Field.

primary controlling mechanism for hydrocarbon recovery under water injection.

2 Imbibition has been shown to be a function of the porosity, temperature and surface chemistry (wettability) of the matrix rock.

3 A field experiment was performed that supported the results of laboratory studies on hydrocarbon recovery by water injection into a fractured chalk field.

ACKNOWLEDGMENTS: The author wishes to thank the staff of the Phillips Petroleum Company, Bartlesville, and the Stavanger Geology and Reservoir Engineering staff for supplying material for this paper. The various co-ventures in the Ekofisk Field are also thanked for their permission to publish this paper. The views and conclusions presented in this paper are those of the author and may not necessarily represent those of the operator or the co-ventures of the Ekofisk Field.

References

BALDWIN, B. A. 1985. Characterization of North Sea chalk reservoir surfaces. *Proc. NPF Chalk Symp., May 1985*, to be published.

BREWSTER, J., DANGERFIELD, J. & FARRELL, H. (1986). Geology and geophysics of the Ekofisk Field waterflood. *J. Mar. petrol. Geol.* **3**, 139–69.

BYRD, W. D. 1975. Geology of the Ekofisk Field, offshore Norway. *In*: WOODLAND, A. W. (ed.) *Petroleum and the Continental Shelf of Northwest Europe. 1. Geology*, pp. 413–27, Applied Science, Barking, Essex.

CRAIG, F. F. 1971. *The Reservoir Engineering Aspects of Waterflooding, SPE Monogr.*, Vol. 3, Society of Petroleum Engineers, Tulsa, OK.

D'HEUR, M. 1984. Porosity and hydrocarbon distribution in the North Sea chalk reservoirs, *Mar. petrol. Geol.* **1**, 211–38.

FARREL, H. E. 1981. Ekofisk Fracture Study. *Internal Report*, Phillips Petroleum Company.

KAZEMI, H., MERRILL, L. S., PORTERFIELD, K. L. & ZEMAN, P. R. 1976. Numerical simulation of water–oil flow in naturally fractured reservoirs, *SPE J.*, December 1976, 317–26.

REISS, L. H. 1980. *The Reservoir Engineering Aspects of Fractured Formations*, Gulf.

SKOVBRO, B., HØY, T. & ZANG, R. D. 1982. Ekofisk Field, defining a reworked Maastrichtian zone in the lower section of the Ekofisk formation (Lower Danian), *Internal Report*, Norwegian Petroleum Directorate and Phillips Petroleum Company, Norway.

THOMAS, L. K., DIXON, T. N., EVANS, C. E. & VIENOT, M. E. 1984. Ekofisk waterflood pilot. *SPE Ann. Tech. Conf. Exhib., Houston, TX, September 1984, Rep. SPE 13120*, Society of Petroleum Engineers, Tulsa, OK.

——, —— & PIERSON, R. G. 1983. Fractured reservoir simulation, *SPEJ.*, February 1983, 42–54.

TORSAETER, O. 1984. An experimental study of water imbibition in chalk from the Ekofisk Field. *Rep. SPE/DOE 12688*, Society of Petroleum Engineers, Tulsa, OK.

VAN DER BARK, E. & THOMAS, O. D. 1981. Ekofisk: first of the giant oil fields in Western Europe, *AAPG Bull.* **65**, 2341–62.

WARREN, J. E. & ROOT, P. J. 1963. The behavior of naturally fractured reservoirs, *SPE J.*, September 1963.

D. A. BROWN, Phillips Petroleum Co. Norway, N-4056 Tananger, Norway.

Index

Pages on which figures appear are printed in *italic*, and those with tables in **bold**